D1370454

Successful Proposal Strategies for Small Businesses:

Winning Government, Private Sector, and International Contracts

Second Edition

For a complete listing of the *Artech House Technology Management and Professional Development Library* turn to the back of this book.

Successful Proposal Strategies for Small Businesses:

Winning Government, Private Sector, and International Contracts

Second Edition

Robert S. Frey

Artech House
Boston • London

Library of Congress Cataloging-in-Publication Data
Frey, Robert S..
    Successful proposal strategies for small business : winning
government, private sector, and international contracts / Robert S.
Frey. — 2nd ed.
    p.   cm.— (Artech House professional development and
technology management library)
    Includes bibliographical references and index.
    ISBN 1-58053-001-X (alk. paper)
    1. Proposal writing for grants—United States.   2. Small business—
United States—Finance.   I. Title.   II. Series.
HG177.5.U6F74   1999
658.8'04—dc21   99-28340
                                                                  CIP

British Library Cataloguing in Publication Data
Frey, Robert S.
    Successful proposal strategies for small business : winning government,
    private sector, and international contracts. — 2nd ed.
    1. Proposal writing in business
    I. Title
    808'.066'658

    ISBN 1-58053-001-X

Cover design by Lynda Fishbourne

© 1999 ARTECH HOUSE, INC.
685 Canton Street
Norwood, MA 02062

All rights reserved. Printed and bound in the United States of America.
No part of this book may be reproduced or utilized in any form or by any
means, electronic or mechanical, including photocopying, recording, or by
any information storage and retrieval system, without permission in writing
from the publisher.
    All terms mentioned in this book that are known to be trademarks or
service marks have been appropriately capitalized. Artech House cannot
attest to the accuracy of this information. Use of a term in this book should
not be regarded as affecting the validity of any trademark or service mark.

International Standard Book Number: 1-58053-001-X
Library of Congress Catalog Card Number: 99-28340

10 9 8 7 6 5 4 3 2 1

Contents

Acknowledgments   xi

Introduction   xiii

Chapter 1
Competitive proposals and small business   1

1.1   Overview   3

1.2   From set-asides to full and open competition   6

1.3   Small business constraints   8

1.4   Maximizing small business strengths   9

1.5   SBIR and STTR programs   10

1.6   Organizing your company to acquire new business   13

1.7   Effective strategic and mission planning   17

1.8   Converting knowledge into proposal success   19

   1.8.1   Knowledge management and proposal development   21

   1.8.2   Internal and external clients: Looking at clients in a whole new way   24

Chapter 2
Strategic partnering and subcontracting opportunities   31

2.1   Subcontracting opportunities and pathways to success   32

2.2   Specific strategies for achieving subcontracts   34

2.3   Becoming part of a government-wide acquisition contract (GWAC) team   37

2.4   Streamlined delegation of authority process   38

2.5   How mentor–protégé programs can help your business   39

Chapter 3
Marketing to and with your clients   43

3.1   More than just selling   43

3.2   Transactions are personal   45

3.3   Listen to your client   46

3.4   Infuse the marketing intelligence into your proposal   47

3.5   Intelligence gathering and analysis techniques   49

3.6   Call plans   52

3.7   Maintain management visibility on your contracts   57

3.8   Project managers as client managers   58

3.9   Commercial-off-the-shelf acquisition   60

3.10   Pursuing firm fixed price and invitation for bid opportunities   61

3.11   Using a request for information and request for comment as valuable marketing tools   62

3.12   Standard form 129s and contractor prequalification statements   63

3.13   Ethics in marketing and business development   64

3.14   Advertising, trade shows, and high-impact public relations   65

Chapter 4
Request for proposals   71

4.1   Overview   71

4.2   Part I—the schedule   73

4.3   Part II—contract clauses   73

4.4   Part III—list of documents, exhibits, and other attachments   74

4.5   Part IV—representations and certifications   74

4.6   The importance of Section L (instructions to offerors)   74

4.7   Section M (evaluation criteria): Toward a maximum score   77

4.8   Greatest value approach   77

4.9   Emphasis on performance-based contracting   78

4.10   Influencing the content of an RFP—legitimately   79

4.11   Other types of solicitation documents   80

Chapter 5
Private-sector solicitation requests   83

Chapter 6
The federal acquisition process: new directions   87

6.1   Overview   87

6.2   Statutory and regulatory requirements for competition   88

6.3   The source selection process   89

6.4   Full and open competition   92

6.5   Major contract types   93

6.6   Significant recent paradigm shifts in federal government acquisition   93

6.7   National Partnership for Reinventing Government (NPR)   94

6.8   Understanding the impact of the Federal Acquisition Streamlining Act of 1994   96

6.9   Federal acquisition computer network   99

6.10   Benefits of electronic data interchange   101

6.11   Understanding ANSI X12 standards   101

6.12   Sources of information on EC/EDI and acquisition reform   102

6.13   Department of Defense Electronic Commerce Office   106

6.14   DoD Electronic Commerce Information Center (ECIC)   106

6.15   DoD electronic commerce/electronic data interchange infrastructure   107

6.16   Using value-added networks and value-added services   107

6.17   Electronic Bid Sets (EBS)   109

6.18   Electronic commerce outside of DoD   110

Chapter 7

The proposal life cycle   115

7.1   What is a proposal in the competitive federal and commercial marketplace?   115

7.2   Where does the proposal fit into the total marketing life cycle?   118

7.3   Bid/no bid decision-making process   127

7.4   Planning and organizing   129

7.4.1   *Draft executive summary   129*

7.4.2   *Theme development   130*

7.4.3   *Bullet drafts/storyboards   133*

7.5   Kickoff meeting   135

7.6   Writing   136

7.7   Major contractor review cycles   136

7.7.1   *Blue Team/Pink Team   137*

7.7.2   *Red Team   137*

7.7.3   *Gold Team   139*

7.7.4   *Black Team   139*

7.8   Preparing for orals and best and final offer   139

7.9   Debriefings (refer to FAR 15.1003)   140

Chapter 8

Major proposal components   143

8.1   Overview   143

8.2   Transmittal letter   144

8.3   Technical volume   145

8.3.1   *Front Cover   145*

8.3.2   *Disclosure statement on the title page   147*

8.3.3   *Executive summary   148*

8.3.4   *Building a compliance (cross-reference) matrix   149*

8.3.5   *Narrative body of the technical volume   150*

8.4   Management volume   154

8.5   Cost volume   161

8.6   Government contract requirements   162

Chapter 9

Acquisition and proposal team activities   165

9.1   Formation and function of acquisition teams   165

9.2   Pre-kickoff activities   166

9.3   Proposal kickoff meeting   169

9.4   Post-kickoff activities   172

Chapter 10

The role of the proposal manager   173

10.1   Overview   173

10.2   Generalized job description   175

10.3   The growing importance of oral presentations   182

10.4   Attending to the details   183

10.5   Control of the schedule   184

10.6   Training additional staff in proposal management skills   187

10.7   Finish the job at hand   187

10.8   Successful proposal managers   188

Chapter 11
Structuring international proposals 191

11.1 Overview 191

11.2 The importance of the World Bank group 192

11.3 Your company's participation in United Nations procurements 195

11.4 European Bank for Reconstruction and Development (EBRD) 197

11.5 Asian Development Bank (ADB) 200

11.6 International market planning 200

11.7 In-country partnerships 202

11.8 Host country procurement environments 202

11.9 Import-export considerations and technology transfer 203

11.10 Risk assessment 203

11.11 Terms and conditions 203

11.12 Export-Import Bank of the United States assists small businesses 205

11.13 Helpful Web-based resources and in-country support infrastructures for small businesses 206

MERX: Canada's national electronic tendering service 206

Government Supplies Department (GSD) Hong Kong, China 207

Euro Info Centres (EIC) 209

British Overseas Trade Bureau (BOTB) 210

Department of Trade and Industry (DTI) (UK) 210

Export Market Information Centre (EMIC) 214

European Procurement Information Network (EPIN) (Ireland) 215

Office of Public Works (OPW) (Ireland) 215

SIMAP (European Commission) 216

THEMiS: The System for Regulated Procurement 216

Procurement Information Online (PIO) (Germany) 217

Confederation of German Trade Fair and Exhibition industries (AUMA) 217

Das Gepa-Projekt (Marketing Assistance Programs for eastern German enterprises) 218

French Committee for External Economic Events (CFME) 218

Contracts and Tenders Worldwide (CTW) 219

JETRO (Japan External Trade Organization) 219

Asia-Pacific Economic Cooperation (APEC) 219

Government Electronic Marketplace Service (GEMS) (Australia) 220

Tender I.N.F.O. (Electronic Tendering and Procurement Network) 220

Australian Trade Commission Online (Austrade) 220

Arabnet: Reaching the Arab World in the Middle East and North Africa 221

Stat-USA GLOBUS Information System 222

11.14 The Unisphere Institute, U.S. SBA, NASA, and International Marketing 223

11.15 British-American Business Council (BABC) 224

11.16 U.S. Trade and Development Agency (USTDA) 225

11.17 U.S. Agency for International Development (USAID) 226

Chapter 12
Proposal production/publication 233

12.1 Internal documentation standards 235

12.2 Document configuration management and version control 236

12.3 Freelance and temporary publication staff 237

12.4   Incorporating technical brilliance up to the last minute   238

12.5   Graphics are an integral part of your proposal   239

   12.5.1   Action captions   240

   12.5.2   Configuration control of graphics   241

12.6   Role and structure of your publications group   242

12.7   Software and hardware compatibility, standards, and recommendations   243

12.8   Electronic proposal submittal and evaluation   245

12.9   Important documentation tips   246

12.10   Virtual proposal centers, intranets, and extranets   248

12.11   Using freelance proposal writers to maintain technical productivity   249

Chapter 13

Human and organizational dynamics of the proposal process   253

13.1   Modifying our thinking to win   254

13.2   Building a competitive work ethic   255

13.3   Strong link between project performance and proposal success   256

13.4   Proposals can be fun!   257

13.5   Maximizing human intellect   258

13.6   Proposal professionals as "change agents"   260

13.7   Wellness in your proposal process   261

Chapter 14

Controlling bid and proposal costs   263

14.1   What does it cost to get new business, and how are those costs recovered?   264

14.2   Tracking B&P expenditures   265

14.3   Business development bonus policy   266

14.4   Stretching limited marketing funds   268

Chapter 15

Tried and true proposal writing and editing techniques   271

15.1   Proposals are knowledge-based sales documents   271

15.2   Active voice adds strength and saves space   274

15.3   Guide the client's evaluators through your proposal   275

15.4   Action captions   278

15.5   Methods of enhancing your proposal writing and editing   279

15.6   Government-recognized writing standards   281

15.7   Additional sources of writing guidance   282

Chapter 16

Packaging and managing proposal information effectively   283

16.1   Overview   283

16.2   The all-important résumés   284

16.3   Project descriptions (project summaries)   287

16.4   Proposal boilerplate ("canned" or "reuse" material)   289

16.5   Marketing targets   291

16.6   Corporate library   291

16.7   Proposal "lessons learned" database   293

16.8   Applying IT solutions: Evolutionary informational data systems   295

   16.8.1   Lotus Notes scenarios   295

   16.8.2   CD-ROM scenarios   296

   16.8.3   Intranet scenarios   297

16.9 Leveraging federal performance appraisal systems to your company's benefit   297

Chapter 17
Leveraging business complexity in a knowledge-based economy   299

17.1 Turbulent transition toward knowledge-based business   299

17.2 How to communicate effectively on your knowledge landscape   302

17.3 Envisioning supple business models   305

17.4 Sample application: Tracing complexity and knowledge management through the proposal development process   309

17.5 Summation   310

Chapter 18
Planning and producing SF254/255 responses for architect-engineer services   315

18.1 SF254/255 and the FAR   315

18.2 Understanding the required structure of the response   325

18.3 Overall strategy of response   325

18.4 Build Block 7 first   328

18.5 Block 8: Selling your project experience   329

18.6 Block 10: Structure according to the evaluation criteria   330

18.7 Block 10 outlining   330

18.8 Other Blocks   331

18.9 Subcontractor participation   332

18.10 Building teaming agreements   334

Epilogue
Thinking to win small business competitive proposals   337

Appendix A
One agency's response to acquisition streamlining   341

The Federal Aviation Administration's (FAA) new Acquisition Management System (AMS)   341

Appendix B
Sample Proposal Kickoff Package   349

Appendix C
Template to capture important résumé information   363

Appendix D
Marketing information and intelligence sources: federal, international, and private sector   367

D.1 Sources of federal marketing leads and information   368

D.2 Sources of international marketing leads and information   377

D.3 Sources of U.S. private-sector marketing leads and information   378

Appendix E
Glossary of proposal-related terms   381

Selected list of acronyms and abbreviations   395

Selected bibliography   435

About the author   453

Index   457

Acknowledgments

THIS VOLUME WAS DEVELOPED, expanded, and refined during the past six years. I would like to express appreciation to Ms. Elizabeth Malone, who provided helpful information borne of years of hands-on experience related to federal source selection, Statement of Work preparation, and the Federal Acquisition Regulations. Dr. Richard Isaacman also provided much in the way of support and momentum for the effort, particularly from 1995 forward. He is both a proposal development colleague and enduring friend. Mr. Bruce Elbert, series editor of the Technology Management and Professional Development Library for Artech House, Inc., must also be recognized and thanked for relentlessly yet good-naturedly stretching the book into new areas. This work has been enhanced greatly through his efforts, and I am genuinely appreciative.

The graphic arts talent of Ms. Lisa Richard must also be recognized. Lisa generated the graphics for both editions of *Successful Proposal Strategies for Small Businesses*. She can be contacted through http://members.aol.com/lotslr/lots.htm. Ms. Sarah A. Withers (e-mail: sarah_withers@msn.com) expertly designed the companion CD-ROM for this volume.

Finally, the business and editorial acumen of Mrs. Terry Raezer Frey contributed significantly to the value, consistency, and readability of this work. And for her love and unconditional day-to-day support, I am both a better author and better person.

Introduction

THE LATE VINCE LOMBARDI, legendary coach of the Green Bay Packers, is reputed to have said that he longed to "lie exhausted in victory." That is, to expend the very best effort, to harness the talent and spirit within, and to channel that immense power toward a very specific goal. In Lombardi's thoughts, that goal was victory in the early Super Bowl competitions of the National Football League. In my own thoughts, that goal is to bring all the knowledge, experience, initiative, and positive emotion I can into producing a winning proposal.

Unlike many other professions, proposal preparation in the contractor arena for federal, state, local, private-sector, and international opportunities occurs in very discrete and often overlapping bundles of intense activity. There is a clear beginning, middle, and end to the preparation process. Often in a mere span of 5 to 45 days and nights, a host of technical and programmatic information, cost strategies, and marketing intelligence must be condensed, distilled, and fitted together into a set of polished documents. Considering the length of time required to bring journals and books to press, it is astounding that such a choreographed process of information retrieval and management, assembly, and packaging must unfold in the space of only a week or several weeks! And yet for those small and large businesses that compete in the contracting marketplace, it is a matter of survival.

Successful proposal preparation is built largely upon a winning attitude, attention to detail, teamwork at all levels, communication, emotional and physical endurance, and adequate and well-timed allocation of company human and material resources. To be sure, success also depends upon marketing intelligence about the customer and your competition, informed and timely bid/no bid decisions, planning, scheduling, and superior information management. But my experience has suggested that what makes the difference once a company decides to respond to a *request for proposal* (RFP) or SF254/255 synopsis lies in the area of human and organizational dynamics rather than in technical and strategic excellence alone. Can a diverse group of technical, management, and support people work together effectively for protracted periods of time—including nights, weekends, and holidays—to produce a winning document? Will company management commit the best technical talent, lease or acquire adequate computer/publishing equipment, make dedicated work space available for the proposal team, or allocate bonus monies to reward the above-and-beyond efforts of particular people?

To lie exhausted in victory. Plans and milestone schedules, bullet drafts and storyboards, writing and editorial guidelines, action item lists, internal review cycles, and document configuration management schemas all come down to one thing—getting a winning proposal assembled, out the door, and delivered before the established due date. While I was coordinating a $100M+ Air Force proposal for a Virginia-based contractor, the entire marketing and proposal lifecycle came down to one overcast Saturday in December, not long before the holidays. Thoughts were not on marketing target identification, intelligence gathering, teaming arrangements, RFP analysis, outline development, program pricing, or Red Team review comments. Rather, there were 150 copies of various volumes that had to be photoreproduced and put into three-ring notebooks, with multiple foldout pages inserted in each one, and an overnight carrier office nearby that was scheduled to close promptly at 5 P.M. Just the night before, several members of the proposal team had worked into the early morning hours. People were exhausted from several weeks of grueling schedules, missed meals, and no recreation. Taping boxes shut at break-neck speed, loading them into several cars, and making multiple trips to the shipping office. When that effort was over, I along with several members of my staff felt too tired to move. And yet, there was a palpable feeling of accomplishment, a feeling of victory.

For those full-time professionals in the proposal development business, proposals must become a way of life if we are to survive and grow in our careers. Alternative strategies for time management, stress management, family life, and personal pursuits must be developed and nurtured. In ways analogous to military combat, the proposal professional must adjust quickly despite tiredness, personal and family concerns, time of day or night, and level of pressure. But the possibility of personal satisfaction from performing proposal work well can be second to none.

Chapter 1

Competitive proposals and small business

Successful Proposal Strategies for Small Businesses: Winning Government, Private-Sector, and International Contracts is designed to provide entrepreneurs as well as beginner and moderately experienced proposal managers, proposal writers, proposal specialists and coordinators, and business development staff with a useful resource for planning, organizing, managing, and preparing effective responses to U.S. federal government *requests for proposals* (RFPs) and *architect-engineer* (A-E) *Standard Form* (SF) 254/255 synopses. (Architectural and engineering firms submit SF254/255s routinely to establish their credentials with client organizations.) There is also attention devoted to responding to U.S. private-sector solicitations and international tenders.

This book illustrates the close relationship between the federal acquisition process and the response lifecycle that unfolds within the contractor community. The specialized statutory and regulatory structure that currently governs and dominates the federal acquisition process and the contractor proposal process is summarized. Important and exciting new directions in federal *electronic commerce* (EC) and *electronic data interchange* (EDI) following the passage of the *Federal Acquisition Streamlining Act* (FASA) in 1994 and the *Federal Acquisition Reform Act* (FARA) of 1996 are highlighted. Ethical business acquisition practices are emphasized, and effective long-term marketing approaches are presented.

Small businesses are confronted with distinctive opportunities and constraints in the federal marketplace. *Successful Proposal Strategies for Small Businesses* focuses specifically upon small business enterprises, exploring the important human and organizational dynamics related to the proposal lifecycle that can facilitate success in acquiring new business. Thinking to win is a crucial aspect in the world of federal, private-sector, and international procurement.

Salient points in the contractor proposal response lifecycle are discussed in detail as are the major components of the proposal documents and the client's RFPs. The role of a small company's proposal manager is explored at length, and valuable information management activities in support of the proposal process are described. Effective proposal-writing techniques are provided along with successful proposal publication and production scenarios. Proposal and marketing cost-tracking, control, and recovery strategies are reviewed; and select client and competitor information and intelligence sources for the U.S. government, U.S. private-sector, and international opportunities are enumerated (Appendix D). Guidance for planning and producing compliant and responsive SF254/255s is presented. And structuring proposals for international and U.S. private-sector clients is discussed as well. Finally, to support the users of *Successful Proposal Strategies for Small Businesses*, a lengthy listing of proposal, business, and acquisition-related acronyms is provided as are definitions of select terminology (Appendix E).

No one person or methodology can offer absolutely definitive step-by-step instructions to win federal, private-sector, or international proposals. There are no shortcuts to building and growing an entire business development infrastructure to market clients and acquire new business. In recognition of the hard work, right thinking, informed decisions, careful planning, and exacting execution of proper proposal techniques, this book is offered as a starting point in proposal literacy. We hope that it serves as a "users' manual," consulted frequently for suggestions and guidance throughout the proposal planning and response process. Best wishes for successful proposals in your company's future!

1.1 Overview

Winning. The federal competitive procurement process [1] is absolutely binary—contractors either win or lose with their proposals. With the exception of multiple-award situations, there are no rewards for coming in second. To allocate your company's *bid and proposal* (B&P), marketing, and *internal research and development* (IR&D) funds to pursue procurements for which there is only a marginal probability of winning is, at best, questionable business planning. Federal agencies often have a variety of domestic as well as overseas[1] contractor/vendor firms from which to select a specific supplier of goods or services. At a minimum, you have to know your potential client and his or her requirements; and in turn, your client must be made aware of your company's particular technical capabilities, relevant contractual experience, available human talent, and financial stability in the context of an ongoing marketing relationship. One or two briefings from your company to top-level government agency administrators will most likely be insufficient to secure new business in the competitive federal marketplace. This applies to the state, municipality, and U.S. private-sector marketplaces as well. Organizations, in general, procure goods and services from companies that they have come to know and trust and that have demonstrated an ongoing interest in an organization's technical, operational, programmatic, and profitability issues.

Many small contracting firms that provide goods and services to the federal government are primarily or even solely dependent upon federal contracts for their survival and growth. Consequently, proposal development, management, design, and preparation are the most important business activities that your company performs. Proposal development and writing are more than just full-time jobs. It can be a 12- to 16-hour-a-day 6- or 7-day-a-week effort just to keep from falling behind hopelessly [2]. Proper, intelligent planning and preparation will certainly make proposal development more manageable. Your company should not start developing a proposal unless it intends to win. An exception to this guideline is if your company wants to submit a proposal on a particular procurement in order to gain experience in assembling proposals or to gain recognition from the government as a potential supplier [3]. The American Graduate University suggests that as many as three-quarters of the proposals received by government procuring agencies are deemed to be nonresponsive or inadequate [4]. If your company competes heavily in the federal marketplace, then proposals are your most important product.

---

1   Competition is growing from Japanese, Canadian, Western European, and emerging Eastern European nations for U.S. government contracts.

It does not matter how large your company is. For example, let us assume that yours is a company with $12M posted in revenue during the last fiscal year. To simply maintain revenues at that level during the next fiscal year, you will "burn" $1M each month in contract backlog, as shown in Figure 1.1. That means that you must win $1M each month in new or recompete business just to keep the revenue "pipeline" full. Yet winning $1M per month in new or recompete business will not allow your company to grow revenue-wise at all! To put that $1M of business per month in appropriate context—your company would have to bid $3M per month in proposals and have a win ratio of 33% to bring in that level of revenue. And $3M worth of proposals translates into identifying 2 to 3 times that amount in potential marketing opportunities that then have to be qualified and pursued. Many times, release schedules for procurement opportunities slip, or funding is withheld, or the specific requirements get rolled into a larger procurement. As a result, what appears to be a solid lead in January has evaporated by June. See Figure 1.2 for an illustration of this pipeline process. The same applies for a company with $1.2B of posted revenue.

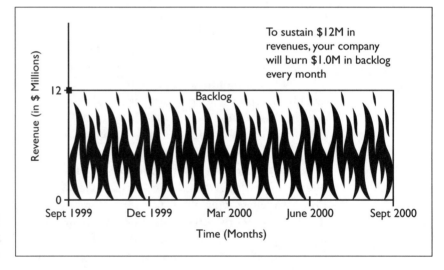

Figure 1.1
Contract backlog
burn rate.

"Without a plan, the proposal process will be chaotic and the product, at best, will be inferior" [5]. Gone is the time of last-minute, haphazard proposal preparation by a few individuals working in isolation from a senior, in-house review and other corporate or divisional guidance. Your company simply cannot compete effectively with the many U.S.-based and overseas contracting firms if every proposal you submit is not your finest effort. Your company will, of course, not win every procurement— 25% to 40% is a reasonable win ratio, although certain firms have been

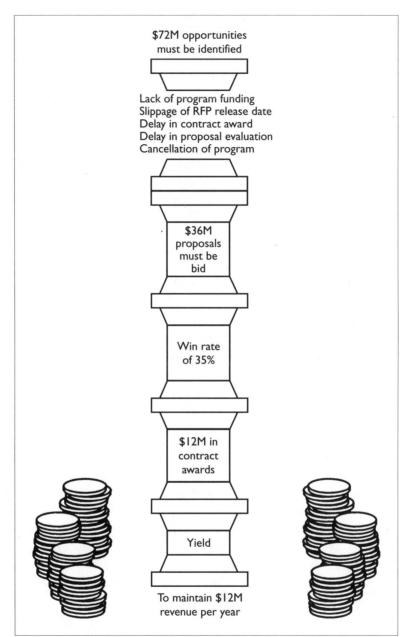

$72M opportunities must be identified

Lack of program funding
Slippage of RFP release date
Delay in contract award
Delay in proposal evaluation
Cancellation of program

$36M proposals must be bid

Win rate of 35%

$12M in contract awards

Yield

To maintain $12M revenue per year

Figure 1.2
Your business pipeline must remain full to maintain and grow your revenue base.

documented to win 60%+ of their proposals consistently—but you must strive to have each and every proposal be in the competitive range [6] from a technical, management, and cost standpoint.

It is important to note that a technically sound, programmatically effective, and competitively priced proposal is not enough. With content and cost must come readability, appearance, and format. And these elements require dedicated time to accomplish. Cover design, page formatting, editing, generating graphics, wordprocessing/publishing, proofreading, photoreproducing, collating, and assembling are all vital steps in the overall proposal preparation cycle. Put yourself in the role of a government evaluator. That person, along with his or her colleagues, has to look at many proposals for each procurement. "Why should I do business with you?" is the question they are asking themselves. Would you enjoy struggling through a poorly written, amateurishly prepared document in the evening or on the weekend? Indeed, there are increasing numbers of small and large businesses chasing fewer and fewer federal dollars. Even relatively minor procurements are resulting in 50 or more proposals that are submitted. Debriefings across a wide variety of agencies suggest that evaluators are spending 15 mins to 30 mins on each company's proposal during the preliminary round of evaluation. There simply is no more time available to them. As a result, it is more imperative than ever that your company's proposal stand out in a positive way. Create difference! Section 52.215-7 of the *Federal Acquisition Regulations* (FAR), "Unnecessarily Elaborate Proposals or Quotations" (April 1984), cautions contractors not to submit proposals that contain elaborate artwork, expensive paper and bindings, and expensive visual and other presentation aids. To be sure, certain federal agencies, as well as state and local organizations that follow the FAR, will look unfavorably upon any proposal documents that go beyond basic typewriter-level presentation values. Yet your competitors are spending tens of thousands of dollars both in-house and through professional proposal consulting firms to prepare full-color, graphics-intensive, high-impact proposal documents. The challenge is to know your client well enough to sense what level of proposal media and presentation style they will respond to favorably. Some clients stipulate that proposals should demonstrate the quality of deliverable document that your company is capable of producing. You will have to balance the perception of cost consciousness in your proposal documents with the genuine need to make your volumes establish the standard of presentation excellence among those submitted for a particular procurement.

1.2 From set-asides to full and open competition

There are no "bluebirds" in full and open competition. The gravy train—to the extent that it still persists in the business world—is over! A reorientation and fundamental transformation of your company's collective thinking and attitude will be required to begin the challenging shift from 8(a) [7] set-aside procurement to competitive federal business acquisition. The term "8(a)" refers to the U.S. *Small Business Administration*'s (SBA) program to assist qualified small and woman- and minority-owned businesses during the early years of their operation. Established in July 1953 under the Eisenhower administration, the SBA provides financial, technical, and programmatic assistance to help Americans start, run, and grow their businesses. The 8(a) program, named from Section 8(a) of the Small Business Act, is the most well-known element of the Minority Enterprise Development (MED) program. In 1997, the SBA provided assistance to more than 1 million small business owners. The Agency's proposed FY2000 budget, as reported in PR Newswire on March 19,1999, stands at $994.5M.

The SBA is administered by Ms. Aida Alvarez, who suggests that diversity, technology, and globalization will be critical to small business success in the 21st century. The agency maintains the Internet site ACE-Net (http://sbaonline.sba.gov/ADVO/acenet.html), which links investors with newly emerging small businesses. Now in its 36th year, the SBA's national Small Business Week honors the contributions of the nation's small business owners. Winners are selected on the record of stability, growth in employment and sales, sound financial reports, innovation, and the company's response to adversity and community service.

Changing attitudes can be a difficult and lengthy process. The process of change must begin and be fostered on an ongoing basis by senior management, for example, through marketing meetings, acquisition team meetings, proposal kickoff meetings, your company's internal newsletter or communications vehicle, and project management mentoring programs. Precisely the thinking that proved so successful and comfortable during the early 8(a) years of your company's history is often the very thing that thwarts your company's potential for growth in the competitive arena. Entrepreneurial companies are often characterized by informal business organizations and cultures that are functional and effective for small companies only.

However, if companies are successful and grow in terms of revenue and human resources, they will reach a point at which an *informal* culture and organization are inadequate. This is particularly apparent in the areas of planning, management structure, internal communications, and support infrastructure. Successful, growing companies should reorganize, bring in new senior operations management as appropriate, and develop a strategic planning process [8]. And management responsibilities and authority should be delegated downward so that a small company's organizational structure is not so "sharply hierarchical" [9]. Empowerment is critical to the lowest possible level within your organization. Senior management should actively encourage all staff to identify and understand problems and then propose positive, team-oriented solutions. When implemented, this approach helps to leverage everyone's talents and contributes to the collective knowledge base of the company.

Dedicated effort in accordance with a well-defined plan, broad-based and in-depth knowledge of your clients and competitors, superlative performance on past and present projects, and a formalized company organization and communication network all contribute to successful proposals. In an extremely important quantitative study conducted by Price Waterhouse from 1990–92, it was determined that companies that exhibit superior performance as measured by competitive contract awards managed "their business acquisition as a formal, disciplined process. These companies view business acquisition as a structured set of interrelated activities to win contract awards. The superior performers continuously improve the methods they use to pursue opportunities" [10].

In an effort to help ensure the long-term success of participants in the 8(a) program, the SBA has established goals for the percentage of 8(a) and non-8(a) contracts that a company should pursue, with the number of non-8(a) contracts increasing each year. A company is put on a schedule during its final five years in the 8(a) program, with non-8(a) business targeted to be 55% to 75% of company revenues by the final year prior to graduation into the arena of full and open competition [11]. Your company will need to develop an "exit strategy" from the 8(a) program if you are to survive and prosper. That strategy might include developing a balanced albeit aggressive mix of government and commercial contracts, establishing long-term business partnerships with major prime contractors, and mergers and acquisitions. Small businesses that have been successful in leveraging partnerships with major prime contractors into increased revenue point to their flexible pricing strategies, solid past performance record, and sufficient financial backing [12].

1.3 Small business constraints

In terms directly relevant to proposal design, development, and preparation, many small businesses must navigate effectively amidst very limited B&P funds, a lack of depth in human resources, a small business base, a contract backlog deficit, a low level of contractual experience, a lack of name recognition in the marketplace, and line of credit challenges. A small business base, for example, can lead to higher indirect costs, which in turn can place a company at a competitive disadvantage during procurement efforts. And insufficient staff can translate to few or no people dedicated to the tasks of advanced and strategic planning, marketing to particular client organizations, proposal operations, proposal reviews, proposal editing and proofreading, proposal publication, and post-proposal marketing. Staffing challenges emerge quickly as full-time project managers work 40 billable hours each week for their client and then additional time to serve as proposal managers or proposal reviewers. (In predominately service-oriented contracting firms, the company's overall profitability is affected by the degree to which its personnel are fully billable. Transfer ratios, that is, billable time versus total time worked, must remain very high.) And thin contractual experience can lead to low scores received for "Past Performance" or "Relevant Experience" sections of the RFP or Blocks 8 and 10 of the SF254/255.

1.4 Maximizing small business strengths

America's 23 million small businesses employ more than 50% of the private workforce [13]. According to President Clinton's "Annual Report on the State of Small Business" (June 6, 1996), a record 807,000 new small businesses recorded initial employment during fiscal year 1995 [14]. And women are creating new businesses and new jobs at double the national rate, and own nearly 40% of all firms in the United States. In 1994, small businesses were awarded 28% of the $160 billion worth of available federal government contract awards [15].

"Small businesses (corporations that employ 500 people or less) constitute 99 percent of U.S. enterprises. These companies have certain competitive advantages: they are lean in terms of administration, they can position themselves in a market niche that large corporations cannot fill, and they can offer superior service to customers" [16]. On June 24, 1998, the White House announced new federal rules to give minority firms an edge when bidding for federal government contracts while respecting a 1995 U.S. Supreme Court ruling that limited affirmative action programs. Under the new rules, small firms that are certified to be disadvantaged by the U.S. SBA will receive a price break of 10% in calculating the lowest bidder for government contracts [17].

Small businesses have the potential to respond rapidly to emerging business opportunities because they have fewer layers of management approval in the decision-making chain. Company policies can be modified quickly to meet client requests and requirements [18]. Small businesses can carefully control their growth in terms of acquiring technical talent and penetrating new market sectors. The opportunity for excellent in-house communications throughout the network of authority exists with small businesses. Small companies are ideally positioned to develop, right from the start, open-architecture internal automated information systems for maintaining and searching staff résumés, project summaries, proposal modules, lessons learned, success stories, and marketing opportunities. And because of the staffing deficit, people tend to become cross-trained and proficient in a wide variety of tasks. More people are given the opportunity to understand the "big picture" of the proposal life cycle and of specific business targets. Conversely, in large, multidivision corporations, very few staff fully understand the multidimensional complexities of massive procurement targets. On large procurements, for example, some major firms devote one or more staff exclusively to handling subcontractor résumés. In a small firm, that level of work breakdown is simply not possible.

1.5 SBIR and STTR programs

Important mechanisms for generating revenue in the small business community for those firms with strong scientific or engineering capabilities include the Small Business Innovation Research (SBIR) program (see http://www.sba.gov/SBIR/sbir.html) and the Small Business Technology Transfer (STTR) (see http://www.sba.gov/SBIR/sttr.html) pilot program. Enacted on July 22, 1982, as part of the Small Business Innovation Development Act (P.L. 97-219), the SBIR program encourages small businesses to explore their technological potential and provides the incentive to profit from its commercialization. Worldwide commercial rights to any patents normally will go to the small company. Public Law 102-564, the Small Business Research and Development Enhancement Act, signed into law on October 28, 1992, continues the SBIR program through September 30, 2000. The SBIR program is highly competitive and merit based—it is in no way an assistance program for small businesses. It is, however, open only to small businesses and is not intended for non-profit organizations.

The risks associated with conducting significant research and development (R&D) efforts are often beyond the economic and resource capabilities of small businesses. SBIR, in effect, protects small businesses and enables them to compete on the same level as larger businesses. In 1995, $900 million was awarded to small businesses under the SBIR program. And in FY96, more than $916 million was distributed across more than 4,000 awards. Of note is that companies with 10 or fewer employees have won more than one-third of all SBIR awards to date. Five of the 11 SBIR federal agencies required to participate in the SBIR program make more than 90% of the awards annually—DoD, Department of Health and Human Services (DHHS), DOE, NASA, and the National Science Foundation (NSF). Other agencies include the Department of Agriculture (USDA), Department of Education, Department of Transportation (DOT), EPA, and the Nuclear Regulatory Commission (NRC).

The Small Business Administration has the responsibility for coordinating the SBIR and STTR programs. Upon request, SBA will mail you a quarterly *Pre-Solicitation Announcement* (PSA) that lists the agencies that will make SBIR offerings in the next fiscal quarter, along with the release, closing, and award announcement dates. The quarterly PSA also provides a one-line statement for each SBIR topic. However, this PSA is not a substitute for the agency SBIR solicitations themselves. DoD will automatically send its solicitations to all companies on the SBA PSA mailing list, but no other agency does. You will have to request SBIR solicitations from each agency in which your company is interested each year. Some agencies, such as DoD and DHHS, issue more than one pamphlet or bulletin each year in which SBIR solicitations are published. Each participating federal agency publishes an extremely helpful volume of abstracts that summarizes the SBIR proposals that it funds each year. To obtain a copy, you can contact the SBIR program manager in each agency in which your company has interest.

The SBIR program has three phases: Phase I, Feasibility Study; Phase II, Full-scale Research; and Phase III, Commercialization. More than 15,000 small businesses have received at least one Phase I SBIR award. Funding at Phase I extends up to $100,000, and finances up to 6 months of research. Phase II funding can reach $750,000, and research can span 2 years. Only Phase I winners are considered for Phase II. With the Department of Energy, for example, success ratios have been about 12% in Phase I and 45% in Phase II. Some states, such as Kentucky through its SBIR Bridge Grant program established in 1988, assist small firms to continue product development research projects begun under federal Phase I SBIR awards.

In most cases, Phase I and II provide full allowable costs and a negotiated fee or profit for the small company. Phase III involves no SBIR funding, although the SBA has developed the Commercialization Matching System to help SBIR awardees locate funding sources for finalizing their innovations. The small business might procure funding for Phase III of the SBIR process from commercial banks (e.g., the Small Business Investment Corporation, the equity investment arm of NationsBank); venture capitalists; private-sector non-profit small business lending groups such as the Development Credit Fund (DCF) of Baltimore, Maryland, or the Maryland-based Council for Economic and Business

Opportunity, Inc. (CEBO); large companies; or other non-SBIR federal agency funding. Other help is available for companies entering the commercialization phase of the SBIR program. For example, Dawnbreaker (http://www.dawnbreaker.com), a professional services firm located in Rochester, New York, has assisted more than 400 DOE SBIR awardees since 1989. Dawnbreaker staff work with DOE SBIR Phase II grantees to develop a strategic business plan and to prepare the firm for commercialization. Dawnbreaker then hosts the Commercialization Opportunity Forum in Washington, D.C., to showcase firms that are ready to enter the third SBIR phase [19].

More than 30% of Phase II SBIR projects will result in a commercialized product. Successfully commercialized SBIR projects include DR-LINK, a natural language retrieval system based on linguistic technology that processes ongoing news streams or information from databases. Patent examiners at the U.S. Patent and Trademark Office now use the DR-LINK system daily to search for prior art in more than 4,000 published databases. This SBIR effort was funded through DARPA, the Defense Advanced Research Projects Agency. SBIR support from the National Science Foundation, Department of Energy, and Department of Defense allowed AstroPower, Inc., of Newark, Delaware, to develop superior thin-layer silicon- and gallium-arsenide technology with optical and speed advantages in photovoltaic devices. The NSF and National Institutes of Health (NIH) have made major SBIR contributions to research conducted by Martek Corporation in Columbia, Maryland. This biotechnology firm discovered how to make a critical ingredient needed by infants that is normally found in mothers' milk. Medical and health organizations in Europe and the World Health Organization (WHO) have recommended that this critical ingredient be added to infant formulas to offset nutrient deficiencies in babies that are not breast fed. And Scientific Computing Components, Inc., in New Haven, Connecticut, has leveraged SBIR funds to produce a number of breakthroughs in commercial software related to high-performance computing.

Although the format for SBIR Phase I proposals varies across the participating federal agencies, a typical format is as follows:

- Cover sheet;
- Project abstract/anticipated benefits/key words;
- Identification of the problem/opportunity;
- Background information;
- Phase I technical objectives;
- Phase I work plan;

- Related work;
- Relationship to future research and development;
- Potential post-applications;
- Key personnel;
- Facilities and equipment;
- Consultants;
- Current and pending support;
- Budget;
- Previous SBIR awards.

The STTR pilot program's role is to foster the innovation necessary to meet America's scientific and technical challenges in the 21st century. Central to the program is the expansion of the public- and private-sector partnership to include joint venture opportunities for small businesses. Each year, DoD, DOE, DHHS, NASA, and NSF are required under the STTR to reserve a portion of their R&D funds for award to small businesses and non-profit research institution partnerships. The non-profit research institution must be based in the United States, and can be a non-profit college or university, domestic non-profit research organization, or Federally Funded R&D Center (FFRDC). Like SBIR, STTR is a three-phase program coordinated by the SBA and announced through the PSA process. Award thresholds are $100,000 for Phase I and $500,000 for Phase II. In FY96, the STTR program distributed $64.5 million across more than 300 Phase I and Phase II awards.

1.6 Organizing your company to acquire new business

To support your company's efforts to acquire new and follow-on business, consider forming, and then actively supporting, a business development or advanced planning group.[2] In many smaller firms, marketing and proposal efforts are handled exclusively through each division or line organization. One division may or may not be aware of duplication of marketing efforts with other divisions, related contractual experience performed by another

---

2   The name of these business and planning groups varies from company to company. Some are called *business development groups* (BDGs), advanced planning, *strategic planning and business development groups* (SP&BDGs), special programs groups, and marketing departments, for example.

division, or human talent in another division, for example. The formation of a centralized corporate BDG should not preclude a given division's involvement in its own business planning and proposal development. Rather, the BDG can serve to focus, channel, and support divisional business-related activities in accordance with your company's formalized mission statement [20] and strategic plan. Because of its corporate vantage, the BDG can help identify and make available the appropriate human talent, material resources, and information resident throughout your entire company in order to pursue a business opportunity. The functional charter of the BDG can also extend to include the following closely related activities:

- Strategic, business, and marketing planning;
- Business opportunity/*Commerce Business Daily* and other opportunity tracking and reporting;
- Intelligence gathering: marketing support/client contacts;
- Formalizing the process of establishing business objectives, gathering data, analyzing data, prioritizing, and action planning;
- Acquisition team formation and guidance;
- Bid/no bid decision-making coordination;
- Teaming agreement coordination;
- Cost strategizing for proposals;
- Company information management, distribution, and archiving (including proposal data center);
- Proposal and documentation standards development and dissemination;
- Proposal coordination and production;
- Company image development and public relations;
- Corporate communications (such as newsletters, trade shows, and advertising);
- Marketing and proposal management training.

To ensure adequate connection with and visibility from senior management, a full-scale BDG should be under the leadership of a vice president (VP) for business development. The most appropriate candidate for this pivotal position is an individual with an advanced and relevant technical degree coupled with at least five years of demonstrated competitive business development experience and success in the federal government arena. Contacts alone are a necessary but insufficient gauge of a business developer's successful performance. This individual must also possess and/or develop a solid knowledge of your company's internal operations and capabilities.

Even under the constraints faced by very small companies (less than $5M in annual revenues), this VP functions most effectively when he or she is not obligated to be an "on-the-road" company marketeer as well as an internal business-development planner, organizer, and administrator. (In addition, an administrative assistant seems absolutely essential to enhance the functionality of the VP position.) Under this VP's guidance would be two primary functional groups: external and internal sales support (see Figure 1.3). The external sales support element might include a full-time corporate marketeer(s) as well as a key division manager(s). External sales efforts would involve direct client contact and interaction at multiple technical and administrative levels. It would also extend to relationship building with potential major teaming partners and specialty subcontractors appropriate to your company's *lines of business* (LOBs) or product lines.

Figure 1.3
A cooperative sales approach ensures information exchange.

The internal sales support element might be subdivided further into proposal development/production and information management, as depicted in Figure 1.4. Proposal development/production should logically include publication of the proposal documents. Proposal design and development become extremely challenging if they do not include oversight of the priorities and resources of the publications group. And finally, information management might include corporate communications, public relations, and corporate library activities. The function and focus of both the external as well as the internal sales support elements are to project a professional, client-oriented corporate image. Understanding your client's

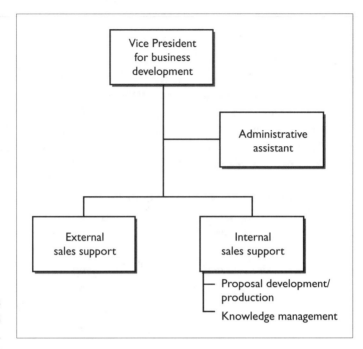

Figure 1.4
Suggested BDG
organization.

business and demonstrating that understanding in every proposal you prepare is absolutely critical to your success in the federal marketplace in 2000 and beyond (see Figure 1.5). Superior proposals should complement the public relations and advertising activities of your firm.

Effective interaction and cooperative information exchange between the external and internal sales support staff are absolutely critical to proposal success. Neither group succeeds on its own. Marketing intelligence learned from client and competitor interaction must be infused into the proposal during the important bid/no bid decision-making process, prekickoff meeting planning sessions, the kickoff meeting itself, the formal internal review steps, and post-proposal activities such as preparation for orals and *best and final offer* (BAFO). Clear, formalized, well-supported internal sales support processes contribute to winning new business. Too many times, companies view the identification of marketing opportunities as paramount to their success. As important as these leads can be, unless there is a well-defined set of processes internal to your company that harness the necessary human talent, computer resources, dedicated space, and B&P monies, the leads do not result in contract awards. They result in also-rans (proposals submitted that do not win).

It has been this author's observation that small companies tend to undergo oscillations, and even convulsions, between *centralized* business development and control, and *decentralized* divisional business activities.

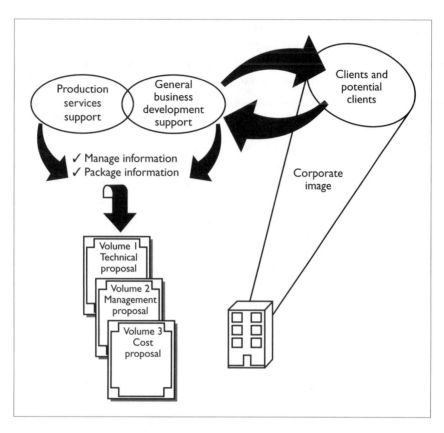

Figure 1.5
Projecting an
appropriate
corporate image.

Instead of using the BDG as a vital corporate support structure, some companies prefer to dissolve or de-scope the BDG, despite the one- to two-year extraordinary level of effort generally required to establish the BDG in the first place. Full-time corporate business development staff can be significantly more effective than employees assigned to BDG support on a part-time basis only. A certain level of centralized oversight and control is highly beneficial. Continual restructuring and refocusing of marketing activities result in the diversion of valuable time and energy into counter-productive internal organizational challenges. Building an effective and responsive business development infrastructure takes time—on the order of two or three years. Senior management must be attentive, yet patient, with the process.

1.7 Effective strategic and mission planning

Strategic and mission planning are crucial considerations to ensure a *planned pattern* of growth for your company and to focus your bid/no bid decision making. According to Thompson and Strickland, a strategic plan is "a comprehensive statement about the organization's mission and future direction, near-term and long-term performance targets, and how management intends to produce the desired results and fulfill the mission, given the organization's overall situation" [21]. Strategic planning works best (that is, is more widely accepted and more easily implemented) when it is interactive to a certain extent, not merely formulated by a separate group who creates the plan with everyone else then told to follow it. The starting point for the strategic planning process is understanding the fundamental goal of your company. No matter how small your company may be, this goal should be identified, understood, and developed. Important questions to keep in mind when formulating a strategic plan for your company are listed below [22]:

- Where is your company now in terms of market sector/line of business (that is, defining your business)?
- What is your company's mission in the marketplace?
- What is your company's distinctive (core) competence?
- Who are your company's principal clients?
- What are your principal services, products, or knowledge lines?
- What is your company's brand?
- Where does your company want to be businesswise in five years? Develop a vision of the future state of your business.
- What are your company's strengths and weaknesses (for example, business, technical, knowledge base, internal business processes, public relations, and fiscal capability)?
- What business/economic opportunities and challenges face your company now (challenges include the entry of new competitors into the marketplace, adverse government policies, and adverse demographic changes, for example) [23]?
- Who are your primary competitors? Understanding how your competition thinks and acts is critical to your success (competitor analysis) [24].
- How are your company's knowledge assets being monitored, developed, and applied?
- How do you identify the domains of knowledge that matter to the competitive strategy of your company? In effect, how do you find the seeds of innovation?

Strategic planning is not a static process; rather, it should be a dynamic, living activity. Business conditions and priorities change over time, and your strategic plan should reflect this evolution. Provide the necessary resources and steering mechanisms to implement your strategic plan, and revisit and revise the plan regularly. The mission statement is a close adjunct of the strategic plan.

Formulating, articulating, and implementing a meaningful strategic plan is critical for obtaining lines of credit through banking institutions and demonstrating to government auditing agencies that planned business growth over a period of time justifies an expansion in business base and therefore a long-term reduction in company overhead costs. Strategic planning also serves as the overarching guidance for your company's marketing activities, which is the focus of our next chapter. In addition, strategic planning provides formalized, informed guidance for making bid/no bid decisions on specific proposal opportunities. If responding to an RFP and winning that piece of business does not mesh with your company's principal lines of business and mission in the marketplace, then submitting a proposal is not appropriate. Just because your company has the technical capability to perform most of the scope of work does not mean that you should invest the time, energy, and resources into preparing a proposal.

1.8 Converting knowledge into proposal success

Dr. Karl E. Sveiby—former executive chairman and co-owner of Ekonomi+Teknik Förlag, one of Scandinavia's biggest publishing companies in the trade press and business press sector, and now an international consultant on knowledge management—defines the term "knowledge management" as the art of creating value by leveraging an organization's intangible assets, or intellectual capital [25]. "Intellectual capital is the most important source for sustainable competitive advantages in companies," suggests Professor Johan Roos of the International Institute for Management Development (IMD) in Switzerland [26]. "The knowledge of any organization is found in its people, substantiated in its business processes, products, customer interactions, and lastly its information systems" [27]. In *Working Knowledge: How Organizations Manage What They Know* (Harvard Business School Press, 1997), Tom Davenport and Laurence Prusak emphasize that "[k]nowledge is as much an act or process as an artifact or thing" [28].

Implementing business processes that actively promote knowledge valuation, development, transfer, management, and congealment will help

American companies maximize both intellectual and structural capital. Traditional command-and-control management functions must now give way to mapping and acquiring intellectual capital, communicating a clear vision for the firm, developing and implementing a business lexicon with shared meanings and nuances to foster generalized understanding of strategic goals, and facilitating the rapid assessment and multidirectional flow of knowledge throughout the organization. Caveats and pronouncements issued from senior management will not be dispersed and internalized throughout the knowledge organization or "knowledge enterprise."

Since 1991, major international corporations and organizations have proactively established executive-level positions in knowledge management. These firms include Arthur Andersen, Booz-Allen & Hamilton, Canadian Imperial Bank of Commerce (CIBC), Coopers & Lybrand, Dow Chemical, and IBM Corporation as well as Monsanto, Skandia AFS, Texas Instruments, U S West, and Xerox Corporation [29]. Other companies and organizations at the vanguard of recognizing and acting upon the reality, and the value, of intellectual capital and "intangible assets" (e.g., corporate brainpower, organizational knowledge, client relationships, innovation ability, and employee morale) include General Electric, Hughes Space & Communications, Price Waterhouse, and Philip Morris, which have launched considerable efforts to understand and enhance intellectual capital management.

In addition, Buckman Labs International, a U.S.-based biotechnology firm, has created a Knowledge Sharing Department. Buckman Labs provides incentives in the form of financial rewards and management positions to those employees who contribute to its knowledge-sharing culture. Chevron has engineered a best practice database to capture and make available the company's collective experience with drilling conditions and innovative solutions to technical problems on site. And Sweden's Celemì company published the world's first audit of intangible assets in its 1995 Annual Report. Celemì's Intangible Assets Monitor focuses on its customers (image-enhancing customers, from whom testimonials are valuable; brand names; trademarks); their internal organization (patents, computer systems, management infrastructure); and their staff (competencies, flexibility). Skandia published the first-ever annual report supplement on intellectual capital. Pfizer of Switzerland has created competence models for recruiting executives that include knowledge building and sharing as important criteria. WM-data of Sweden, a fast-growing information technology (IT) company links non-financial indicators to strategy, and considers financial ratios of little use for management. And Sweden's leading telecommunications company, Telia, has published an annual Statement of Human Resources for each of the past 7 years. This statement includes a profit-and-loss account that visualizes human resources costs and a balance sheet that shows investments in human resources [30].

During a speech before the American Accounting Association's annual meeting held in Dallas, Texas, in August 1997, Michael Sutton—chief accountant of the U.S. Securities and Exchange Commission (SEC)—noted that "historically, accounting has been strongly influenced by the reporting needs of a manufacturing-based economy" [31]. In April 1996, the SEC had convened a symposium on intangible assets in Washington, D.C., during which invited participants from prestigious business, academic, and government organizations discussed issues related to the measurement of intangible assets by preparers of financial reports, concerns about disclosures related to intangible assets, and the experience of U.S. and overseas trendsetters with regard to the accounting and disclosure of intangible assets. However, the SEC has not yet provided any guidelines or issued any directives vis-à-vis intangible assets for direct application in American corporations. Baruch Lev, professor of accounting and finance at New York University's Stern School of Business, concludes that "[n]early 40% of the market valuation of the average company is missing from its balance sheet" [32]. Clearly, your company's success in the new millennium will depend increasingly upon the quality of your people in terms of the level of knowledge, information, and know-how at their fingertips [33] as well as your management's ability to systematically foster highly adaptive and openly collaborative business processes that leverage this collective knowledge and enable intelligent decision making (see Figure 1.6).

1.8.1 Knowledge management and proposal development

Now how does this relate to proposals and proposal management? Just as knowledge management is concerned with which knowledge should be available where, when, and in which form within an organization, proposal management—at its core—is concerned with assembling, synthesizing, and packaging knowledge within a very limited time frame. With small companies in particular, leveraging the collective intellect is integral to business development achievement and superlative, long-term client support. Improving the processes and systems used to transfer and manage knowledge is an ongoing, critical function.

Proposals are, first and foremost, *knowledge products* that include a host of marketing, technical, programmatic, institutional, pricing, and certification information. Through a choreographed process of knowledge generation, transfer, and congealment, a proposal is designed to sell both technical and programmatic capabilities of a company to accomplish all

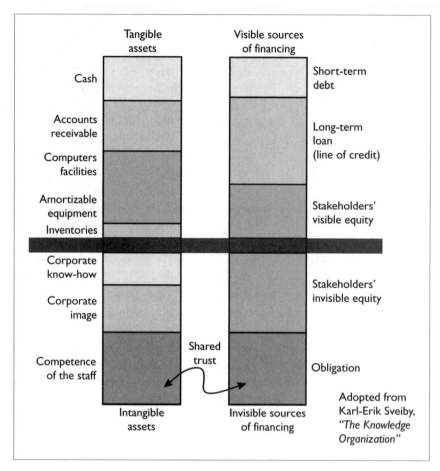

Figure 1.6
Balance sheet of
the knowledge
organization.

required activities on time and at a reasonable cost to meet client needs and requirements (see Figure 1.7). In addition, your proposal must often convey more intangible values such as "peace of mind," problem-solving acumen, fiscal stability, and so forth. Your company's proposal document(s) is scored, literally, by the client's evaluators against formalized, specific standards in a process analogous to pattern recognition. A proposal is the tangible result of knowledge-building processes, supported in turn by hard and directed work and buoyed by a positive collective attitude within your company's proposal team of knowledge workers, as Peter Drucker termed employees in the new American economy in 1988.

In actuality, résumés and project summaries reflect in one form a significant percentage of intellectual capital of an organization, small or large. Relevant knowledge resident in the employees of the company as gained through education, training, and professional experience coupled with

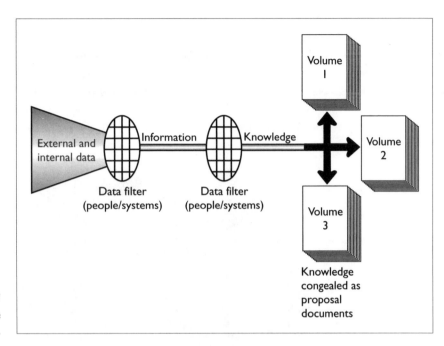

External and internal data | Information | Knowledge

Data filter (people/systems)

Data filter (people/systems)

Volume 1

Volume 2

Volume 3

Knowledge congealed as proposal documents

Figure 1.7
The knowledge
transfer process.

initiative, dedication, and innovation is highlighted in each résumé. Similarly, effective project summaries reflect the organizational knowledge of the firm from technical, managerial, and geographic perspectives.

Crucial to proposal success is the transfer of relevant marketing information, lessons learned from current and past contracts, and technically and programmatically relevant "best practices" to appropriate members of the Proposal Team. And from there, that information must be interpreted, assigned value, and translated into knowledge that will ultimately be congealed into the actual proposal documents—into narrative, theme statements, graphics, captions, compliance matrices, tabs, and so forth. If marketing and other data and information do not find their way into the knowledge product called a proposal, then a company's proposal will not receive maximal point score. Therefore, the marketing data are valueless. In effect, if intellectual capital measured in client, project, competitor, and political knowledge is not transferred to appropriate staff (knowledge workers) and converted efficiently into hardcopy or electronic proposal documents via established, articulated business processes, your company's proposal win ratios in terms of *dollars* (X dollars of contracts awarded for every $100 of contract value pursued) and *number* (X number of proposals won) will be suboptimal.

With information and knowledge becoming the "currency" of American business in the late 20th century, the organizations that can share this new currency "ecologically" in an effort to arrive at effective and appropriate solutions for both their external and internal clients in the shortest possible time and with the least amount of resources expended will be those firms that occupy positions, or *optima*, above the knowledge landscape (see Figure 1.8). From these vantage points, companies will be less likely to be displaced by new technologies, reconfigured marketing paradigms, social and demographic shifts, and political and legislative climate changes.

Figure 1.8
Strive for optima above the knowledge landscape.

1.8.2 Internal and external clients: Looking at clients in a whole new way

Let us reconsider the notion of *clients*. Within a traditional company, clients are perceived as the buyers or procurers of products or services being offered. In a knowledge-oriented company, clients include both *external elements,* such as buyers, key subcontractors, regulatory agencies, and government institutions, as well as *internal clients.* The latter might include professional, project management, and support staff; operational units; sales and marketing elements; financial and accounting resources; senior management; and so forth. The currency of knowledge must flow in meaningful, multidirectional pathways among all of these "clients." Each client is critical to the sustainability and performance (i.e., positive energetics) of the company, and each requires, and in turn provides, vital information and knowledge to the other much as enzymes are shared within a living cell to perform a host of vital maintenance, replicative, and growth functions.

END NOTES

1. See Federal Acquisition Regulation (FAR) Part 6.

2. "Do-It-Yourself Proposal Plan," TRW Space & Defense Proposal Operations, Rev. 3, February 1989, p. iii.

3. Stewart, Rodney D., and Ann L. Stewart, *Proposal Preparation*, New York: John Wiley & Sons, 1984.

4. *Proposal Preparation Manual, Vol II*, Covina, CA: Procurement Associates, Inc., 1989, pp. 1–2.

5. "Do-It-Yourself Proposal Plan," p. ii.

6. See FAR 15.609. Competitive range consists of those proposals that have a reasonable chance of being selected for contract award. Win percentages can be presented in two very distinct ways: in terms of numbers of proposals won versus total number submitted and also in terms of dollars awarded versus total potential dollars for all procurements on which your company proposed. For example, if your company has won 35% of its proposals in terms of number and only 12% in terms of dollars, that means that you are winning proposals of low dollar value.

7. 8(a) refers to Section 8(a) of the Small Business Act (15 U.S.C. 637(a)), which established a program that authorizes the SBA to enter into all types of contracts with other agencies and let subcontracts for performing those contracts to firms eligible for program participation based upon the criteria established in 13 CFR 124.101-113. The SBA's subcontractors are referred to as "8(a) contractors." (See FAR Subpart 19.8.) Eligibility in the program focuses on those groups that have historically been denied access to capital, educational resources, and markets.

    In 1995, the SBA listed 5,500 8(a) firms throughout the United States. Of those firms, 45.7% were owned by African-Americans, 22.9% were owned by Spanish-speaking people, and 20.7% were owned by Asians. The remaining companies were owned by Puerto

Ricans, native Americans, and Aleuts. Slightly more than 20% of the 8(a) companies are currently based in the Washington, D.C. area. The SBA allocated $5.2B in 1995 in set-aside contracts for minority-owned businesses. (See Munro, Neil, "Clinton Set-Aside Plan Becomes Election-Year Pawn," *Washington Technology*, 11 April 1996, pp. 5, 93; and Munro, Neil, "8(a) Program Survives Republican Attack," *Washington Technology's 8(a) and Small Business Report*, 12 Sept. 1996, p. S-6.)

8. Cranston, H. S., and Eric G. Flamholtz, "The Problems of Success," *Management Decision*, Vol. 26, Sept. 1988, p. 17.

9. Hartman, Curtis, and Steven Pearlstein, "The Joy of Working," *INC.*, Nov. 1987, p. 62.

10. O'Guin, Michael, "Competitive Intelligence and Superior Business Performance: A Strategic Benchmarking Study," *Competitive Intelligence Review*, Vol. 5, 1994, p. 8.

11. Murray, Bill, "Looking Beyond Graduation," *Washington Technology's 8(a) and Small Business Report*, 7 March 1996, p. S-4.

12. Andersen, Tania, "8(a) Companies Learn the Secret of Partnering," *Washington Technology*, 10 Oct. 1996, p. 22.

13. "Vice President Gore Praises SBA for Expanding Economic Opportunity; Congratulates Agency on 45th Birthday," U.S. Small Business Administration Press Release 98-68, 30 July 1998; http://www.sba.gov/news/.

14. "Annual Report on the State of Small Business," The White House to the Congress of the United States, 6 June 1996.

15. Kelleher, Kevin, "Feds, State Go On Line with Contracts," *San Francisco Business Times*, Vol. 9, April 1995, p. 3.

16. Fox, Harold W., "Strategic Superiorities of Small Size," *Advanced Management Journal*, Vol. 51, 1986, p. 14.

17. "Disadvantaged Firms to Get Edge in Bids for Federal Contracts," *Baltimore Sun*, 25 June 1998, p. 2A.

18. Hartman and Pearlstein, "The Joy of Working," p. 67.

19. Frank, C., "Uncle Sam Helps Small Business Get High Tech to Market," *Missouri TechNet News*, 4 November 1997; http://www.umr.edu/~tscsbdc/sbirsumm.html.

20. A mission statement is a concise, written expression of a company's long-term business, technical, and programmatic goals. It is a narrative outline of "who we are, what we do, and where we are headed." (Thompson, Arthur A., and A. J. Strickland, *Strategy Formulation and Implementation: Tasks of the General Manager*, Fourth ed., Homewood, IL and Boston, MA: BPI/IRWIN, 1989, p. 23.) Once formulated, this Statement should be communicated publicly to employees, clients, and vendors, among others. One effective mechanism is to add a "pull quote" from your company's mission statement to your marketing brochures, annual reports, company letterhead, or newsletter. Some companies have poster-sized copies of their mission statement displayed prominently in their facilities to clearly convey their identity.

21. Thompson and Strickland, *Strategy Formulation and Implementation*, p. 19.

22. Certain elements in this listing were adopted and modified from Piper, Thomas S., "A Corporate Strategic Plan for General Sciences Corporation," 1989, p. 1.

23. For additional business challenges, as well as opportunities, strengths, and weaknesses, see Thompson and Strickland, *Strategy Formulation and Implementation*, pp. 109–111.

24. Benchmarking is a rigorous process for linking competitive analysis to your company's strategy development. Benchmarking is a method that measures the performance of your "best-in-class" competitors relative to your industry's key success factors. It also is a mechanism for identifying, learning, and adapting the business processes by which the best-in-class achieve those performance levels to your company.

    "Benchmarking results should provide insight into how well we are doing and specific process changes that will improve our operations." See Leibfried, Kate H. J., and Joe Oebbecke, "Benchmarking: Gaining a New Perspective on How You Are Doing," *Enterprise Integration Services Supplement*, Oct. 1994, pp. 8–9.

25. Community KM Glossary, http://knowledgecreators.com/km/kes/glossary.htm and Karl E. Sveiby, "What is Knowledge Management," p. 2; http://www.sveiby.com....owledgeManagement.html.

    The term "intellectual capital" is attributed to economist John Kenneth Galbraith (1969). Intellectual capital includes individual talents and knowledge as well as documented knowledge that spans reports, research papers, patents, software source code, books, articles, and manuscripts. See also Touraj Nasseri, "Knowledge Leverage: The Ultimate Advantage," *Kognos: The E-Journal of Knowledge Issues* (Summer 1996); http://www.magi.com/~godbout/Kbase/kognos11.htm.

26. Johan Roos, "Intellectual Performance: Exploring an Intellectual Capital System in Small Companies," 30 October 1996; http://www.imd.ch/fac/roos/paper_lr.html.

27. Brad Hoyt, "What Is KM?," *Knowledge Management News,* 1998; http://www.kmnews.com/Editorial/whatmk.htm.

28. Tom Davenport and Larry Prusak, "Know What You Know," http://www.brint.com/km/davenbport/cio/know.htm.
    Tom Davenport is a professor and director of the Information Management Program at the University of Texas at Austin. Larry Prusak is managing principal of the IBM Consulting Group in Boston.

29. Michel Grundstein, "Companies & Executives In Knowledge Management," http://www.brint.com/km/cko.htm.

30. Information in this paragraph was drawn from Karl E. Svieby, "What Is Knowledge Management?," pp. 5–6; http://www.sveiby.com....owledgeManagement.html.
    See also Karl E. Sveiby, "Celemì's Intangible Assets Monitor," http://www.sveiby.com....Ass/CelemiMonitor.html. Celemì employs metrics such as efficiency (sales per customer, staff turnover), stability (repeat orders, growth in sales per administrative staff), and growth/ renewal (average years of professional competence). Celemì is headquartered in Malmö, Sweden, and is dedicated to creating processes that help companies leverage the power of learning.

31. Michael H. Sutton, "Dangerous Ideas: A Sequel," p. 3; http://www.sec.gov/news/speeches/spch175.txt. (Remarks delivered during the American Accounting Association 1997 Annual Meeting in Dallas, Texas, on 18 August 1997.)

32. http://www.martech.co.nz/kbl.html#intro.
    Professor Lev notes, "How ironic that accounting is the last vestige of those who believe that things are assets and that ideas are expendable." See Baruch Lev, "Accounting Needs New Standards for Capitalizing Intangibles," *ASAP: Forbes Supplement on the Information Age,* 7 April 1997; http://www.forbes.com/asap/97/ 0407/034.htm.

33. Uri Merry, "Postings #1–7 in Complexity & Management," Posting # 4, "Knowledge Landscapes," http://pw2.netcom.com/~nmerry/post1.htm#post4.

Chapter 2

Strategic partnering and subcontracting opportunities

R ALPH PETERSON, president and CEO of CH2M Hill Companies, Ltd., a 7,000-person, 50-year-old firm that provides professional services worldwide in water, environmental, transportation, and industrial facilities, has remarked that "[p]artnerships and alliances are being forged that just a short time ago could not have been imagined" [1]. Strategic alliances can contribute synergy that results in sustained competitive advantage in a rapidly changing global business environment. *The stakes are high and the opportunities are real.* Many major corporations now maintain their own internal procurement and small business centers or agencies. For example, the Bethesda, Maryland-based space and aeronautics megacorporation, Lockheed Martin, staffs five internal procurement agencies nationwide. "Twenty-two to 28 percent of the company's contracting dollars go to small, disadvantaged and women-owned businesses." And computer manufacturer Gateway, Inc., based in South

Dakota, awarded several million dollars worth of contracts to more than 1,000 small businesses in 1997 [2].

Hamilton Standard, a division of United Technologies in Connecticut; The National Renewable Energy Laboratory (NREL) in Golden, Colorado; and the Jet Propulsion Laboratory (JPL) in Pasadena, California, are all recent winners of the Dwight D. Eisenhower Award for Excellence from the U.S. Small Business Administration. The Eisenhower Award is a national program for large prime contractors in the manufacturing, service, research and development (R&D), and construction arenas that have excelled in their use of small businesses as subcontractors and suppliers. Divisions of Lockheed Martin in Orlando, Florida, and Cherry Hill, New Jersey, were among the winners of the Eisenhower Award in the manufacturing and service categories announced in June 1998 [3].

2.1 Subcontracting opportunities and pathways to success

Evidence of new directions in subcontracting is seen in the February 1998 public/private partnership among the SBA, General Motors Corp., Ford Motor Co., and Chrysler Corp. that is designed to improve opportunities for minority-owned small businesses in a major industry. This landmark memorandum of understanding (MOU) marks the first time that a private-sector industrial group of this magnitude and importance has engaged in partnership with the SBA. The MOU represents a sharp increase in the contract dollars flowing to small disadvantaged businesses (SDBs). Subcontracting awards of $3 billion during the next 3 years are anticipated [4].

In general, proactive small business entrepreneurs are significantly leveraging their limited marketing, financial, and technical resources as well as augmenting their revenue stream by attracting and forming strategic partnerships, as subcontractors, with large prime contracting companies to pursue new business opportunities and expand work with existing clients. Many procurements let by the U.S. federal government as well as by the public sector at the state and local levels stipulate small, small disadvantaged, minority-owned, and/or women-owned business target thresholds for the resulting contracts. For example, a procurement related to the Everglades in 1998 by the state of Florida required that a minimum of 16% of the contract revenue be allocated to minority/woman business enterprises (M/WBE).

Critical to a small firm's subcontracting success are the following elements: (1) superlative contractual performance on existing or past projects (being the best-in-class); (2) flexible, audited, and approved pricing strategies and structures; (3) the allocation of top-flight professional staff to proposal teams with the prime contractor; (4) fiscal strength as measured by such parameters as operating profit, net income, positive cash flow, operating history, and future assumptions; (5) well-defined and articulated core competencies and product lines; and (6) fair, equitable, and ethical teaming agreements that are honored over the long term. Strategic alliances or partnerships must be managed as enduring business relationships—with mutually compatible objectives, shared risks and resources, real-time knowledge transfer, and trust.

Understand that the prime contractor will have to *sell* your small company's human talent and knowledge base, contractual experience, and fiscal solvency in their proposal to a given public- or private-sector client. Make that task extremely easy for them by having a great story to tell, and make that information available to them in a timely manner during the proposal response life cycle. Too many small businesses submit poorly written, incomplete materials to the prime contractor for integration into the prime's proposal. This inappropriate practice adds to the prime contractor's Bid & Proposal (B&P) costs, and detracts from your working relationship with that prime in the future. You may need to contact freelance or temporary staff to prepare required proposal modules and other materials in an electronic form compatible with the prime contractor's publishing systems. Above all, adhere to the proposal schedule as developed by the prime contractor. These aspects of working with prime contractors are reviewed in Chapter 7.

Whenever possible, small businesses should strive to forge and negotiate exclusive teaming agreements with prime contractors which are active and successful within your company's major lines of business (LOBs). Cultivate successful experience within the prime's and your target client's technical areas of interest. Understand how to conduct contractual business with the prime and the ultimate client. Demonstrate your company's ability to staff the project with stellar professional staff (through incumbent capture, for example) who are committed to the life of the project and not merely senior staff whose résumés help to win the job, but who leave the project after a minimal amount of time. Build camaraderie in the relationship, and establish a strategic partnership for the long term.

2.2 Specific strategies for achieving subcontracts

The following strategies can help you achieve subcontracts with prime contractors:

- *Register your small business with the SBA's new Internet tool called Pro-Net.* Pro-*Net* has taken the place of the PASS, or Procurement Automated Source System, database, which is now defunct. As of August 1998, in excess of 177,000 small, disadvantaged, and women-owned businesses—including the more than 6,000 firms certified under the SBA's 8(a) business development program— were listed on Pro-*Net*. Small businesses with Internet home pages can hyperlink their Web site to a Pro-*Net* profile of their firm. Business profiles of the Pro-*Net* system can be searched by such user-friendly and meaningful data parameters as SIC codes, keywords, geographic location, business type, ownership race and gender, and electronic data interchange (EDI) capability. The profiles are structured like executive business summaries, and are kept up-to-date by the individual small business through password-controlled access.

  Pro-*Net* gives small companies increased visibility and exposure to government contracting officers in charge of the 200 billion dollar-per-year federal market. (In total, the federal government will spend more than $1.7 trillion in 1999, which includes funding for Social Security, Medicare/Medicaid, means-tested entitlement programs, and interest payments.) Pro-*Net* can be found on the SBA's Web site at http://pro-net.sba.gov. This electronic gateway is a *free service* to federal and state government agencies as well as prime and other contractors seeking small business contractors, subcontractors, and partnership opportunities. It also provides easy access to the electronic version of the *Commerce Business Daily* as well as government agency home pages on the Internet.

- *Register and prequalify your small business with the SADBUS, or Small and Disadvantaged Business Utilization Specialist, in specific federal agencies.* At the state level, the analogous office may be called the "Office of Supplier Diversity and Outreach." And many large corporations now have small business centers, staffed with Small Business Liaison Officers (SBLOs). Every federal contracting agency and prime contractor doing substantial business with the federal government is obliged to designate an SBLO or SADBUS within the agency or firm. This person is the first point of contact for your small business wishing to make your products and services known to the contracting entity.

- *Create, maintain, and register an easily accessible Web site for your small company.* Prime contractors often perform Internet searches to locate potential teaming partners. Registration refers to listing your site with various Internet-based search engines, such as Submit It!, Add Me!, Multi-Submit, Add It!, The Postmaster, and Submit All.

- *Interact with Small Business Development Centers (SBDCs) in your state.* The 900+ SBDCs nationwide are educational and research resources for small businesses. The Association of Small Business Development Centers (ASBDC) links each SBDC program into a national network.

- *As fiscally appropriate for your small business, employ reputable consulting firms to open the door to strategic partnership discussions.* Reputable consulting firms include INPUT, Inc., and Federal Sources, Inc. (http://www.fedsources.com), both located in Virginia.

- *List your company in the TRY US National Minority Business Directory.* Maintained by TRY US Resources, Inc., in Minneapolis, Minnesota, the 1998 directory lists more than 7,000 minority suppliers with regional or national sales capabilities. Available in hardcopy and electronic form, the directory can be searched by such parameters as keyword, SIC code, minority type, and commodity/service. You might also consult the *Purchasing People in Major Corporations* directory, which lists more than 750 corporate purchasers by name, title, address, and telephone. You can contact TRY US Resources at http://www.tryusdir.com or by calling 612-781-6819.

- *Attend and participate in appropriate conferences and trade shows for your industry.* For example, the U.S. Department of Defense Office of Small and Disadvantaged Business Utilization (OSADBU) is conducting a series of conferences on the industry THRUST program. The DoD program focuses on the following industries: environmental, manufacturing, health care, telecommunications, and management information including simulation. Each conference features a specific THRUST industry and is hosted by a designated military department or agency. Consider also the series of SBA-sponsored national satellite conferences designed to help women entrepreneurs develop winning small business strategies. The conferences provide information on leading-edge technology, doing business on the Internet, protecting your business from the Year 2000 computer issue, and financial resources available via the Internet.

The New England Small Business Outreach Program (Nashua, New Hampshire; Tel. 1-800-861-5037) acts as an advocate for firms in the northeast United States with regard to NASA business opportunities. Each year, the Outreach Center coordinates a NASA conference for small businesses where NASA Center representatives and prime contractors are present.

During the first week of March annually, NASA and the Jet Propulsion Laboratory (JPL) in California host a national high-tech conference for all small businesses. The conference is well-attended by NASA Centers, major prime contractors, and other federal government agencies. Each year, NASA convenes the Space Science Symposium for Small Businesses in Washington, D.C. Small businesses are provided the opportunity to meet senior executive-level NASA Associate Administrators and managers as well as major prime contractors. And the NASA Conference of SDB Trade Associations was established to inform SDB trade associations about the latest NASA initiatives to increase the agency's utilization of minority and women-owned businesses.

- *Secure your company's membership and participate in trade and business associations.* This includes interaction with such associations as the NAPM, the National Association of Purchasing Management. With more than 44,000 members, NAPM is a not-for-profit association based in Tempe, Arizona, that provides national and international leadership in purchasing and supply management research and education. NAPM meetings offer forums for developing important business relationships with procurement representatives of large corporations. NAPM can be contacted on the Web at http://www.napm.org or by calling 1-800-888-6276.

- *Consider listing your company in the Thomas Register of American Manufacturers.* Major corporations and government agencies often consult this resource for suppliers that operate nationally. Visit www.thomasregister.com or call 1-800-699-9822, ext. 444.

- *Visit Internet sites to learn more about potential prime contractors.* For example, http://www.hoovers.com provides detailed business and financial information about major corporations.

2.3 Becoming part of a government-wide acquisition contract (GWAC) team

Government-wide acquisition contracts (GWACs) are multiple award task order contractual vehicles for various resources and services hosted, or "owned," by one agency of the federal government, but which other specified federal agencies can use as well. GWACs were given statutory weight through the Federal Acquisition Streaming Act (FASA) of 1994. There is a limitation on the total percentage of the GWAC that one agency can use, and each GWAC has its own specific terms and conditions. Despite a small administrative and contracting fee levied by the host federal agency, the direct benefit is that other government agencies do not need to incur major acquisition and procurement costs when using the GWAC vehicle. The General Services Administration (GSA) strongly supports this type of contracting. As part of a GWAC team, your small business can proactively sell the fiscal and efficiency benefits of GWACs to a broad group of constituencies, including government agency chief information officers (CIOs), as well as establish government-funded venture capital programs to support R&D within the IT sector.

The Department of Energy's (DOE) Telecommunications Integrator Service (TELIS) telecommunications contract is an example of a GWAC vehicle open to all federal agencies including DOE headquarters, DOE field offices, and DOE contractors. The TELIS Team includes 8(a) small businesses such as Native American Sales, Inc., headquartered in Englewood, Colorado. Other active GWACs include the Defense Information Systems Agency (DISA) DEIS-II systems integration vehicle and the Department of Transportation's (DOT) Information Technology Omnibus Procurement (ITOP) contract, which provides information systems engineering, systems and facilities management and maintenance, and information systems security support services. As one of the most successful GWACs, DOT's ITOP program has been saluted for its stellar contribution to the National Performance Review (NPR, now the National Partnership for Reinventing Government) in making the federal government a service-oriented, cost-conscious source of value for IT customers. The ITOP customer base extended throughout the DOT, DoD, Department of Justice, DOE, Department of Commerce, U.S. Department of Agriculture (USDA), and the U.S. Small Business Administration. And in mid-1998, DOT let the ITOP-II RFP, a performance-based omnibus contract with both a small business and full-and-open competition component.

Other active GWACs include the Department of the Navy's New Technologies for Office and Portable Systems (NTOPS) office automation GWAC; the Department of the Air Force's Desktop (DTOP)-V microcomputer, operating system, and application software GWAC; and the National Institutes of Health (NIH) Chief Information Officer-Solutions and Partners (CIO-SP) GWAC. Additionally, there is NASA Goddard Space Flight Center's (NASA/GSFC) Scientific and Engineering Workstation Procurement (SWEP) II GWAC. Currently used by 24 federal agencies, NASA's SWEP has done $40 million worth of business using electronic commerce (EC). Astrox, Inc., of Rockville, Maryland, an 8(a) minority-owned small business, stands among the SEWP team of contractors [5].

2.4 Streamlined delegation of authority process

In mid-1998, SBA Administrator Aida Alvarez signed delegation of authority agreements with 25 federal agencies. In the past, SBA functioned as a "middleman" when federal agencies contracted with 8(a) firms. Now, following a successful pilot project between the Department of Transportation (DOT) and SBA, these 25 agencies work more closely and directly with small businesses. The greatest benefit is accelerated approval time. In FY97, the DOT obligated $28 million using this streamlined process. Approval time for the sole source awards was reduced from several weeks to 5 days.

Federal agencies that have signed delegation of authority agreements with the SBA include USDA, DOC, DoD, DOE, DHHS, the Department of Housing and Urban Development (HUD), DOI, DOJ, DOL, and the Department of State. In addition, the Treasury Department, VA, EPA, GSA, the Federal Emergency Management Agency (FEMA), NASA, NRC, and the Social Security Administration are among the other signatories to these agreements [6].

In 1998, the SBA released a plan that affects the pool of contracts for services ranging from $25,000 to $100,000. This plan offers federal con tracting officers the chance to trim weeks off normal contracting procedures. Agencies are currently required to publish their contracting opportunities in the *Commerce Business Daily*, wait 30 days to receive bids, and then evaluate all bids. Basic contract awards typically take a minimum of 45 days. More detailed contracts can take much longer. Under this new plan, the SBA will grant a waiver that permits use of a simplified process if a contracting officer solicits bid information from five small businesses— including one woman-owned firm and one small disadvantaged business. These solicitations will be done quickly, cutting the time it takes to identify and award the contract to just a few days. The new process should save time and money in the contracting process. The SBA is also encouraging contracting officers to use the agency's Pro-*Net* database [7].

2.5 How mentor–protégé programs can help your business

Created by Congress in 1991, the DoD Pilot Mentor–Protégé Program (PMPP) encourages major DoD prime contractors (mentors) to develop the technical and business capabilities of small disadvantaged businesses and other eligible protégés. Through credit toward subcontracting goals or some direct reimbursement of costs, the MPP provides tangible incentives for the major prime contractors to establish and implement a developmental assistance plan designed to enable the small business to compete more successfully for DoD prime contract and subcontract awards. Mentors represented in the MPP encompass a broad range of industries: environmental remediation, manufacturing, telecommunications, and health care. Protégés receive assistance in marketing and business planning, management, training, needs assessment, bid and proposal preparation, human resources practices, quality assurance (including ISO 9000 certification), environmental protection, emerging technologies, and health and safety procedures. NASA's Office of Small and Disadvantaged Business Utilization in Washington, D.C., coordinates a similar mentor–protégé program. San Diego-based Science Applications International Corporation (SAIC), under its Atmospheric Science Research and Technology Support Services contract with NASA Langley Research Center in Virginia, was awarded the first Goldin-Stokes Mentor–Protégé Award for its relationship with Analytical Services & Materials of Hampton, Virginia [8]. And companies such as TRW participate in both the DoD and NASA programs.

Recent data suggest that small firms participating in the mentor–protégé programs have experienced significant growth. Specifically, among 226 agreements reviewed there was a net gain of 3,342 jobs within protégé firms; there was a net revenue gain in excess of $276 million within the protégé firms; and mentors reported an additional $695 million in subcontract awards to small disadvantaged business firms. Mentor firms reported value accrued to them as the direct result of developing a technically qualified and more competitively priced supplier base for DoD requirements. As mentor firms restructured and right-sized, they often formed strategic alliances with protégé firms for the specific purpose of outsourcing functions previously performed in-house. Similarly, the DoD gained by having an increased number of cost-effective, technically qualified small business sources for defense prime contracting as well as subcontracting requirements. To obtain an information package and sample agreement for DoD's mentor–protégé program, call 1-800-553-1858 or review the Web site at http://www.acq.osd.mil/sadbu/mentor_protege/.

Your reasons to enter and participate in the mentor–protégé programs should extend far beyond subcontracting dollars alone. Your company should be willing to invest its nonreimbursable funds and your employees' billable time for training in order to develop a long-term teaming relationship premised upon mutual trust, respect, communication, and flexibility. The process of being selected by a mentor corporation can take up to 1 year. The mentor–protégé programs present opportunities for realizing diverse technical capabilities, efficient contract management, exposure to new technologies and business practices, expanded work potential, and exposure to new federal clients. The CEO of one protégé firm indicated that "We measured success by what we learned rather than what was earned."

Minco Technology Labs, Precise Hydraulics and Machining Center, TN&A Associates, Kuchera Defense Systems, SUMMA Technology, Omega Environmental Systems, Fuentez Systems Concepts, and Vista Technologies are several of the 174 small disadvantaged businesses that have participated in the mentor–protégé programs. The 102 mentors span corporate giants such as SAIC, Boeing, Raytheon, Lockheed Martin, Texas Instruments Defense Systems & Electronics, IT Corporation, Bell Helicopter Textron, Earth Tech, and Hughes Aircraft as well as Booz-Allen & Hamilton, McDonnell Douglas, and Northrup Grumman.

END NOTES

1. Ralph R. Peterson, "Plotting a Safe Passage to the Millennium," http://bstconsulta.../whatshot/ch2masce.htm.

2. Charlotte Mulhern, "Round 'Em Up," *Entrepreneur*, August 1998, p. 120.

3. "Lockheed Martin Division, Nova Group Earn SBA's Top Awards for Subcontracting," SBA Number 98-43, 4 June 1998; http://www.sba.gov/news/current/98-43.html.

4. "Vice President and SBA Administrator Announce Pact with Big Three Automakers," SBA Number 98-09, 19 February 1998; http://www.sba.gov/news/current/98-09.html.

    Public–private partnership initiatives of value and interest to small businesses also include the comprehensive joint effort of IBM, the U.S.

Chamber of Commerce, and the SBA called the "Small Office Solutions" or "SOS." Announced in 1998, this nationwide initiative is designed to promote new opportunities for small business enterprises by educating them about the benefits of information technology. Expert guidance will be delivered free of charge to small businesses through the SBA's Business Information Centers (BICs) in Albany, New York, El Paso, Texas, and Spokane, Washington; the U.S. Chamber of Commerce and local chambers; and IBM. Small businesses seeking support on technology issues can contact IBM at http://www.ibm.com/businesscenter or by calling 1-800-426-5800. See "IBM, U.S. Chamber of Commerce Announce Results of New Study on U.S Small Business and Technology," http://www.uschamber.org/NEWS/sb980601a.htm.

5. Doug Hanson, "Electronic Commerce and NASA's Scientific and Engineering Workstation Procurement (SEWP)," http://www.sewp.nasa.gov/edidoc/hanson9711.html.

6. "SBA Streamlines 8(a) Contracting," SBA Number 98-24, 6 May 1998; http://www.sba.gov/news/current/98-24.html.

7. "Vice President Announces Plan to Help Women-Owned Firms Win Federal Contracts," SBA Number 98-67, 31 July 1998; http://www.sba.gov/news/current/98-67.html.

8. "NASA Mentor–Protégé Program Award Presented to SAIC," *SAIC Magazine*, Fall 1996; http://www.saic.com/publications/.

   The award is named in honor of Congressman Louis Stokes and NASA Administrator Daniel Goldin.

Chapter 3

Marketing to and with your clients

[G]et in the path of their judgment…
to make your product part of an already existing code.
—*Adcult USA* (1997)

3.1 More than just selling

Demonstrate how you can help meet your client's requirements, provide solutions, and minimize associated technical, contractual, and fiscal risks.

Marketing involves far more than just short-term or one-time *selling* in response to immediate issues. It is a long-term process of dedicated commitment to and cultivation of your client's success as well as your specific business and product lines as defined in your strategic plan. Your client wants to know that you'll be around next year, and five years down the road, to offer him high-quality products and superior services. Marketing means learning about your client's technical requirements, contractual and fiscal constraints, and programmatic as well as regulatory concerns. It involves developing an in-depth understanding of your client's business strategy and culture, organizational structure and dynamics, and procurement decision-making processes and buying influences. You will want to help your client appreciate the value of not just buying a product or service, but rather *buying a solution* [1].

Functionally, marketing is about understanding your client, your competition, your potential teaming partners, and the specific technical and programmatic benefits that your company can provide (see Figure 3.1).

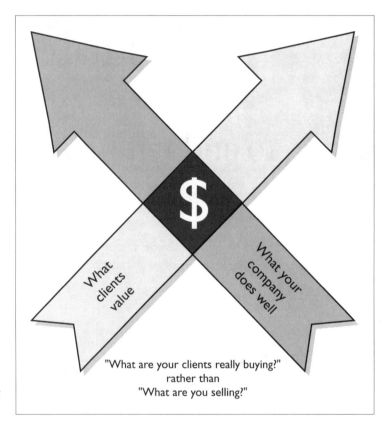

Figure 3.1
The point of intersection.

What clients value

What your company does well

"What are your clients really buying?"
rather than
"What are you selling?"

- What other companies have provided goods and services to your client?

- How are those companies perceived by your client?

- Because potential teaming partners are part of the knowledge equation as well, does your client favor the formation of teams? How are your potential teaming partners perceived by your client?

- How can your company contribute in positive ways to your client's bottom line (private sector), or program budget (federal and state), or hard currency deficit (international clients)?

Fundamentally, marketing means "championing the customer" [2]. In the case of federal procurements, your company must determine:

- That the particular target program is real and will be funded;
- That the RFP will be issued;
- That political conditions within the specific government agency will allow for genuine competition in the case of a recompetition.

Learn which civil servants will serve on the evaluation or selection board for your procurement. Does the client expect to get the most cost-effective solution? In effect, is *price* the real driver of the acquisition? Or is *best value* important to your client [3]?

Over time and in a variety of ways, inform the client about your company's distinctive human talents, technical capabilities and knowledge base, past and present contractual experience, and financial stability in ways that demonstrate how you can help meet his or her requirements and minimize associated technical, contractual, and fiscal risks. Be sure to meet your client's emotional or intangible needs as well—help them feel secure knowing you are looking out for them. Provide them with peace of mind [4].

If your potential client does not know your firm, he or she is less likely to buy from you. You must differentiate your company from every other company in a positive, credible way. Marketing is the process of persistently helping your client understand and believe that your "brand" of product or service will contribute substantively to his or her success. Build your differentiated suite of solutions into the decision pathway of your client's infrastructure and buying influences.

3.2 Transactions are personal

Interaction with your client should be conducted in a concerted manner by a variety of levels within your company—from marketing and senior management staff to mid-level technical personnel and project managers. Client visits by your marketing staff are necessary to open doors and establish new relationships with a potential client. These visits should be ongoing and augmented by additional information-sharing visits by members of your senior management as well as appropriate, select staff from among your technical and programmatic ranks. For your company's president to invest energy and time in focused direct client contact is essential to demonstrate and authenticate your corporate commitment to a given program or project. The only way to develop a robust, contextual understanding of your client and his mission, overall program goals, funding processes, and technical concerns is to personally interact with him and his staff as well as cognizant contracting, procurement, and management decision-making personnel.

Given the *decentralized* marketing activities of many firms—both small and large, technical and line managers are often in a position to make their own decisions about pursuing marketing opportunities. In my experience, this has resulted frequently in the pursuit of marginal targets that may have been appropriate from a technical standpoint but were not considered thoroughly from programmatic, cost, or strategic planning perspectives. In fact, many times no marketing has taken place—the announcement in the *Commerce Business Daily* (*CBD*), on the Internet, or a governmental agency *bulletin board system* (BBS) is the first time that the project has been identified and considered for a bid/no bid decision. If your company does not market the client, you cannot expect to win contracts. There are, at minimum, several firms that qualify technically for most any government opportunity. Unless you can set your company apart in a positive way from your competition, you will not acquire new business.

Paul Lombardi, chief operating officer of DynCorp, a Reston, Virginia-based services company, has noted that "[u]ltimately, the government is buying people. The transactions are personal" [5]. A sales representative for a mid-Atlantic analytical chemistry laboratory remarked to me that a proposal that he had submitted in 1995 received extremely high marks from the prime contractor and the ultimate government client. Why? Because this rep had marketed the client and was able to answer all of his relevant questions and concerns in the context of responding to the RFP. And in competitions between experienced contracting firms, relationships assume an even larger role in the final decision-making process. Plan to attend association gatherings, meeting and talking with your clients at conferences, and going out to lunch (with each party paying for their own check), for example.

3.3 Listen to your client

First and foremost—listen to your client. And listen to understand, not to reply [6]. Take every opportunity to learn about his or her requirements, budget concerns, technical problems, staffing goals, scheduling issues, experiences with your competitors, and professional and personal likes and dislikes. Too many times, a company attempts to force-fit its clients into the company's own technical or programmatic compartments. Give your prospective client a chance to participate in your presentations. Get your technical people in front of your client to have meaningful discussions on a variety of important technical, programmatic, and financial issues well in advance of the release of the RFP or *CBD* SF254/255 synopsis. Your company should be marketing an existing federal opportunity for the life of the current contract. Introduce and reinforce innovative, solution-oriented technical and management approaches that will appear later in your own proposal. Indeed, a compelling case can be made that the *sale* is made pre-RFP and that the proposal only closes the deal. A reasonably well-constructed proposal will not make up for inferior technical homework, poor management plans, and/or high costs [7]. And inferior marketing intelligence about your client, teaming partners, and competitors will most certainly result in a proposal loss.

3.4 Infuse the marketing intelligence into your proposal

To be useful at all, the marketing information that your company collects must be analyzed, distributed, and archived so that it can be retrieved and updated easily. There is no point in marketing your client if you do not apply what has been learned into designing and producing your proposal. All information gathered that is relevant to a given procurement should be provided to the specific proposal manager. Appropriate marketing information must shape and also be fitted directly into your proposal at several critical stages, as illustrated in Figure 3.2. During the proposal response lifecycle, there is a minimum of three critical inflection points at which marketing information should be infused into the process. These include: (1) pre-kickoff activities, (2) the formal kickoff meeting, and (3) the formal review processes such as the Red Team review. These topics are discussed in Chapters 9 and 7, respectively. Current marketing information should shape post-proposal activities as well. Procurement-specific, client-specific, and competition-specific information collected and analyzed by your company's external sales staff must be shared (as appropriate from a competition sensitivity and need-to-know standpoint) with the proposal manager and proposal team/internal sales staff at these times. During the pre-kickoff phase, accurate, current marketing intelligence helps drive (in addition to the RFP requirements found in Sections L, M, C, H, and others) the architecture of the proposal outline, the thematic structure of the

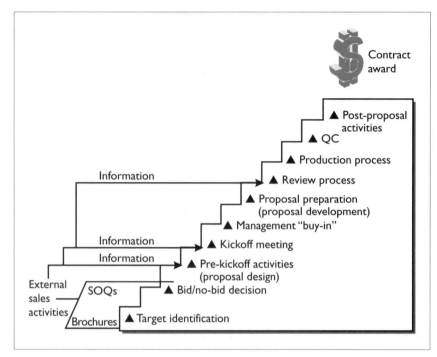

Figure 3.2 Marketing information is critical to proposal success.

proposal narrative, the project management approach, and the costing and staffing strategies. Without this marketing intelligence, your company would be like any other firm that was responding to the RFP only, with no further guidance from the client than the words on the pages (or bytes in the electronic file) of the RFP.

At the formal kickoff meeting, in-depth marketing intelligence as shared by a member of your company's external sales group will assist members of the proposal team—particularly those charged with writing significant sections of the proposal narrative—with a reasonably full-life "big picture" of the procurement opportunity, the client's sentiments and concerns and "hot buttons," and the competition.

Again, during the formal review cycles, marketing intelligence should be used as a benchmark against which to measure the thematic success of the proposal documents at that stage in the response lifecycle. In the terminology of remote sensing, marketing should supply the "ground truth" that authenticates your company's proposal for the particular client and project at hand.

As depicted in Figure 3.2, the flow of information and support moves in both directions between the internal sales staff to the external sales staff. For example, the internal sales staff will assist their external sales counterparts in the development of effective marketing brochures and *statements of qualification* (SOQs) that are valuable tools in helping your client to better understand your company's capabilities and talents.

3.5 Intelligence gathering and analysis techniques

One potentially effective technique of intelligence gathering and analysis is for your company's external sales support staff and division managers to visit the contracting offices of your primary federal client agencies in order to obtain specific information on existing contracts. Learn the contract title and numbers of existing contracts, the incumbent contractor(s), the projected contract renewal date, and the nature of the work, as well as the staffing size, contract dollar amount, contract duration, and contract type. Ask for the names and telephone numbers of the *contracting officer's technical representative* (COTR) and *contracting officer* (CO) for each contract. Supplemental information can be gathered by contacting specific federal agencies and requesting copies of their acquisition forecast lists. These lists are called different names by various agencies, for example, procurement forecast, active contracts list, advanced procurement plan, and active contracts register-contractor sequence.

Many federal agencies now maintain their acquisition forecasts online on the Internet. For example, the USEPA provides its procurement forecasts at http://www.epa.gov/oam/main/forecast/; the Defense Logistics Agency (DLA) posts its upcoming procurements two times each year at http://www.dscc.dla.mil/programs/acqforecast/fore_3.html; and the NASA Acquisition Internet Service (NAIS) site at http://ec.msfc.gov/hq/forecast/index.html provides NASA Center-specific procurement projections.

When you obtain this marketing intelligence from in-person contacts, agency procurement briefings, and budget documents; Internet postings; and published sources, record it carefully in a table, such as the one presented in Table 3.1. Someone within your organization must be responsible for maintaining the tracking tables and keeping them secure. It is critical that the information is accurate, current, and relevant. Try to collect information on both short- and long-term opportunities. One to two years is not too long a lead time for advanced business planning (see Figure 3.3). Once client target tracking tables are compiled for a given federal agency, preliminary bid/no bid decisions, proposal resource allocation, and revenue stream projections can be made in a planned and rational manner. This initial downselecting process, if done in a structured way, will allow your company to focus scarce resources in the direction of viable targets

#	Project Title	Acquisition Manager	Incumbent(s)	Contract #	Anticipated RFP/CBD Synopsis Release Date	Duration of Contract	Prime /Sub	Estimated Level of Effort (LOE) Per Year	Potential Contract Dollar Value
1	A-E Services to *perform Hazardous Toxic & Radioactive Waste (HTRW) &* environmental compliance services USACE-Mobile District	Ms. Jennifer Brothers	AABC Corp.	DACA01-99-R-0115	7/99	1 base year + 4 option years	Prime	35 *full-time* equivalents (FTEs) per year	
2	Investigation and design for remediation of misc. HTRW sites USACE-Omaha District	Mr. Tim Grayson	AAAC Corp.	DACA45-99-R-0011	10/99	Indefinite delivery	Sub	8 FTEs	
3	Environmental contract for Ft. George Meade, Maryland	Mr. Greg Yost	AACB Corp.	DACA31-99-R-0039	4/99	Indefinite delivery	Sub	5 FTEs	
4									
5									

Table 3.1 Sample Client Tracking Table (U.S. Army Corps of Engineers)

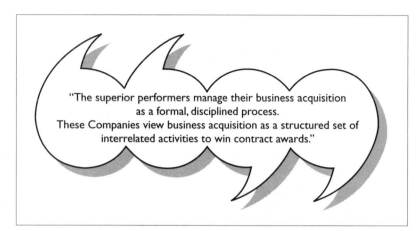

Figure 3.3
Formalized
business
acquisition
processes produce
tangible results.

"The superior performers manage their business acquisition as a formal, disciplined process.
These Companies view business acquisition as a structured set of interrelated activities to win contract awards."

and prevent your internal sales group and publications staff from becoming totally overburdened. There is nothing more debilitating to morale company-wide than to produce losing proposal after losing proposal because of too few resources chasing too many targets. Be selective, and then produce a stellar proposal every time you submit one to your client.

There are automated tools available to support business opportunity tracking and decision making. The WinAward marketing management system from Bayesian Systems, Inc. (http://bayes.com/mainpage.html), is one example of a Windows-based, robust software product. WinAward retains and manages data characterizing specific marketing leads, retains costs associated with the bid process, determines award probabilities, and supports a rigorous bid decision process.

Once a preliminary downselecting procedure has been completed, capture plans and associated call plans (direct contact plans) can then be generated to pursue the qualified list of targets. The type of information your company will want to collect in its capture plans includes:

- Project name (according to government sources, that is, use the client's name for this project);
- Client name (to the level of the line organization) and organizational chart;
- Client mission statement and recent developments in the client environment;
- Incumbent contractor (if applicable);
- Potential competitors;
- Specific services or products being procured;
- Key evaluation factors for award/client success criteria;
- Anticipated RFP release date;

- Anticipated contract award date;
- Anticipated contract dollar value to your company;
- Source evaluation board membership;
- Relevant contractual experience (for your company);
- Proposed project manager (for your company);
- Other proposed key staff (from your company);
- Potential teaming partners.

Take the time to know your teaming partners, their technical expertise, human talent, cost strategies, contractual performance, financial stability, business processes, contract-performance liabilities, and pending litigation. Ask your client what they think of your prospective teaming partner. Consider the benefits and disadvantages of exclusive versus nonexclusive teaming arrangements. (See Chapter 18 for more discussion regarding teaming agreements.)

3.6 Call plans

A call plan should be designed to clearly indicate whom within your company will be visiting the client, as well as when, why, and at what level of the client organization. Information gathered from direct client contact should include the following.

Government or commercial buying organization and key personnel.
Obtain the official organizational charts for the client organization, and learn the full names and functions of the important technical and contractual decision makers.

Currently, what is the client doing technically and programmatically?
Is the client planning to migrate their automated process from one computer platform to another? Have regulatory changes imposed new demands on the client's programs? Will "service-to-the-customer"-style government necessitate a telecommunications center capable of handling significantly more queries per unit time? How will the impending century change (Y2K) impact your client's accounting systems? *Bottom line:* Learn the technical and managerial issues that are important to your client now and over the life of the upcoming contract.

What is planned for the future?

Through interaction with the client's staff, understand the evolutionary changes anticipated and planned for the period of performance of the contract.

What relevant documents/articles have the client and incumbent contractor personnel published?

Has the client established a public-access reading room or library that archives documents and articles relevant to the particular contract? If so, plan to visit and copy or take notes from appropriate materials. Sometimes copying has to be scheduled, so call ahead.

Strengths and weaknesses of the incumbent contractor.

Incumbent contractor refers to the company performing on the current contract.

Can your company hire incumbent personnel?

Incumbent personnel refer to those staff members employed by the incumbent contractor on the specific project that is being recompeted. Placing these people under contingent hiring agreements and including their résumés in your proposal can be highly beneficial to your prospects of winning the contract. Incumbent personnel are "known quantities" to the client. Those staff are knowledgeable about and familiar with such elements as the client's technical directions, organizational and reporting structures, and procedures and protocol. There is a reduced learning curve associated with contract phase-in and startup given a staffing mix that includes a significant number of key incumbent staff. *Contingent hiring* refers to the establishment of a signed agreement between an individual and a company that stipulates, upon contract award, the person will be hired to perform certain duties on the particular contract. Ensure that the client holds any incumbent staff to whom your company extends a contingent hiring agreement in high regard professionally and personally.

Which client personnel will sit on the source evaluation board?

Attempt to learn who the probable key candidates are for the SEB, including technical, contracting, and administrative staff. Learning their sentiments, concerns, goals, and professional backgrounds can be very advantageous when composing your proposal narrative.

Who is the competition? What does your company know about them?

The client's contracts office will tell you the name of the incumbent contractor, as will the published and Internet-resident agency Acquisition Forecast lists. Advance knowledge of the identity and technical competence of additional competitors can be gleaned from conversations with other companies in the same industry, watching newspaper and professional journal classified ads,[1] preproposal conference attendance lists, and *Freedom of Information Act* (FOIA) requests for previously submitted proposals to the client and monthly technical progress reports for the specific program or project.

Is your company qualified to bid alone, or must you team as the prime contractor or as a subcontractor?

There are circumstances under which teaming to win a contract is very appropriate. Talk with the client about his or her predispositions about and experiences with teaming arrangements. And sometimes your company may be too small to prepare a credible management plan as the prime contractor. For example, the contract staffing requirement may be 35 *full-time equivalents* (FTEs) per year. If yours is a 60-person company, you may want to consider establishing a teaming relationship with a larger prime.

What are your company's technical, programmatic, costing, and staffing strengths and weaknesses regarding the particular opportunity? Can you overcome your shortcomings in time?

After preparing a realistic inventory of your strengths and weaknesses, your Acquisition Team will have to determine if ongoing marketing efforts combined with hiring success rate (to address staffing deficits), cost restructuring (in accordance with *Defense Contract Audit Agency* (DCAA) stipulations), and exploring specialty subcontracting arrangements (to mitigate technical deficits and programmatic risks) will result in a favorable bid position for your company.

It is critical that the information collected during interaction with your client be recorded and then distributed on a need-to-know basis within your company. You may want to develop "Client Contact Reports," which include such information as:

- Client;
- Contact name and title;

---

[1]   This practice can be very beneficial to learn what projects your competition is pursuing as well as their staffing deficits.

- Contact date;
- Company representatives;
- Contact type (such as visit, call, and conference);
- Mission (your company's objectives for the contact);
- Summary of results;
- Next action.

Support services contracts are often characterized by very diverse, sometimes even conflicting, requirements, because they frequently take the form of umbrella task-order, level-of-effort (LOE) contracts serving multiple government representatives. The challenges associated with marketing and preparing the proposal for such bids center on creating as complete and detailed a picture as possible of various decision makers' needs within the client's infrastructure. To this end, it very useful to build a detailed "technical marketing" activity into the proposal development life cycle, using domain experts to interview as many of the cognizant government staff as possible. The purpose of this activity is to bring your company's understanding of the requirements to the next level of technical detail after the initial round of marketing visits. The pre-proposal interview form presented in Figure 3.4 is very useful for this purpose. Armed with it, your company's "technical marketers"—usually working-level support services staff—can hold technical discussions with many government task leaders in a manner that yields fruitful management approaches and proposal themes. Note that the questions center on three areas: (1) "hot buttons," that is, critical issues and their likely evolution; (2) "ghosts," that is, possible problems with the current support (if there is an incumbent); and (3) "discriminators," that is, approaches that allow you to propose a superior, well-tailored solution to your client's requirements.

As a practical matter, it is *not* a good idea to walk into a client interview and start reading from a list of questions; the form should be completed by the interviewer as soon as possible after the interview. When feasible, it is useful to send two company staff members to the interview; they can alternate asking questions and taking notes, and address any gaps in each other's recollections when completing the interview form later. The completed interview writeups should be reviewed by the proposal manager and key proposal writers before actual writing begins in order to identify common issues out of which proposal themes can emerge. These writeups also represent a wealth of technical detail that can be mined by the proposal writers in order to address client concerns in depth. Figure 3.5 provides another potential interviewing tool when interacting with your private-sector client or potential client.

Pre-Proposal Interview Form

Date: _____

Interviewer(s): _____   _____

Interviewee(s): _____   _____

1. Please provide a brief description of the work that you expect contractors to be performing under this contract.

   Note to interviewer: This should be a ½- to 1-page description. Attach a separate writeup to accommodate it. It is very helpful to get important reprints and other documents from the interviewee, though not so much that it will be impossible to assimilate. Concentrate on the main points! Also, be on the lookout for material for graphics.

2. What are the most critical areas of the work, i.e., the technical or management issues that are driving the contractor?

   Provide some key sentences or bullets. This is very critical, as they will drive the proposal themes and technical response.

3. How many contractors are currently supporting this work? Do you expect this level to change?

4. Are there any particularly important contractor personnel whose names you can give us, who are key to your effort?

5. What kind of skill mix do you require for this work?

   Does the client want more high-level people ("partners"), or lower-level support? On technical jobs, what kind of proportion of senior programmers, analysts, junior programmers, scientists, engineers, etc.?

6. What kind of management structure best supports this work?

   How does the client think management would best support him, e.g., what kind of organizational structure?

7. Are there any areas in which you think substantial change or innovation is required? "Innovation" could be either technical or managerial.

   Probe for areas where we can suggest innovation! Examples: Does the client feel that she needs short-term support with highly specialized skills? Is she looking for some particular kind of management structure?

8. If you had the chance, are there any changes you would make in the nature of your contractor support?

   Tie this to the previous question. Get her to tell you how you can look like a hero!

9. How do you see your contractor support requirements evolving in the future?

   Get information about future skill mix and staffing levels. What are her hopes/fears/biases about her budget, and how does she see these affecting her need for support? For example, does she see a larger or smaller role for her support contractor in the areas of science, operations, instrument engineering, etc.?

10. What do you think your future critical issues are likely to be?

    Extremely critical question! Get as much information as possible regarding projected future problems and drivers, and the directions from the client thinks solutions might emerge. This will be a dominant part of the proposal!

11. What would you be looking for in a winning proposal for this effort? In particular, what do you see as the highest priorities that would be given the most attention in a winning proposal?

    Try and draw the client out about his/her "hot buttons" and the likely selection discriminators. That is, what could distinguish you from your competitors in a strong, positive way?

12. Can you suggest anyone else that we might talk to to get more information on these topics?

Figure 3.4   Pre-proposal interview form. (Adapted with permission from Dr. Richard Isaacman.)

Client:_____ **Address:**_____ **Telephone No.: (__)_____**

1. Are you meeting with a representative of corporate management, or facility management?

2. What are the specific types of needs or activities of the client?

3. Are these needs managed through an "in-house" technical organization, professional service contracts, or both?

4. What is the total company or facility budget for all goods and services for the current and upcoming year? (Is the budget published and can we have a copy?)

5. How much of this budget will be spent on professional services or consulting?

6. Who is responsible for management and procurement of services in each specific requirement areas?

7. Is there any organizational chart available that can help us understand the client? Can we have a copy?

8. Which of our competitors are under contract or have recently performed work for the client? Are any of the competitors under term or national service contracts? When are these contracts up for re-bid?

9. Are there any opportunities that are evident from the discussions for which our company may provide benefit to the client? What is the timeline for these opportunities?

10. Is there a requirement that our company be listed as an "approved" firm on a procurement list? Are we on it? Who in the client's organization is responsible for maintaining the list?

11. Does the client have Minority Business Enterprise/Woman-owned Business Enterprise (MBE/WBE) goals for their procurements? Are there specific MBE/WBE firms that the client recommends?

12. How does the client conduct the procurement of service contracts? Who is (are) the decision maker(s)?

Figure 3.5
Client information
interviewing tool.

3.7 Maintain management visibility on your contracts

A very practical and relatively easy marketing technique applicable to contracts on which your company is the incumbent (contractor performing on the current contract) is for members of your company's executive and division-level management to visit the client staff as well as your on-site personnel on a regular basis. Build and continue to cultivate communication networks with client staff at both professional and appropriate social levels, including professional association meetings and government-industry council meetings. Let your client know that his contract is important to your company by showing a real physical presence at his job site during the entire contract lifecycle and not merely at contract renewal. A discussion of your management's commitment to a particular project in the proposal narrative is far more persuasive when it has been substantiated by regular site visits by your senior managers.

Senior management can communicate on an ongoing basis with, for example, on-site company staff through electronic mail (e-mail), informal briefings, friendly on-site luncheons ("brown bag," bring-your-own lunches are fine), and performance recognition meetings on-site. There is an enhanced sense of loyalty among your on-site staff to your company that can follow from ongoing management presence and genuine interest. Too often, on-site staff, due to sheer proximity to the client and isolation from contact with your company's management, develop greater loyalty to the project than to your company. When recompetition comes around, your staff may not be so concerned as to which company they work for but rather that they stay associated with the project. And your competition may find them easy targets to lure away with contingent hire agreements.

It is critical that senior management take an active role in client interaction, relationship building, and other marketing activities. "A senior executive's instinctive capacity to empathize with and gain insights from customers is the single most important skill he or she can use to direct a company's strategic posture" [8]. Know when to have your senior management visit your client, and make sure that your management know what is necessary to convey as well as learn.

3.8 Project managers as client managers

Your company's project managers (PMs) should develop self-recognition as client managers and first-tier marketing generalists. These managers are in the enviable position of being in day-to-day contact with your clients. PMs should be taught and mentored how to cross-sell your company's services and products. They can help your client better understand and appreciate the direct benefits of your firm's core competencies. PMs can appropriately, yet persistently, inquire as to what additional services your company can provide to this client. Expanding services to existing clients is the most cost-effective form of marketing activity. You need to recognize your PMs as an integral part of your public relations, or reputation management, initiatives. PMs put a face on your company; they need to be your local advocate. Your project managers can also be instrumental in obtaining letters of commendation from your clients, which in turn can be used in proposals to authenticate your technical and programmatic expertise. Letters of commendation should be filed in your company's proposal library, and scanned into electronic format for easy placement into a proposal document.

Your company should also regularly assess performance levels on each of your current contracts. That assessment must necessarily include an evaluation of your performance by your client. Figure 3.6 offers one such performance scorecard that might be used.

Please rate "how we are doing" for the following activities:

	(Not doing well at all)			(Doing very well)		
	1	2	3	4	5	N/A
Clear/Effective Planning (schedules, updates)						
Support for Client Decision Making						
Regular Communication and Feedback						
Innovative Ideas						
Appropriate Experienced Staffing Applied						
Appropriate Technical Support Provided						
Cost-Effective Project Performance						
Issues Resolved Appropriately						
General Client Satisfaction						
Meeting Project/Task Objectives						
On-time Deliverables						

Other comments:

Thank you.

Figure 3.6   Performance scorecard.

3.9 Commercial-off-the-shelf acquisition

There is a clearly discernible trend toward *commercial-off-the-shelf* (COTS) procurement within the federal arena, particularly in the area of software products. COTS products represent industry's best—they are tested and piloted before deployment in the marketplace. They are also readily available off of the GSA schedule. And Congress and the Office of Management and Budget (OMB) within the Executive Office of the President have indicated a preference for COTS solutions, which constitutes an important consideration in obtaining funding for a federal agency.

Established in 1961, the U.S. Agency for International Development (USAID) is the independent government agency that provides economic development and humanitarian assistance to advance U.S. economic and political interests overseas. Through the Principal Resource for Information Management Enterprise-wide, or PRIME, contract, USAID seeks to form an alliance with a world-class information technology contractor and its subcontractors for the purpose of consolidating current operations and implementing a comprehensive approach to the acquisition, integration, total life cycle management, and operation of USAID's information technology (IT) resources within a framework of standards and consistent methodologies. In an exhaustive study, a full COTS alternative was considered carefully and deemed to present the least risk as measured by factors such as schedule, cost, technology obsolescence, and architecture dependency.

A variety of federal agencies have already implemented COTS solutions from GSA Financial Management Systems Software (FMSS) vendors. The U.S. Department of Labor installed a COTS system package that was deployed ahead of schedule. And the U.S. Department of Agriculture's COTS system was installed within the planned budget. The U.S. Army National Guard and Army Reserve made extensive use of COTS products in the areas of forms processing and data access and retrieval for its worldwide Reserve Component Automation System (RCAS). NASA is one of the federal agencies that maintains a Web-based list of COTS products that offer Year 2000 compliance (http://www.lerc.nasa.gov/www/yr2000pub/).

As software systems become increasingly complex to build, developers are turning more and more to integrating pre-built components from third-party developers into their systems. This extensive use of COTS software components in system construction presents new hurdles to system architects and designers. Some federal agencies are discovering, however, that a practice that was intended to streamline acquisition—and that does so in many cases—also raises some challenging issues associated with data conversion and systems integration. For example, government requirements may have to be "shoehorned" to fit the features of COTS products. In specific cases, software "bridges" and interfaces can be designed and developed to address this issue. There are also the matters of intellectual property and licensing that must be addressed. When COTS products are modified, ownership of the new product becomes open to question. And "[f]ormerly, government officials who were in a hurry to buy off-the-shelf equipment such as personal computers could sidestep the normal, time-consuming bidding practices by awarding a sole-source contract directly to a minority-owned firm that could fill the order" [5]. According to the National Federation of 8(a) Companies based in Arlington, Virginia, the 1996 federal acquisition reform legislation allows "agency officials to make purchases of selected equipment from a large list of preapproved contractors, large and small" [5].

The *Federal Acquisition Streamlining Act* (FASA) of 1994 stipulates that a commercial item can still be a commercial item if it has one of two modifications. The first is a modification, any kind of modification, that is customarily available in the marketplace. This is a type of modification that the contractor has done in the commercial marketplace. But the other kind of modification allowed under FASA is a modification for government use. Even if it has a modification for government use, it is still considered to be a commercial product. Commercial item services are services that are customarily sold in the marketplace for a set price for a particular task. This is a great opportunity for small business.

3.10 Pursuing firm fixed price and invitation for bid opportunities

In negotiated services contracts, relationship building through marketing is paramount to success. However, selling computers or reprographic paper—in effect, products—under *firm fixed price* (FFP) contracts often involves less in the way of personal interaction. FFP contracts require that "the contractor delivers the product or performs the services at a specific time at a price that is not subject to any adjustment of actual costs incurred during contract performance. It places 100% of the risk on the contractor and places the least amount of administrative burden on the Government" [9]. Maximum profit can be earned by managing FFP contracts aggressively and identifying, understanding, and controlling risk. When marketing, your company will want to establish an appropriate mix of targets between cost-reimbursement and FFP types of opportunities.

Invitations for bid (IFB), or sealed bids, result in a FFP contract. Bids are opened publicly, and the award is made immediately to the lowest priced, responsible, responsive bidder. No discussions are allowed, and no exceptions may be taken by the contractor. IFBs are suitable for COTS purchases as well as for small support services contracts. Understanding your company's cost structure and your vendors' costs as well as being able to operate at maximum efficiency during contract performance are critical to making profit on IFB/FFP opportunities. Profit on cost-reimbursement type contracts is termed "fee."

3.11 Using a request for information and request for comment as valuable marketing tools

In circumstances under which the government is seeking to establish that adequate competition exists within industry to release a solicitation or to determine if industry has developed specific technologies that the government requires (assess project feasibility), an agency will issue a *request for information* (RFI), or *request for comment* (RFC) on a *statement of work* (SOW). In fact, the Federal Acquisition Regulations (FAR) encourage government agencies to promote early exchanges of information about future acquisitions (see FAR, Part 15.201). This practice is intended to identify and resolve concerns regarding such things as proposed contract type, requirement feasibility, suitability of evaluation criteria, and availability of reference documents.

Your company's response to an RFI or RFC should be prepared in the same professional manner as a proposal. Too many times in the case of an RFI, companies will submit marketing brochures or other prepared, "canned" materials rather than develop a customized, client-oriented package. Responding to an RFI presents a significant opportunity for your company to demonstrate in writing that you are interested in supporting the client and in developing solutions for your client's issues and challenges. Also, you will learn about the client and possibly the competition. As with an RFP, respond specifically to the client's requests. Authenticate your technical capabilities with references to specific contractual experience. Help your client to better understand your company's particular strengths. Submit the RFI on time and to the correct location. Hand delivery can also be beneficial.

3.12 Standard form 129s and contractor prequalification statements

Certain client organizations require that a company be *prequalified* or *pre-approved* prior to the company being able to receive RFPs, RFQs, RFIs, and other solicitation documents. An example of such a prequalification vehicle is the federal Standard Form 129 (Revised 10-83), which is a "Solicitation Mailing List Application." The U.S. Army Corps of Engineers, Forest Service, *General Services Administration* (GSA), and various port administrations are among the many organizations that use this application. The SF129 contains 21 blocks for information ranging from the type of organization your company is and the names of officers, owners, or partners to tax identification numbers, applicable *Standard Industrial Classification* (SIC) codes,[2] and net worth. The application must be signed by the principal of the company. Certain federal agencies attach a Supplemental Commodity list along with instructions to the SF129. The *Standard Army Automated Contracting System* (SAACONS) Vendor Information Program pamphlet (Revised 4/16/89) is one such Supplemental Commodity list. Your company will have the opportunity to identify up to 12 *Federal Supply Class* (FSC) categories of supplies and services on this particular list. Accurate and timely completion of the SF129 application and supplemental documents will result in your company's name being added to a particular agency's automated bidder's mailing list.

State and local governments as well as private-sector clients may issue contractor prequalification applications and/or contractor qualification questionnaires[3] in order to prequalify a listing of potential contractors. Examples of the wide-ranging list of clients that request such information are the State of Georgia, MCI, Southwestern Bell, Philadelphia Electric Company, Shell, and Mobil Oil. The information required can be broad-ranging in scope but often includes such information as the names of principals of the company, financial information, years of contractual experience, debarment status (has your company currently been debarred from submitting bids?), minority business status, professional registrations by state, and equipment lists. Again, timely and accurate completion is important.

---

2   SIC codes will be changed to North American Industry Classification System (NAICS) codes.

3   These questionnaires can be called bidder's list applications, general subcontractor surveys, potential supplier profile questionnaires (PSPQs), preliminary qualification questionnaires, qualified suppliers lists (QSLs), and vendor forms.

3.13 Ethics in marketing and business development

Ethical conduct in all business activities, whether domestic or international, should be your company's corporate standard at all times [10]. The development and implementation of a formal set of ethics rules that prohibits any type of procurement fraud by employees, agents, or subcontractors would be a prudent step. Indeed, *DoD FAR Supplement* (DFARS) 203.7000, entitled "Contractor Responsibility to Avoid Improper Business Practices," notes that a contractor's management system should include a written code of business ethics and conduct as well as an ethics training program for all employees. According to an article published in the *Harvard Business Review* in 1994, new federal guidelines now recognize the organizational and managerial roots of unlawful conduct and base fines partly on the extent to which companies have taken steps to prevent misconduct [11].

In their important work entitled *Formation of Government Contracts*, George Washington University law professors John Cibinic, Jr., and Ralph Nash, Jr., discuss standards of conduct in two broad categories, namely, "those dealing with improper influence on Government decisions and those requiring honesty and disclosure of relevant facts in dealing with the Government" [12].

Part 3 of the FAR presents guidance regarding "Improper Business Practices and Personal Conflicts of Interest." One example, the Anti-Kickback Act of 1986 (codified at 41 U.S.C. 51–58 (1988)) discussed in FAR 3.502-2, is legislation designed to deter subcontractors from making payments and contractors from accepting payments for the purpose of improperly obtaining or rewarding *favorable treatment* [13] in connection with a prime contract or a subcontract relating to a prime contract. The term "kickback" means any money, fee, commission, credit, gift, gratuity, thing of value (for example, promise of employment), or compensation of any kind. Unlike its precursor in 1946, the 1986 Act includes all types of government contracts. Finally, the 1986 Act places new reporting requirements on prime contractors and subcontractors. "Possible" violations of kickback laws must be reported in writing to the *inspector general* (IG) of the appropriate federal agency, the head of the contracting agency if the agency has no IG, or the United States Department of Justice.

The Procurement Integrity Act, about which there is considerable confusion on the part of both civil servants and contractors, provides that the government cannot impart information to one bidder or proposer without making that information public to all potential bidders or proposers.

And the False Claims Act Amendment of 1986 facilitated prosecution of complex defense procurement fraud cases.

Many private-sector corporations—your potential prime contractors or clients—have enacted written policy codes that govern gifts and entertainment. One major East Coast manufacturing firm stipulates that its employees are forbidden to accept any incentives in exchange for increased purchases through a given supplier or subcontractor. Gifts are defined very strictly, to include dinners (unless incidental to a business meeting), money, discounts on personal purchases, holiday presents, tickets to theaters and sporting events, airline tickets, hotel accommodations, and repairs or improvements at prices other than for fair value.

3.14 Advertising, trade shows, and high-impact public relations

Like "brand name" recognition in consumer advertising and marketing, increasing the baseline level of name recognition for your company is important to your overall business development success. Well-targeted public relations (PR), trade show, and advertising efforts will probably not result directly in sales or contracts but will help to build an image of legitimacy and solid foundation for your company that can and should be leveraged by your external and internal sales teams and public relations or reputation management efforts (see Figure 3.7).

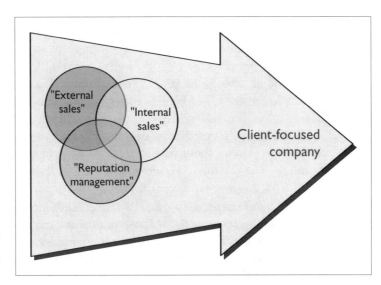

Figure 3.7
One common
vision.

There are many freelance graphics artists, for example, who can develop high-impact ads on a cost-per-ad basis for placement in key business and trade publications without the significant fees associated with retaining an advertising and design agency. Figure 3.8 provides an example of an ad developed for a small company. Watch for special issues of publications important to your lines of business that may focus directly on market segments that are part of your strategic plan. Cases in point are *Washington Technology; World Dredging, Mining, and Construction;* and *Chemical Week.* The latter publication, for instance, ran a special supplement devoted to the State of New Jersey. Companies that did chemical or environmental business or were positioning themselves to do such business in New Jersey could then place ads in that particular supplement.

Increasing the baseline of name recognition for your company is important to your overall business development success.

Professional-looking tabletop trade show booths can be designed and populated with attractive, powerful photographs and other imagery for only a few thousand dollars total. If your company does work for the Department of Defense, for example, that agency maintains public domain photographs that are available for nominal fees at its Anacostia Naval Air Station near Washington, D.C. Professional stock imagery may be procured from photo houses or from the Internet (e.g., http://www.photodisc.com) for limited usage instead of taking your own photographs. Appearance at selective trade shows will increase your company's visibility in the specific business community you serve.

Other public relations efforts can take the form of news releases to appropriate business, financial, and professional/trade journals. In the Baltimore-Washington area, companies can purchase and consult *The Media Directory* (P.O. Box 2200, Annapolis, MD 21404). News releases will, of course, need some kind of "news hook." That is, you will need to provide a financial or business editor with something around which to build a headline and weave a story line.

Your company's credibility depends in part upon the efficient distribution of news about your success. You might consider employing the services of a media relations firm, such as *BusinessWire* (http://www.businesswire.com), which has headquarters offices in New York and San Francisco. *BusinessWire* provides simultaneous electronic delivery of full-text news releases and photos and graphics to the domestic U.S. and international media, financial communities, and targeted industry trade publications. Your news releases will also be delivered automatically to 150 online services and databases worldwide, including Reuters Business Alert, Pointcasts, Wall Street Journal Interactive, New York Times,

Open Systems

T R A I N I N G

NeXTSTE

The best training available? Don't take our word for it. Talk to our customers. They've consistently rated our courses superior to vendor-supplied training. Just ask the USDA, who named Synex **Small Business Contractor of the Year**.

The NeXT Challenge.

Unfortunately, learning to program in Objective C under NeXTStep 3.0 is no easy task. Many developers found that NeXT's 'Dev Camp' didn't meet their needs.

That's why customers like **Mobil Corporation** and **Fannie Mae** came to us for help. These companies found out that Synex had designed a two-week training program on Objective C, AppKit, and Interface Builder. They learned that

Synex's courses were designed to transfer programming skills to the students, and that the courses included measurable objectives—in the form of lab exercises—that required the students to write solutions from scratch.

The Courses.

Objective C Programming Under NeXTStep. Part I of a two-week NeXTStep training program. A hands-on intensive on Objective C and the AppKit. Learn to design and code Objective C classes, and to allocate, initialize, and manipulate NeXT interface objects programmatically.

NeXTStep 3.0 Development Tools. Part II of our two-week NeXTStep training program. Advanced training on Interface Builder, Project Builder, and AppKit. Hands-on exercises permit students to

build meaningful applications in cl

ANSI C Programming. Thorough cover age of the ANSI C language definition. Training on proper use of all C constructs. Realistic, hands-on examples give students the confidence they need to program effectively in C.

Introduction To Unix. In-depth coverage of the Unix filesystem, the Shell, process execution, Unix commands, and I/O management. Plenty of supervised, hands-on exercises, and complete integration of topics.

We're Ready.

Call us today for information on our customized on-site training. We're sure you'll agree with Fannie Mae, Mobil and the USDA: Synex Training is the answer for your organization's productivity.

UNIX/C

I N T E G R A T I O N

Department of Labor, Department of the Army, USDA, Mobil, Fannie Mae...they know Synex as the rightsizing experts. Our customers have discovered that our open systems expertise and service orientation makes the tough transition from the mainframe to client/server architectures just a little easier. They like to work

with our friendly, knowledgeable, and experienced Unix, C, C++, NeXT, Sybase, Oracle, Networking, and project management experts.

They've also found that Synex's broad experience in managing the complete life-cycle of downsizing projects translates to a reliable and predictable timeframe and budget.

Open Systems Integration Services

- **Unix-based MIS**
- **Unix and C Training**
- **Distributed DBMS**
- **OLTP**
- **LAN/WAN**
- **Secu**

Client/Server

Synex inc.

ARCHITECTS OF OPEN SYSTEMS

NeXTSTEP is a trademark of NeXT, Inc.
Unix is a trademark of USL

Figure 3.8 Advertisement for a small company that appeared in *Washington Technology's 8(a) Report*. [Ad developed by Robert S. Frey for Synex, Inc., Columbia, Maryland. Reprinted with permission from Robert S. Conner, president, Synex, Inc.]

Dow Jones News, MSNBC News, Bloomberg Financial Markets, and Yahoo! (Internet). Of note is that Pointcasts is the engine of choice for many U.S. corporate executives. For no additional cost, you can also specify that *BusinessWire* send your news releases to U.S.-based trade journals that focus on specific industries.

Another effective PR technique related to trade and professional journals is to generate and place articles in these publications [14]. This process is highly cost-effective, and reprints of articles that have been published by your company's staff can be used in proposals to authenticate the firm's and the proposed staff's expertise. It is advantageous to contact the editorial staff of appropriate trade journals to obtain the agendas and themes of upcoming issues as well as submission guidelines. In addition, your company can create and maintain its own Web page on the Internet, replete with logo, easy-to-use icons, and engaging content, and accessible for "data mining" with well-known search engines such as www.hotbot.com; www.metacrawler.com, www.yahoo.com; www.altavista.com; www.web crawler.com; www.lycos.com; and www.infoseek.com as well as www.excite.com; www.dejanews.com; and www.inktomi.com. Your Web page should clearly identify your company's lines of business upfront, and provide user-friendly, quick-loading access to information about how your firm can support clients successfully. Include an electronic fill-in form that Web site "visitors" can complete and e-mail or mail to your marketing staff. You might build in the text from recent external news releases, information about your office/project locations and client base, corporate awards, core competencies, and upcoming trade show appearances. Add your Web site address to business cards, marketing brochures and fliers, and promotional items such as golf balls, keyrings, or refrigerator magnets. Finally, include electronic point-and-click hyperlinks on your Web pages to other pertinent Internet sites that provide additional regulatory, technological, business, or contractual information. Register your Web site with appropriate search engine providers on an ongoing basis.

Remember that whatever the level of funding that can be allocated for these types of important activities, the critical thing is to do them on a *consistent* basis. Clients are more likely to buy goods and services from companies they know or have heard about in the press or trade literature. Public relations helps bring your company and its capabilities to life and keeps your company's name embedded in decision makers' minds.

We have now introduced the federal competitive proposal environment and the strengths and constraints of small businesses operating in that arena. A company organizational scenario was offered along with a presentation of strategic and mission planning. Specific, effective marketing techniques as well as ethical conduct for business development were discussed.

We will now examine the RFP, the solicitation document issued by both government and private-sector client organizations. The RFP should not be a call to action, as it many times is, but rather a continuation of the ongoing process of marketing your client.

END NOTES

1. Graham, John R., "Getting to the Top, And Staying There," *The New Daily Record* (Baltimore, MD), 26 July 1997, p. 7B.

2. This metaphor was borrowed from Marshall, Colin, "Competing on Customer Service: An Interview with British Airways' Sir Colin Marshall," *Harvard Business Review*, Nov./Dec. 1995.

3. "Any time the government makes an award to a contractor who is neither the highest technical offeror nor the lowest price offeror, then the government has made a best-value determination; it has made a trade-off between technical and cost features." See Hackeman, Calvin L., "Best Value Procurements: Hitting the Moving Target," McLean, Va.: Grant Thornton, 1993, p. 3.

4. Ray, Dana, "Filling Your Funnel: Six Steps to Effective Prospecting and Customer Retention," *Selling Power*, July/August 1998, p. 44.

5. As quoted by Behr, Peter, "Just Say Know," *Washington Post*, 1 April 1996.

6. Mim Goldberg, "Listen to Me," *Selling Power*, July/August 1998, p. 58.

7. "Selected Viewgraphs from Judson LaFlash Seminars on Government Marketing and Proposals," Government Marketing Consultants, July 1980.

8. Gouillart, Francis J., and Frederick D. Sturdivant, "Spend a Day in the Life of You Customers," *Harvard Business Review*, Jan./Feb. 1994.

9. McVay, Barry L., *Proposals That Win Federal Contracts*, Woodbridge, VA: Panoptic Enterprises, 1989, p. 29.

10. For a variety of articles that all focus on business and professional ethics (including international marketing ethics), see *BRIDGES: An Interdisciplinary Journal of Theology, Philosophy, History, and Science*, Vol. 2, 1990.

11. See Paine, Lynn Sharp, "Managing for Organizational Integrity," *Harvard Business Review*, Mar./Apr. 1994.

12. Cibinic, John, and Ralph C. Nash, *Formation of Government Contracts*, Second ed., Washington, D.C.: The George Washington University, 1986, p. 107.

13. According to Ropes & Gray, examples of "favorable treatment" that are illegal include receiving confidential information on competitors' bids, obtaining placement on a bidders' list without meeting requisite qualifications, obtaining unwarranted waivers of deadlines, obtaining unwarranted price increases, and recovering improper expenses. Contractors should also be aware of the "Fraud Awareness Letter" issued in September 1987 by the DoD Council on Integrity and Management Improvements, which identified "'indicators of potential subcontractor kickbacks.'" See "Complying With The Anti-Kickback Act: *Guidelines and Procedures*," *Developments in Government Contract Law* No. 10, Sept. 1990.

14. Schillaci, William C., "A Management Approach to Placing Articles in Engineering Trade Journals," *Journal of Management in Engineering*, Sept./Oct. 1995, pp. 17–20.

Chapter 4

Request for proposals

4.1 Overview

In addition to federal government RFPs, there are RFPs released by state, municipality, and city governments as well as by private-sector companies and international governments.

The federal government RFP that your company requests and receives is the culmination of a lengthy planning, budgeting, and approval process. It is a solicitation document issued to obtain offers from contractors who propose to provide products or services under a contract to be awarded using the process of negotiation. The RFP is a complex document, often prepared under proposal-like conditions. That is to say, it is written and reviewed by a variety of civil servants under tight schedule constraints and is subject to delay caused by, for example, late inputs, protracted legal reviews, program modifications, funding issues, or changes in contractor support of the SOW development.[1] RFPs often contain conflicting or ambiguous requirements as well as incorrect cross-references, particularly in Section L, "Instructions, Conditions, and Notices to Offerors." (An "offeror" is any company or organization that responds with a proposal to the RFP.) This may result in part because RFP documents are often assembled using "boilerplate" materials from previous or similar RFPs. When there are conflicting guidelines, submit questions for clarification in writing to the appropriate government point of contact by the date specified in the RFP. Your questions,[2] along with those from other firms, may result in the government issuing an amendment to the RFP. Within the DoD, Integrated Product Teams have been employed to ensure the proper integration of the various parts of the RFP.

Contractors should not assume that the RFP reveals the full intent of the client's preferences, sentiments, or requirements. By the same token, care must be exercised to respond precisely to the RFP requirements in the order in which they are provided. For example, if the RFP calls for a discussion of technical understanding and approach, past performance, and key personnel in that order, *your proposal should be structured to respond in exactly that order*. Do not build in additional levels of quality or unrequested services that will inflate your company's costs when compared with your competitors. Do not propose a Mercedes when the client wants a Buick! A general rule is to take no deviations or exceptions to the stated RFP requirements, and do not submit alternate proposals, even though they may be allowed. The government has spent considerable time, energy, and resources in performing functional and data requirements analyses; conceptualizing measurable evaluation factors (e.g., past performance), subfactors (e.g., project profiles), elements (e.g., management of complex contracts), and standards; and developing the RFP and its SOW[3] to reflect its concept of a technical solution for a given project or program. It can, therefore, be perceived as presumptuous on a company's part to propose an alternative. Your pre-RFP marketing efforts should have provided intelligence as to the feasibility and acceptability of alternate proposals.

---

1   Many SOWs and specifications are researched, written, and prepared by contractor personnel under separate contract to the government procuring agency. Companies that have contracts to prepare SOWs cannot legally compete on particular procurements for which they have prepared the SOW.

2   Because they will become public record, be certain that your questions do not reveal a weakness or lack of understanding about the project on the part of your company.

3   A Statement of Objectives (SOO) is sometimes used instead of a government-written SOW to maximize the flexibility afforded to offerors to propose innovative and cost-effective approaches.

RFPs can range from a few pages to thousands of pages, replete with attachments, for major aerospace and defense procurements. Select RFPs are now being distributed via electronic media (diskettes, Internet, *Federal Acquisition Computer Network* (FACNET), agency electronic BBSs) instead of in hardcopy form. When hardcopy RFP documents are mailed, some contractors elect to scan the RFP into electronic files using some type of *optical character recognition* (OCR) or *intelligent character recognition* (ICR) technology. Scanning can facilitate electronic searches for RFP requirements, which in turn can be transferred to electronically stored "story board" templates. It also facilitates searches for hard-to-locate requirements or duplications. A word of caution regarding scanning— time must be allotted to spell check electronically, review, and correct the scanned file. Scanning accuracy levels vary widely depending upon the scanner technology itself and the physical quality of the hardcopy document.

In accordance with the Uniform Contract Format established at FAR 15.406-1, government COs must prepare and assemble RFPs in a specific manner as enumerated in the following sections.

4.2 Part I—the schedule

Section A: Solicitation/Contract Form (Standard Form 33);

Section B: Supplies or Services and Prices/Costs;

Section C: Description/Specifications/Work Statement (The SOW may include system specifications, contractor tasks and services, products, contract end items, data requirements, and schedules, for example. It is an essential part of the RFP.)

Section D: Packaging and Marking;

Section E: Inspection and Acceptance;

Section F: Deliveries or Performance;

Section G: Contract Administration Data;

Section H: Special Contract Requirements.

4.3 Part II—contract clauses

Section I: Contract Clauses.

4.4 Part III—list of documents, exhibits, and other attachments

Section J: List of Attachments.

4.5 Part IV—representations and certifications

Section K: Representations, Certifications, and Other Statements of Offerors ("Reps and Certs");

Section L: Instructions, Conditions, and Notices to Offerors;

Section M: Evaluation Factors for Award (used to determine proposal page allocations, writing emphases, and thematic structure).

Presented a slightly different way, the RFP:

- Describes the requirement in Sections B, C, D, E, F, and (J);
- States the government agency's terms in Sections A, B, G, H, I, K, and (J);
- Describes the evaluation criteria in Section M;
- Prescribes the proposal format and content in Section L;
- Provides process information in Sections A and L.

Figure 4.1 presents the table of contents (Section 11) from an RFP issued by the U.S. Army Corps of Engineers. This form is a Standard Form 33, "Solicitation, Offer and Award," which appears in federal RFP documents.

The contractor's response to the government's RFP is called the *proposal*, the focus of Chapters 7 and 8 of this work.

4.6 The importance of Section L (instructions to offerors)

Section L of the RFP provides specific instructions for preparing and structuring the proposal document. (Note, however, that on occasion these instructions are found in Section M. RFPs from the Sacramento District of the U.S. Army Corps of Engineers' are cases in point.)

SOLICITATION, OFFER AND AWARD	1. THIS CONTRACT IS A RATED ORDER UNDER DPAS(15 CFR 700)	RATING		PAGE OF PAGES
	▶			1

2. CONTRACT NO.	3. SOLICITATION NO	4. TYPE OF SOLICITATION	5. DATE ISSUED	6. REQUISITION/PURCHASE NO.
	DTTS59-98-R-00001	☐ SEALED BID(IFB)		
		☒ NEGOTIATED(RFP)	07 July 98	

7. ISSUED BY   CODE   SVC-180   **8. ADDRESS OFFER TO (If other than Item 7)**

U.S. Department of Transportation
TASC Acquisition Services
400 7th Street, SW, Room 5106
Washington, DC 20590

NOTE: In sealed bid solicitations "offer" and "offeror" mean "bid" and "bidder".

SOLICITATION

9. Sealed offers in original and   **See L.8**   copies for furnishing the supplies or services in the Schedule will be received at the place specified in Item 8, or if handcarried,

in the depository located in   **U.S. Department of Transportation Room 2126**   until   **1:00 P. M.**   local time   **14 Aug 1998**
                                                                            (Hour)   (Date)

CAUTION-LATE Submissions, Modifications, and Withdrawals: See Section L, Provision No. 52.214-7 or 52.215-01. All offers are subject to all terms and conditions contained in this solicitation.

10. FOR INFORMATION CALL:	▶ A. NAME Nancy Martus	B. TELEPHONE NO. (include area code)(NO COLLECT CALLS) (202) 366-6962

11. TABLE OF CONTENTS

(X)	SEC	DESCRIPTION	PAGE(S)	(X)	SEC.	DESCRIPTION	PAGE(S)
		PART I - THE SCHEDULE				PART II – CONTRACT CLAUSES	
X	A	SOLICITATION/CONTRACT FORM	1	X	I	CONTRACT CLAUSES	13
X	B	SUPPLIES OR SERVICES AND PRICES/COSTS	5			PART III - LIST OF DOCUMENTS, EXHIBITS AND OTHER ATTACH.	
X	C	DESCRIPTION/SPECS./WORKSTATEMENT	28	X	J	LIST OF ATTACHMENTS	48
X	D	PACKAGING AND MARKING	1			PART IV- REPRESENTATIONS AND INSTRUCTIONS	
X	E	INSPECTION AND ACCEPTANCE	1	X		REPRESENTATIONS, CERTIFICATIONS AND	
X	F	DELIVERIES OR PERFORMANCE	1		K	OTHER STATEMENTS OF OFFERORS	13
X	G	CONTRACT ADMINISTRATION DATA	4	X	L	INSTRS., CONDS., AND NOTICES TO OFFERORS	18
X	H	SPECIAL CONTRACT REQUIREMENTS	21	X	M	EVALUATION FACTORS FOR AWARD	6

OFFER (Must be fully completed by offeror)

NOTE: Item 12 does not apply if the solicitation includes the provisions at 52.214-16, Minimum Bid Acceptance Period.

12. In compliance with the above, the undersigned agrees, if this offer is accepted within 120 calendar days (60 calendar days unless a different

period is inserted by the offeror) from the date for receipt of offers specified above, to furnish any or all items upon which prices are offered at the price set

opposite each item, delivered at the designated point(s), within the time specified in the schedule.

13. DISCOUNT FOR PROMPT PAYMENT (See Section I. Clause No. 52-232-8)	▶ 10 CALENDAR DAYS %	20 CALENDAR DAYS %	30 CALENDAR DAYS %	CALENDAR DAYS %

14. ACKNOWLEDGMENT OF AMENDMENTS (The offeror acknowledges receipt of amend-Ments to the SOLICITATION for offerors and Related documents numbered and dated:	AMENDMENT NO.	DATE	AMENDMENT NO.	DATE

15A. NAME AND ADDRESS OF OFFEROR	CODE	FACILITY	16. NAME AND TITLE OF PERSON AUTHORIZED TO SIGN DIFFER (Type or print)	
15B. TELEPHONE NO. (include area Code)	15C. CHECK IF REMITTANCE ADDRESS ☐ IS DIFFERENT FROM ABOVE. ENTER SUCH ADDRESS IN SCHEDULE.		17. SIGNATURE	18. OFFER DATE

AWARD (To be completed by Government)

19. ACCEPTED AS TO ITEMS NUMBERED	20. Amount	21. ACCOUNTING AND APPROPRIATION	
22. AUTHORITY FOR USING OTHER THAN FULL AND OPEN COMPETITION: ☐ 10 U.S.C. 2304(c)( ) ☐ 41 U.S. C. 253 (c)( )		23. SUBMIT INVOICES TO ADDRESS SHOWN IN (4 copies unless otherwise specified) ▶	ITEM
24. ADMINISTERED BY (If other than Item 7) CODE		25. PAYMENT WILL BE MADE BY CODE	
26. NAME OF CONTRACTING OFFICER (Type or Print) Richard A. Lieber		27. UNITED STATES OF AMERICA (Signature of Contracting Officer)	28. AWARD DATE

IMPORTANT - Award will be made on this Form, or on Standard Form 26, or by other authorized official written notice.

Figure 4.1   Table of contents from an RFP.

Turn to Section L first when evaluating a new RFP. Section L drives the structure and form of your company's proposal response. Together with Sections C (SOW), M (Evaluation Criteria), and H (Special Contracts Requirements). Section L provides the basis for the "architecture" of the proposal documents. Of note is that recent versions of Section L within the U.S. Department of Transportation include an element titled "Why Should We Contract With You?"

Margin requirements, font family and size, number of foldout (11 × 17 inches) pages permitted, page count, double-sided photocopying, binding and packaging specifications, résumé formats, project experience formats, number of copies of each volume, and oral presentation slide specifications are among the publication parameters that may be addressed in this important section of the RFP. Here, you may also find specific guidance as to how your proposal should be structured in terms of the outline and numbering conventions. For example, you may see that your technical volume should consist of six major sections, each numbered according to the convention I.A, I.B, I.C. Outlining the proposal volumes should take into account guidance from Section L as well as Sections M and C and other parts of the RFP as appropriate. Remember to use the verbiage from the RFP itself, particularly Section C, to build your proposal outline. Evaluators, their administrative support staffs, and government-hired consultants/evaluators[4] will be looking for those same words to ensure that you have a *compliant* submittal.

The proposal manager for a given RFP should ensure that your internal publication group receives a copy of all of the instructions from Section L during the pre-kickoff planning activities. Your publication group can then develop RFP-specific page templates and provide the technical writers with word counts per page. Words-per-page guidelines are important so that the writers adhere to the page allocations established either within the RFP itself or by the proposal manager. To provide vague instructions such as, "Write 10 pages of text on systems engineering," could result in 2,500 words or 7,500 words, depending upon font, margin, and line spacing settings. Instead, provide written instructions to each writer, indicating that one should provide, for example, 1,500 words for Section 2.2, "Effective Software Engineering Approaches."

---

4   Certain government agencies use consultants to verify past performance client references.

4.7 Section M (evaluation criteria): Toward a maximum score

The general criteria by which the government will evaluate your proposal are presented in Section M of each RFP. The Competition in Contracting Act (CICA), as implemented in the FAR, requires that price or cost to the government be included as an evaluation factor in every source selection. In some cases (such as *National Aeronautics and Space Administration* [NASA] RFPs), points are clearly allocated for each *scored* portion of the proposal. However, many times the evaluation criteria must be derived from a somewhat vague narrative, as in Table 4.1.

Table 4.1
Example of
Nebulous
Evaluation
Weightings

Criteria	Weight
1. Offeror's understanding the problem and proposed technical approach	Most important
2. Offeror's applications-related experience	Less important
3. Experience and training of individuals who will work under the contract	Same importance as 2
4. Offeror's general experience in developing software of comparable size, complexity, and content to this project	Same importance as 2
5. Offeror's general experience in *maintaining and operating* (M&O) a computer-based system of comparable size, complexity, and content to this project	Less important than 2
6. Offeror's proposed management plan	Less important than 5

4.8 Greatest value approach

Although it tends to increase the cost of acquisitions, federal agencies have been encouraged proactively by the U.S. General Services Administration (GSA) to develop and issue source selection procedures that include the "greatest value approach," especially as it applies to the acquisition of Federal Information Processing (FIP) resources. Among the strengths of the greatest value approach from the federal government's perspective are that it (1) allows greater flexibility to balance technical and total cost factors on the basis of the proposals actually received rather than on the basis of a predefined formula that combines technical and total cost ratings or rankings; (2) leverages the experience and personal judgment of senior governmental decision makers; and (3) permits the evaluation to take place just before the start of actual contract performance [1].

4.9 Emphasis on performance-based contracting

Executive Order 12931, issued on October 13, 1994, directs executive agencies of the U.S. government to "place more emphasis on past performance and promote best value rather than simply low cost in selecting sources for supplies and services." Additionally, in accordance with the Office of Federal Procurement Policy (OFPP) Best Practices Guide of May 1995, "the use of past performance as an evaluation factor in the contract award process makes the award (de facto) 'best value' selections." The U.S. Navy led the way in this acquisition innovation in 1989 when they employed a methodology for "greatest value source selection" of firm-fixed-price supplies in which cost and past performance were the only factors for award.

It is a central principle of acquisition reform in the federal arena that source selections must place greater weight on past performance empirical data and commensurably less on the technical and management proposal writing skills in determining best value. A performance-based statement of work focuses on contractor accountability for required outcomes and results rather than on the details of how the contractor is expected to accomplish the tasks. Performance-based SOWs include methods and standards that will be used for performance measurement, clearly established deliverables and other reporting requirements, and well-defined task completion criteria.

Contract Data Requirements Lists (CDRLs) have also migrated toward a performance-based structure within Department of Defense procurements. Performance-based specification of contract data requirements defines the government's need for data (in hardcopy, electronic, e-mail, or other format) and gives the contractor latitude to propose the content and format of data to be provided to meet the requirement.

Because severe competitive pressure has encouraged some contractors to promise more than they can deliver, the government has found that it is particularly useful to compare actual performance with promised or bid performance on earlier or similar contracts. Did the contractor deliver the quality of work promised? Was the contract executed in a timely manner?

Was cost adequately controlled? With respect to the proposed effort, are proposed costs realistically estimated? Are salaries consistent with experience for the area? Do they raise questions about the ability of the contractor to perform as proposed and, thus, put the government at risk? Using a best value approach, these criteria can be compared and contrasted for each offeror to determine the overall greatest value for the government.

Your company's strategy for responding to an RFP should definitely take into account the evaluation criteria. For example, if "Key and Other Résuméd Personnel" are stipulated to count 50 out of 100 total points, then résumés and biographies should receive significant emphasis in your proposal response. Evaluation criteria should also serve as a guide for page allocations. Heavily weighted items should have an appropriately high number of pages allocated to them. Section M should also serve as a guide in your company's final bid/no bid decision for the particular procurement. Let us suppose that the evaluation criteria place significant scoring weight (50%) on past performance with the particular agency. Your firm, although it has marketed the client well, has only performed on one or two small contracts for this client. When you evaluate Section M and determine this, you may want to rethink the decision to expend the time, resources, and money to prepare a credible proposal.

4.10 Influencing the content of an RFP—legitimately

RFPs are generated by client staff who have problems to solve and needs to meet within certain budgetary and time constraints. Your company's marketing efforts should be directed toward understanding those problems and constraints and introducing the government staff to innovative solutions that can be applied better, faster, and more cost effectively than your competition. By talking with your client's staff at a variety of technical, programmatic, and contractual levels well in advance of the release of an RFP, your company may be successful in helping to shape the requirements of that RFP. Help your client understand his or her problem in terms that your company can meet. Consider the following: a government agency has a need to archive important hardcopy documents. Your company supplies storage space, file boxes, and warehousing staff. Your goal is to help the government staff understand their requirements in terms of physical warehousing. Now let's say that your company specializes in microfilming services. Your goal is to help that same government agency understand their requirement in terms of microfiche and microfilm processing. And finally, let us consider that your company's focus is on high-resolution document-scanning technologies and optical storage media. Your marketing goal is to define the requirement in terms of optical image processing and storage. Then when the RFP is written, the requirements stipulated will reflect the industry or technology that the government believes will provide the best solution. Influencing the RFP legitimately is a skillful craft, but it can be and is accomplished all the time. You will often hear of RFPs being "wired" for this or that company. A case could be made that with effective marketing, every RFP is predisposed in terms of technical requirements and position descriptions, for example, for one vendor or another, or at least one class of vendor over another. It is your company's task to predispose RFPs in your direction!

Influencing the RFP legitimately is the product of relationship building, careful listening, and idea sharing. Let us suppose that your company employs Ph.D.-level scientists with combined specialties in water resources management and advanced software design. Convincing your client that his problem requires the analytical depth that sophisticated computer models can provide might very well result in RFP position descriptions that require dual training in science and computer modeling for key staff. This capability may be one that your competition does not have.

In the course of marketing your client, you may determine that rapid, on-site response to task orders is perceived to be critical to the upcoming contract's success. You may be able to convince the client to stipulate a distance requirement in the RFP. That stipulation might require all offerors to have offices within 15 miles of the client's site—something which you have but which will cost your competition time and money to establish.

4.11 Other types of solicitation documents

In addition to federal government RFPs, there are RFPs released by state, municipality, and city governments as well as by private-sector companies and international governments (see Chapter 11).[5] Many of the marketing and proposal planning, designing, scheduling, writing, and publication guidelines and processes that apply to federal procurements can be used effectively with these other important types of proposals. Central to any proposal effort is developing a response that carefully, concisely, and completely meets your client's needs—whether that client is the U.S. Air Force Space Command, PEMEX in Mexico (petroleum organization), the

---

5   International RFPs may be called "tenders" or "Terms of Reference."

Environmental Protection Administration in Taiwan, Chrysler Corporation, or the Port Authority of New York and New Jersey.

Because private-sector companies are not obligated to adhere to the federal Uniform Contract Format, their solicitation documents can vary substantially. General Electric's or Public Service Electric and Gas's RFPs, for example, can be quite different than DuPont's or British Telecom's RFPs. Many private-sector solicitations, however, address the following major elements:

- Understanding the requirements;
- Technical approach;
- Project organization and management approach;
- Project schedule;
- Health and safety;
- Qualifications and experience: résumés and project summaries;
- Terms and conditions;
- Subcontractors;
- Cost and other financial terms (supplier financing, countertrade, and letters of credit).

In addition, some companies require the completion of a complex set of forms as an addendum to the proposal itself. These forms can request a variety of institutional, financial, and contractual information and can be very time-consuming to prepare.

With international proposals, efforts must be made to remove references to U.S.-centric references and regulations from project summaries and the main proposal narrative. In many cases, time must be allocated in the proposal response milestone schedule to have the proposal and its associated documents (such as corporate certifications and articles of incorporation) translated into another language, such as Spanish or Mandarin Chinese. Your company's ranking by prestigious business journals or organizations is often important to include in international proposals.

Let's take a closer look at private-sector solicitations.

END NOTE

1. See *Source Selection: Greatest Value Approach*, Document #KMP-92-5-P, U.S. General Services Administration, May 1995.

Chapter 5

Private-sector solicitation requests

IN ADDITION TO FAR-driven RFP-based proposals submitted in
response to federal government procurement solicitations, companies
should consider other important types of proposal documents such
as private-sector proposals, product proposals, R&D internal proposals,
grant proposals for education, international proposals, and health care
proposals. Many times, nongovernment proposals are most analogous to
bids in that *price* is the primary determinant for award. In the commercial
sector, communication with the client organization during the proposal
process is not precluded by FAR-type regulations, so your company's
marketing efforts can be more aggressive and protracted. In private-sector
solicitation documents, there may be no instructions to offerors (i.e.,
Section L) or published evaluation criteria (i.e., Section M).

A product proposal tends to be built upon a point-by-point format.
There is no management plan or other response required wherein you
have to take into consideration client preferences that go well beyond what
is explicitly in the RFP. Health care proposals are similar to product

proposals, and lend themselves to boilerplate response far more easily than solution-driven RFP-based proposal responses. In the grant field, private sources (e.g., John Simon Guggenheim Memorial Foundation) or government funders (e.g., National Endowment for the Humanities) are much more inclined to request a one- to three-page letter or completed form that explains your entire project and need.

Private-sector solicitation requests may be issued in hardcopy form as Requests for Proposal (RFPs) or Inquiries documents, and might also be presented as an informal telephone call or FAX from the client organization. Whatever the form or format of the request from the client, the response on the part of your company must be customized and targeted, reflect the relevant marketing intelligence you have obtained, and convey the benefits of doing business with your firm. Private-sector proposal documents that you will prepare can range from a very focused letter (1 to 4 pages) to volumes that are well in excess of 100 pages. In many instances, private-sector clients will include specific Vendor Qualification and Data Questionnaires, Bid Forms, and/or Professional Services Agreement (PSAs) that must be completed fully and accurately and included with your proposal submittal package.

Most private-sector proposals must be submitted to a named location and point of contact by a specific date and time. Although corporations and other private entities are not bound by the stringent proposal acceptance parameters of the Federal Acquisition Regulations (FAR 52.215-1) when conducting business with other companies for private projects, timely submittals are becoming increasingly important. Plan to have your private-sector proposal arrive on time and with the appropriate number of copies in your submittal package. And be sensitive to your client: refrain from sending a proposal to United Parcel Service via Federal Express, for example.

The following listing represents the variety of elements that *may* be required in a private-sector proposal. Note the similarity to federal proposal responses. Boldface items are among the ones encountered most frequently. Some elements may be presented as appendixes in your proposal document.

- **Introduction or Executive Summary**
- **Scope of Work**
- Understanding of the Project
- **Technical Approach**
  Alternative Approaches
  Bibliographic References

- **Experience and Qualifications**
  - Corporate contractual experience
  - Experience of key personnel
- **Team Organizational Structure**
- **Project Management Approach**
  - Roles and responsibilities
  - Project control
  - Ongoing communications with client
  - Availability of staff
- **Proposed Project/Task Milestone Schedules**
- **Task Subcontracting Approach**
- **Project Deliverables**

  Quality Control and Quality Assurance Programs

  Future Action Plans
- **Client References**
- **Résumés**
- **Project Summaries**
- **Sample Work Products and Deliverables**
- **Annual Reports/10Ks/Financial Statements/Dun & Bradstreet (D&B) Reports**

  Equal Opportunity Employer Program and Goals

  Training Certificates

  Licenses/Registrations/Certifications
- **Certificates of Liability Insurance**
- Affirmative Action Plan

  Existing Contract Disclosure

  Complete List of Contractor Affiliates and Their Locations
- **Representations and Certifications**
  - Type of business organization
  - Taxpayer Identification Number (TIN)
  - Corporate status (sole proprietorship, partnership, etc.)
  - Certification of Independent Price Determination (Non-Collusive Bidding and Code of Ethics Certification)
  - Preference for Labor Surplus Area (LSA)
  - Certification of nonsegregated facilities

- Certification regarding a drug-free workplace
- Cost accounting practices and certifications
- Certification of no investigation (criminal or civil anti-trust)
- **Indemnification** (protection for the client against the risk of legal claims)
  - Damage to persons or property
  - Negligence on the part of the contractor
- **Non-Conflict of Interest Statement**
- **Confidentiality Agreement**

  Health and Safety Plan

  Commercial Terms and Conditions

  Equipment Billing Rate Schedules
- **Project Costs**
  - Names, titles, and hourly rates for specific job classifications
- Sample Invoice
- Bid Bond
- Comprehensive General Liability Insurance and Worker's Compensation Insurance

As part of your proposal library, your company should have a collection (in both hardcopy and electronic form as applicable) of various licenses, certifications, insurance coverage, TIN numbers, training certificates, and so forth. Ready, albeit *controlled*, access to this information will make building your responses to private-sector solicitations much easier.

Next we will examine the dynamic arena of federal government acquisition in an era of *electronic commerce* (EC) and *electronic data interchange* (EDI). There are many new dimensions to consider. We will begin by focusing on how RFPs emerge from the federal acquisition process.

Chapter 6

The federal acquisition process: new directions

6.1 Overview

How do RFPs originate? The answer is that RFPs are extraordinarily complex contractual documents prepared by a variety of government employees in accordance with an intricate, formalized sequence of regulated procedures. It was during World War II that the federal government began to carefully regulate its procurement processes. From that time until the mid 1990s, United States procurement law evolved toward increased *complexity* (multiphased contracts and elaborate evaluation methods) and increased *competition*, even in subcontracting. The focus was on accountability, that is, objective and defensible acquisition decisions [1]. Contracting officials and technical staff alike were extremely concerned that the competitive efforts for which they had oversight responsibility were conducted in full accordance with appropriate agency-specific procurement regulations. However, federal acquisition streamlining and reform legislation enacted in 1994 and 1996 has placed more decision-making and discretionary authority back in the hands of government contracting officials.

6.2 Statutory and regulatory requirements for competition

The FAR system is part of the *Code of Federal Regulations* (*CFR*) [2]. It is the FAR that is the primary source of procurement regulations used by all federal agencies in their acquisition of supplies [3], construction, services [4], and research and development with appropriated funds. All of the provisions and clauses that are used in government contracting are found in the FAR, which is interpreted and applied in areas of dispute through a complex process of litigation and court and special governmental board decisions [5]. The DoD FAR Supplement, NASA FAR Supplement, Air Force Supplement, and Department of Energy Supplement all augment and amplify the FAR and should be used in conjunction with the FAR when determining acquisition regulations relevant to DoD and Air Force contracts, and NASA and DOE policies and procedures. Your company can order copies of the FAR and its supplements directly through the Superintendent of Documents U.S. Government Printing Office Washington, DC 20402 FAX (202) 275-0019. Agency supplements to the FAR and services regulations, such as *Air Force Regulation* (AFR) 70-15 and the *Naval Sea Systems Command* (NAVSEA) *Source Selection Guide*, complement the FAR but do not contradict it. You can also review and download the FAR from the Web at http://www.arnet.gov.far.

The FAR is the primary source of procurement regulations used by all federal agencies in their acquisition of goods and services.

The FAR system, which became effective on April 1, 1984, replaced the *Federal Procurement Regulation* (FPR) used by civilian agencies of the Federal Government, the *Defense Acquisition Regulation* (DAR) used by DoD, and the *National Aeronautics and Space Administration Procurement Regulation* (NASAPR) used by NASA. The Commission on Government Procurement, established on November 26, 1969, by Public Law (P.L.) 91-129, was given the charter of studying the statutes, policies, regulations, and practices affecting government procurement and recommending improvements. Following years of effort, the FAR was announced in the *Federal Register* on September 19, 1983 [6]. You can locate the important SFs included in RFPs in FAR Part 53.

The Federal Acquisition Regulations System was established to codify and publish uniform policies and procedures for the acquisition of goods and services by all executive agencies. The FAR record is divided into Subchapters a through h, and Parts 1 through 52. Give particular attention to Subchapter c, "Contracting Methods and Contract Types." The opportunity for your company to submit written comments on proposed significant revisions to the FAR is provided through notification of proposed changes in the *Federal Register*. The current volume of the *Federal Register* can be found online at http://law.house.gov/7.htm.

FAR Part 15—Contracting by Negotiation, establishes that in negotiated procurements the bidder must be responsible; evaluation criteria are flexible, that is, tailored to the procurement; and the acquisition process may be competitive or noncompetitive. Unlike sealed bidding (IFBs), negotiated procurement is not merely a series of steps. Every federal government agency has a somewhat unique pattern of procurement activity. Be sure to visit the contracting offices of your company's target client agencies and obtain copies of the handbooks that govern procurement practices for that particular agency.

6.3 The source selection process

Competitive negotiation is formally called "source selection." This process, which is regulated by the FAR at Subpart 15.3 and designed to select the proposal that represents the best value to the government, normally involves the following steps.

First, an RFP is prepared and publicized by the government. Federal government RFPs all look essentially the same in terms of major sections (A through M) as a result of the application of the *Uniform Contract Format* (UCF), established at FAR Subpart 15.204.1 (Table 15.1). As the Federal Government migrates toward electronic commerce (EC), the application of the UCF has become open to question.

Second, technical and cost proposals are submitted to the client organization by offerors in the contractor community.

Third, proposals are analyzed and evaluated by client staff against stated criteria and unstated standards.

An award can be made at this stage without discussions, based upon the decision of the Source Selection Authority (SSA).

Fourth, potentially successful proposals are identified and included in the "competitive range" (shortlisted, see FAR Subpart 15.306) based upon price and other factors; all others are eliminated from further competition and the offerors notified in writing of such.

Fifth, oral and written discussions are conducted with those offerors in the competitive range for the principal purpose of eliminating deficiencies in their proposals.

Sixth, those offerors are given an opportunity to submit *best and final offers* (BAFOs) (see FAR 15.306).

Seventh, BAFOs are evaluated and a contract award is made to the offeror whose proposal is most advantageous to the client, as determined on the basis of stated evaluation criteria.

Eighth, unsuccessful offerors are notified promptly in writing (per FAR 15.503).

Finally, debriefings are held with unsuccessful offerors.

According to FAR 2.101, *acquisition* means the acquiring by contract with appropriated funds of supplies or services by and for the use of the federal government. There are four primary phases in the acquisition process used by the United States government: (1) needs assessment/ determination; (2) acquisition planning; (3) solicitation release, proposal evaluation, contractor selection, and contract award (source selection phase); and (4) contract administration. Within NASA, for example, there are the following major steps in the acquisition process: (1) a candidate project is selected; (2) a commitment to project planning is made; (3) project planning review; (4) project approval by Deputy Administrator; (5) RFPs; and (6) system design, development, and operation (see Figure 6.1). The federal government *fiscal year* (FY) begins on October 1 each year.

Fundamentally, the source selection involves two processes:

1. Selection of a contractor;

2. Formation of a contract.

The federal government uses a hierarchical source selection organization, the size of which depends upon the complexity, size, and importance of the procurement [7]. The source selection process has traditionally been dominated by weapons procurement. The formal source selection procedures of the U.S. Air Force[1] serve as a frequent model for other federal agencies as well as international governments. All formal U.S. government source selection systems, such as presented in the *NAVSEA Systems Command Source Selection Guide* [15 February 1984], are structured for objectivity, legality, and thoroughness. Most systems use successively

---

1   *Air Force* (AF) Regulation (Reg) 70-15 prescribes Formal Source Selection for Major Acquisitions, and AF Reg 70-30 prescribes Source Selection for Non-Major Acquisitions. AFR 70-30, "Streamlined Source Selection Procedures," promotes page limitations in proposals.

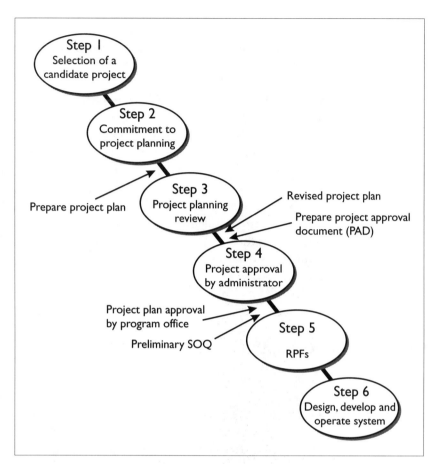

Figure 6.1
NASA Acquisition
Process. (*Source:
Statements of Work
Handbook* NHB
5600.2., p. 3.)

weighted levels of evaluation, which allow the government to assign relative importance to each evaluation criterion. In the Air Force system, a *Source Selection Authority* (SSA) is supported by a *Source Selection Advisory Council* (SSAC), which in turn is supported by a *Source Selection Evaluation Board* (SSEB).[2] The SSEB is actually groups of teams that evaluate the proposal. The SSEB teams are further divided into subpanels, areas, items, factors, and subfactors. It is at the level of "subfactor" that the actual scoring (evaluation) occurs. The overall evaluation is then a *compilation* of subscores. Scoring can take the form of numbers; a plus, check,

---

2   The *U.S. Army Corps of Engineers*' (USACE) source selection organization is the SSEB. In this agency, the SSEB is comprised of separate technical-evaluation and cost-evaluation teams. This organization is designed to ensure active, ongoing involvement of appropriate contracting, technical, logistics, legal, cost analysis, small business, and other functional staff management experience in the procurement process.

minus scheme; colors[3]; or adjectives (pass/fail; outstanding to unsatisfactory). Scoring of proposal responses is weighed against the prescriptions (standards) [8] set forth in the government's highly proprietary "Source Evaluation Guide (or Handbook)." The NASA system includes a *Source Selection Official* (SSO), an SEB, committees, panels, and subpanels. Again, evaluation occurs at the "subpanel" level. Of note is the fact that evaluation scores are *not* binding on source selection officials "as long as the official has a rational basis for the differing evaluation" [9].

In accordance with FAR 1.602-1(a), COs have the authority to enter into, administer, or terminate contracts and make related *determinations and findings* (D&F). The D&F detail the proposals that will be included in the competitive range and the reasons for those decisions. The "determination" is the conclusion or decision supported by the "findings." The findings are statements of fact or rationales essential to support the determination and must cover each requirement of the statute or regulation (FAR 1.701).

6.4 Full and open competition

"Full and open competition" means all responsible sources are permitted to compete. The major requirement of the *Competition in Contracting Act* (CICA), of 1984,[4] was that full and open competition is the required standard for awarding contracts. CICA '84 established the civil service position of *competition advocate* to promote and ensure the use of full and open competition whenever feasible [10]. "The Office of Federal Procurement Policy Act (41 U.S.C. §404) requires that each executive agency of the Federal Government designate an advocate for competition for the agency and for each procuring activity within the agency" [11]. The fact that cost and technical volumes are evaluated separately is not a statute or part of the FAR but is simply part of traditional practice. In full and open competitively negotiated bids, the CO is not bound by the Source Selection Board decision. He or she can, in fact, override the conclusions and recommendations of the selection board. And no BAFO is required.

No matter how objectively the RFP and the proposal evaluation process are structured, or how much the client's contracts office enforces competition in contracting protocol, in the final analysis the evaluation is one of human judgment.

---

3   For example, AF Reg 70-15 stipulates blue, green, yellow, and red, with blue being "exceptional" and red "unacceptable."

4   The CICA '84 was implemented in the FAR through Federal Acquisition Circulars (FACs) 84-5 and 84-6.

6.5 Major contract types

There are two basic types of contracts that result from the negotiated bid process: fixed price and cost reimbursement. A firm-fixed-price contract provides for a price that is not subject to any adjustment on the basis of your company's cost experience in performing the contract. This contract type places full responsibility and maximum risk on your company for all costs and resulting loss or profit. The latter calls for paying the contractor all incurred direct and indirect costs, as defined and specified in the contract, plus some profit [12]. When bidding fixed price opportunities, your company must ensure that it fully understands the scope of work and terms and conditions of the contract in order to ensure an acceptable profit.

6.6 Significant recent paradigm shifts in federal government acquisition

The federal government spends approximately $200B each year in procuring goods and services [13] out of a total budget of $1.7 trillion for FY99. Managing and modifying the complex set of processes by which the government procures goods and services has proven to be daunting.

David Osborne, a senior advisor to the Clinton administration and co-author with Ted Gaebler of the best-selling *Reinventing Government: How the Entrepreneurial Spirit Is Transforming the Public Sector* (1992), foresaw a profound movement away from large, centralized, command-and-control bureaucracies toward decentralized, entrepreneurial organizations. These new governmental organizations will be driven by competition and accountability to their customers for the results they deliver [14]. In a manner that echoes the work of Alvin and Heidi Toffler, Osborne contends that industrial-era bureaucracies must be restructured so that they can handle the problems of the information age. Movement in this direction can be seen in *Business Week* magazine's recent rating of the Social Security Administration's 800 telephone service as the best in the business, surpassing even L.L. Bean and Disney.

Internationally known futurists Alvin and Heidi Toffler have demarcated human history in accordance with three major watershed processes [15]. Each of these three dynamic waves resulted in, and continues to result in, fundamental human social, political, and economic shifts and transitions. The invention of agriculture 8,000 to 10,000 years ago is identified as the first major wave in human organizational and technological dynamics. The *hoe* is the Tofflers' symbol for this first phase. They then point to the Industrial Revolution of approximately 250 years ago, wherein brute-force technologies amplified human and animal muscle power and gave rise to an urbanized, factory-centered way of life. Emphasis was on mass production, mass consumption, mass media, and mass education—each symbolized by the *assembly line*. This Second Wave continues to energize various parts of the world today, and is manifested in emerging nationalist movements as well as existing nation-states.

In addition, unfolding at present is the "Third Wave." Pentium-class computers, electronically accessible satellites, and the ubiquitous Internet with its far-flung backbone of servers and hyperlinked documents, along with customized production, micro or particle markets, and single-issue political groups all characterize this new transition phase. The Tofflers symbolize knowledge-, information-, or mind-based technologies and their associated social changes by the *computer*. Economies of scale—bigness—are being replaced by economies of speed in this digital age [16]. And Swedish Professor Johan Roos of the International Institute for Management Development (IMD) in Switzerland speaks of the emerging "knowledge economy" and "knowledge landscapes" of the late 1990s.

6.7 National Partnership for Reinventing Government (NPR)

Morley Winograd, a former senior policy advisor to Vice President Al Gore, now serves as the director of the National Partnership for Reinventing Government (NPR), formerly the National Performance Review [17]. The mission of this federal task force—"America@OurBest"—is literally to reinvent government to work better, cost less, and get results that Americans care about. For example, toward this objective the NPR designated the Department of Education's Office of Student Financial Assistance as one of 32 High Impact Agencies (HIAs), that is, federal agencies that serve the most Americans. In October 1998, the president signed a law that established the Department of Education as the first Performance-Based Organization (PBO). Executive leadership within the department will be personally responsible for delivering results that center on partnerships, information technology, and customer service. Students and families are already able to apply for federal student aid over the Internet [18].

The former National Performance Review was created by President Clinton on March 3, 1993. Led by Vice President Al Gore, the NPR was an interagency task force of approximately 250 career civil servants from various levels of government. David Osborne served as a key advisor to Mr. Gore during this initiative. Federal agencies were also directed to form their own internal reinvention teams that worked in concert with corresponding NPR teams. The task force's final report, "Creating Government That Works Better and Costs Less," was presented to the president on September 7, 1993. By December of that year, President Clinton had signed 16 directives that implemented specific recommendations from the final report, including reducing the federal workforce by more than 250,000 people, cutting federal internal regulations by 50%, and requiring federal agencies to establish customer service standards.

Following the release of the NPR report, 50 members of the task force remained in place to start the implementation of a series of initiatives. These broad-based efforts spanned training federal employees about customer service, promoting the streamlining of headquarters' functions, staffing NPR-recommended cross-agency councils, and creating reinvention laboratories to pilot innovations. And NPR created an automated tracking system to monitor progress on individual action items. A brief examination of two federal agencies' progress will provide insight into the general trend throughout the federal government.

At the heart of the Air Force's new "Acquisition and Sustainment Reinvention Process" is the concept of Reinvention Teams. Each team, headed by a hand-selected leader and made up of field personnel from multiple disciplines, will study and develop selected ideas in specific high-payoff areas. Their charter is to look beyond tasks, jobs, and organizational structures to focus on processes and process improvement. They will redesign key processes and package the change to become part of the acquisition culture within the USAF. Reinvention Team results will take the form of executable actions, whether allowing for the future test of concepts the team has developed or paving the way for deployment across the Air Force acquisition and sustainment community. Planning for deployment will include creation of the tools, education, and supporting policies needed for successful implementation in the field.

Within the Department of Transportation (DOT), the Procurement Management Council (PMC) has taken highly effective ownership of the acquisition reform initiatives. The Acquisition Re-engineering and Realignment Taskforce (ARRT) has been charged with dramatically reinventing DOT's procurement processes. One direct, very visible result was that the entire agency was designated as a Procurement Reinvention Laboratory—a dynamic, evolving concept that has received the support and active endorsement of DOT's senior management. The objective is to conceptualize and implement innovative strategies within, and actually outside of, the current parameters driven by the Federal Acquisition Regulations (FAR), Transportation Acquisition Regulations (TAR), and Transportation Acquisition Manual (TAM). The ultimate goal is to have the DOT's procurement process work better, cost less, and enable the timely receipt of satisfactory products and services. In addition, DOT will share its success stories and "best practices" with other federal agencies to foster government-wide acquisition process improvements.

In a letter from then-Speaker of the House Newt Gingrich, to Jacob J. Lew, acting director of the Office of Management and Budget (June 9, 1998), congressional evaluators singled out for praise the Department of Transportation along with the personal involvement of both Secretary Rodney Slater and Deputy Secretary Mortimer Downey. "Transportation produced the best strategic plan and the best performance plan of all the agencies we evaluated" [19]. Sharing specific information about innovative procurement thinking and strategies, technology breakthroughs, and reductions in life cycle costs (LCC) constitutes the cornerstone of DOT's Reinvention Laboratory. DOT can already point to specific successes—the formal source selection threshold was increased from $5M to $50M [TAM 1215.612(a)(1)], draft RFPs have been implemented to foster communication and partnership with industry, and oral proposals are beginning to set the trend. Within the Federal Aviation Administration, one of the 12 operating administrations of DOT, all procurements in excess of $5M are required to include a draft RFP to industry.

In 1994, the 103rd Congress enacted about 25% of the NPR recommendations that required legislative action. The Federal Acquisition Streamlining Act (FASA) of 1994 was one important piece of legislation that became law in October of that year [20].

6.8 Understanding the impact of the Federal Acquisition Streamlining Act of 1994

Signed by President William J. Clinton on October 13, 1994, the *Federal Acquisition Streamlining Act* (FASA) of 1994 (P.L. 103-355) "repeals or substantially modifies more than 225 provisions of law to reduce paperwork burdens, facilitate the acquisition of commercial products, enhance the use of simplified procedures for small purchases, transform the acquisition process to electronic commerce, and improve the efficiency of the laws governing the procurement of goods and services" [21]. FASA streamlines the acquisition process by providing greater discretion to contracting officers. It increases the *Truth in Negotiations Act* (TINA) thresholds and streamlines the procurement process for commercial items. "The bill strongly encourages the acquisition by federal agencies of commercial end-items and components, including the acquisition of commercial products that are modified to meet government needs " [22]. Instrumental in the generation of this specific legislation was Executive Order 12862, the

Government Performance and Results Act (1993), and the Chief Financial Officers Act of 1990.

FASA took effect on 1 October 1995. It emphasizes the acquisition of COTS items, streamlines acquisition procedures under an elevated small purchase threshold, implements a governmentwide electronic commerce system, establishes uniformity in the procurement system, improves protest and oversight processes, and authorizes specific pilot programs.

The trend in federal acquisition is toward electronic commerce, purchase of commercial products, and streamlining the procedural framework.

The Federal Acquisition Streamlining Act also established FACNET, which requires "the government to evolve its acquisition process from one driven by paperwork into an expedited process based on electronic data interchange" [23].

EC/EDI should not be confused with the Internet [24]. Currently, the Internet is not as secure as it needs to be for contractual transactions. Electronic data interchange refers to direct computer-to-computer exchange of data that are machine-readable and processable using a public standard. EC/EDI is a "different medium created to process government bids and contracts" [25]. It is the paperless transfer of routine business documents. Government procurement will be 100% accomplished via EC/EDI by the year 2003, in accordance with FASA. The EDI program will allow companies to sort through listings to locate contracts and then submit bids electronically via modem. The U.S. SBA and the California Technical Assistance Network, for example, are assisting small businesses to gain access to the information superhighway for better access to contracts [26]. The SBA is working with the APL Group, a software manufacturer in Wilton, Connecticut, to develop a program for personal computers that will provide EDI access [27].

Centralized contractor registration is required to inform the government that your company is ready to do business with the government using EDI. The *Central Contractor Registry* (CCR) sites are in Columbus, Ohio, and Ogden, Utah. Registration in the CCR database involves preparing and submitting a complete Trading Partner Profile (ANSI X12 838). Your company will register once and have your Trading Partner Profile shared with all government agencies. Your *Value-Added Network* (VAN) will be able to assist you with this process, or visit the Central Contractor Registration home page at http://ccr.edi.disa.mil/ccragent/plsql/ccr.welcome. A trading partner is a business that has agreed to exchange basic business information electronically.

Each trading partner will be assigned a unique identification number. The information required of each trading partner is:

- Company name;
- Address;
- General point of contact (name, telephone, FAX, e-mail);
- Electronic data interchange point of contact;
- Contractor identification numbers [Commercial and Government Entity (CAGE) code, Taxpayer Identification Number (TIN), Data Universal Numbering System (DUNS), Standard Industrial Classifications (SIC)];
- Electronic data interchange and address.

In addition to FASA, further modifications to government procurement occurred with the enactment of the Federal Acquisition Reform Act of 1996, a part of the National Defense Authorization Act. The most significant change introduced by the 1996 procurement rules is that federal officials have been given more latitude to award contracts based upon a contractor's performance or expertise rather than price alone [28]. "Awards will be tied to trust and confidence" [29]. Government COs will be given greater flexibility in limiting competition where advisable. This measure will also reduce bid protests [30]. FARA also permits limiting the number of offerors (contractors or vendors) that must be included in competitive range determinations. In addition, certain contractor certification requirements were eliminated, and specific government employees were authorized to make "micropurchases" (purchases of less than $2,500) without competition.

In light of these federal acquisition reforms, solicitation requirements were reduced greatly for two key military weapons systems—the *Joint Direct Attack Munition* (JDAM) and the Army Tactical Missile System Block II. JDAM, which had been identified under FASA as a pilot program for reform, was released with a seven-page SOW versus the original 100-page version. The contract data requirements were reduced by 70%, and *military specifications* and *standards* (MIL-SPECs and MIL-STDs) were eliminated completely. When the Army's Tactical Missile System *Block I* RFP was released, it contained a 503-page SOW and 112 MIL-SPECS. The *Block II*, on the other hand, had an 11-page SOW and no MIL-SPECS [30].

6.9 Federal acquisition computer network

The Federal Acquisition Streamlining Act of 1994 (Title IX) mandated the establishment of a Federal Acquisition Computer Network (FACNET) architecture to enable federal agencies and vendors to do business electronically in a standard way. FACNET is the federal government's EC/EDI for the acquisition of supplies and services. This pivotal network facilitates the two-way electronic data interchange of acquisition information between the government and the private sector. It employs commercially available hardware and software and nationally and internationally recognized data formats, and provides universal user access. FACNET is the preferred contracting method for purchases between $2,500 and $100,000. The network may be used for products, services, and construction (drawings and wage determinations may have to be supplied in hardcopy form).

FACNET is a universal electronic capability that will permit potential contractors to, at a minimum, obtain information on proposed procurements, submit responses, query the system, and receive awards on a government-wide basis. The system, which should be fully functional in five years, is being designed to inform the public about federal contracting opportunities, outline the details of government solicitations, permit electronic submission of bids and proposals, facilitate responses to questions about solicitations, enhance the quality of data available about the acquisition process, and be accessible to anyone with a personal computer and a modem. The Administrator for Federal Procurement Policy, who heads the Office of Federal Procurement Policy (OFPP) in the Office of Management and Budget, has responsibility for overall policy direction and leadership of the FACNET program. The Electronic Commerce Acquisition Program Management Office (ECA-PMO), co-chaired by the General Services Administration (GSA) and DoD and reporting to the administrator, has been chartered to coordinate and oversee FACNET implementation throughout the federal government. Several agencies have been tasked to lead specific government-wide FACNET projects. In particular, DoD has lead agency responsibility for developing, operating, and supporting the FACNET infrastructure and the Centralized Contractor Registration database used with it [31].

The Naval Sea Systems Command (NAVSEA) no longer mails hardcopy solicitation documents but requires offerors to obtain copies electronically by dialing into its agency BBS. The *U.S. Environmental Protection Agency* (USEPA) has also established its own BBS on which they post solicitations for offerors. USEPA will still mail hardcopies of solicitation documents to interested parties. FACNET differs from agency bulletin boards in that each company will need to subscribe to an on-line service (VAN) operated not by the government but by numerous profit-driven private concerns. Currently, FACNET seems more appropriate for the acquisition of products rather than services, although service-type procurement transactions are increasing in number. Approximately 400 government agencies were distributing their solicitations through FACNET as of late 1997 [32]. The U.S. Army Corps of Engineers, for example, has aggressively implemented FACNET.

The U.S. General Accounting Office (GAO) reviewed the federal government's progress in developing and implementing FACNET, focusing on federal agencies' use of FACNET, problems and benefits of using FACNET, concerns relating to the Federal Acquisition Streamlining Act's requirements for FACNET, and management obstacles to effective government-wide implementation of FACNET. GAO found that overall, the federal government has executed relatively few procurement actions through FACNET. Officials from at least 14 of the 18 agencies GAO contacted rated the lack of a sound FACNET infrastructure, effective engineering, and operational management, and a well-populated and fully functional centralized contractor registration database as significant obstacles to effective FACNET implementation. Many federal agency officials said the current FACNET approach is out of step with new, cost-effective technologies and buying practices. Agencies and vendors have consistently cited leadership and management shortcomings as major reasons for delays and unresolved problems in FACNET implementation [31]. Dr. Steven J. Kelman, administrator of the Office of Federal Procurement Policy, has acknowledged FACNET's flaws and has indicated that OFPP has revised its EC strategy to emphasize all aspects of automated procurement [33].

GAO determined that although FACNET has, in some instances, resulted in lower prices and expanded access to vendors, agency officials and vendors have often said that FACNET is not producing the benefits expected. Of note, however, is that federal organizations with the most success in using electronic data interchange technology for purchasing typically use it to transmit high-volume, routine, and repetitive transactions with a small group of known suppliers [31].

Already, the U.S. Health and Human Services Department is piloting a commercial electronic system that provides contracting firms with a Web-based alternative to FACNET [34]. Significant, recent advances and efficiency gains in hardware platform computing power and speed, software and network applications, signal and telecommunication technologies, data and information processing strategies, and encryption and digital signatures have made a toolbox of alternatives to FACNET—with its reliance upon ANSI X12 EDI transaction set—not only possible, but desirable, secure, and cost effective.

6.10 Benefits of electronic data interchange

EDI is intended to increase business opportunities through wider diffusion of procurement information. There are to be fewer errors in data, reduced processing times, less reliance on human interpretation of data, and reduced unproductive time. Greater competition and reduced prices to the government are the ultimate goals. Savings are to be realized through reductions in inventories, mailroom sorting/distribution time, elimination of lost documents, and reduction in postage and other mailing costs. EDI will also facilitate the flow of better and more up-to-date information for enhanced management decision making. According to one VAN located near Philadelphia, more than 50,000 companies nationwide were using EDI in 1996.

6.11 Understanding ANSI X12 standards

The *American National Standards Institute* (ANSI) is the coordinator and clearinghouse for national standards in the United States. ANSI-chartered organizations called *Accredited Standards Committees* (ASCs) are composed of voluntary representatives from industry, labor, consumer groups, and government to prepare consensus standards. Upon public comment, ANSI ASCs publish national standards.

The ASC X12 was chartered to develop the structure, format, and content of business transactions conducted through the EDI. The ASC X12 was supported by the *Data Interchange Standards Association, Inc.* (DISA). The results of the ASC X12 committee's efforts are the ANSI X12 standards.

An EDI transaction involves the electronic transmission of a business document in the form of a transaction set that is prepared in accordance with an ANSI X12 standard format. In other words, a transaction set is the electronic equivalent of a document, such as a purchase order or request for quotation, enclosed in an "electronic envelope." Just as you can enclose several paper letters in one envelope, you can send several transaction sets (or "functional groups") enclosed in one electronic envelope. There are currently almost 200 transaction sets that support communications and controls, product data, finance, government, materials management, transportation, purchasing, industry standards transition, distribution and warehousing, and insurance. Important ANSI X12 transaction sets include: (1) ANSI 840–*Request for Quotation* (RFQ); (2) ANSI 843–Response to an RFQ; (3) ANSI 850–*Purchase Order* (P.O.); and (4) ANSI 838–Trading Partner Profile (see Figure 6.2). The federal

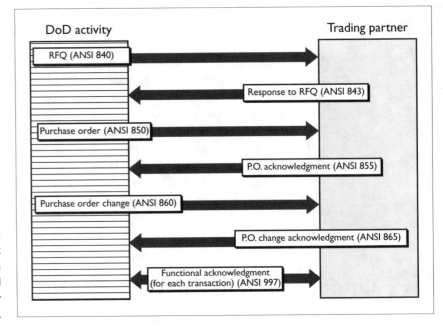

Figure 6.2
Typical document
exchange between
the DoD and
private-sector
trading partners.

government has endorsed the use of ANSI X12 standards for EC/EDI with the U.S. Government through the *Federal Information Processing Standards* (FIPS) Publication 161.

ANSI X12 standards build on the success of private industry in implementing EDI. Trading partners will be able to take advantage of COTS ANSI X12-compliant translation software, an example of which is MarketLink™ 1.0.

In addition to ANSI X12 standards, there is another United Nations-sponsored set of EDI standards. This second set is called EDI for Administration, Commerce and Transport (EDIFACT). The EDIFACT standards are primarily used in Europe and Asia. However, in order for everyone to benefit from a single global EDI standard, ANSI X12 agreed in 1997 to begin a gradual alignment with EDIFACT.

6.12 Sources of information on EC/EDI and acquisition reform

Important resources for information about EC/EDI and acquisition reform are presented in Table 6.1.

U.S. Small Business Administration's primary Web site:

> http://www.sba.gov
> See also http://www.sba.gov/y2k and http://www.dell.com/smallbiz/y2k for a
> free risk assessment of year 2000-related computer issues. SBA loan
> guarantees are available for small businesses that face Y2K challenges.

U.S. Small Business Administration's (SBA) *Pro-Net*:

> http://pro-net.sba.gov

Central Contractor Registration (CCR):

> http://ccr.edi.disa.mil/ccragent/plsql/ccr.welcome

Joint Electronic Commerce Program Office (JECPO):

> http://www.acq.osd.mil/ec/

Procurement Technical Assistance Center (PTACs):

> http://tsd.r3.gsa.gov/bsc/bsc_iiih.htm

Electronic Commerce Resource Centers (ECRCs):

> http://www.ecrc.ctc.com/about.htm

DoD-certified Value Added Network (VAN) list:

> http://www.armyec.sra.com/knowbase/interim/main.htm

Federal government's free *Commerce Business Daily* site (CBDNet):

> http://cbdnet.access.gpo.gov/index.html

SBIR Information:

> http://www.eng.nsf.gov/sbir/about_sbir.htm
> http://www.sbaonline.sba.gov/hotlist/sbir.html

Federal Acquisition Jumpstation:

> http://nais.nasa.gov/fedproc/home.html
>   – Procurement and acquisition sites by contracting activity

FedWorld Information Network:

> http://www.fedworld.gov/detail.htm

U.S. Army Electronic Commerce Center:

> http://www.armyec.sra.com/
> The Army Electronic Commerce Center Web site provides a wide variety
> of services related to the implementation of electronic commerce within the
> Army community. These services include an EC information repository; links
> to EC-related sites; EC news flashes, upcoming events, and success stories;
> and a virtual electronic forum.

Table 6.1
EC/EDI
Information
Resources

Navy Electronic Commerce Online (NECO):

> http://ecic.abm.rda.hq.navy.mil/
> - Direct online access to Navy solicitations
> - Daily announcements of procurement opportunities to registered vendors
> - General Navy information on electronic contracting initiatives

The Federal Marketplace Procurement Data Warehouse (Wood River Technologies, Inc.):

> http://www.fedmarket.com/sales_resources/bids/federal.html
> - Federal procurement
> - Federal information
> - Electronic Data Interchange (EDI)
> - Small Business Innovative Research (SBIR) Program
> - Military Specifications and Standards
> - Federal laboratories
> - Business resources
> - Active federal contracts
> - *Commerce Business Daily*
> - Laws and regulations
> - State and local procurement
> - International procurement

NASA Acquisition Internet Service (NAIS):

> http://ec.msfc.nasa.gov/NAIS/html/naishome.html
> - NASA Center-specific procurement projections

U.S. Army Acquisition Web site:

> http://acqnet.sarda.army.mil/home.htm

Office of the Undersecretary of Defense for Acquisition and Technology:

> http://www.acq.osd.mil/acqweb/jump/acq.html

Office of Small and Disadvantaged Business Utilization (OSDBU) Directory:

> http://www.sbaonline.sba.gov/GC/osdbu.html

Assistant Secretary of the Air Force/Acquisition (SAF/AQ):

> http://www.safaq.hq.af.mil/
> Provides information on Air Force-specific projects and procedures.

Department of Transportation's Office of Small and Disadvantaged Business Utilization (OSDBU):

> http://osdbuweb.dot.gov/main.shtml

General Services Administration (GSA) Electronic Posting System (EPS):

> http://eps.eps.gov

Table 6.1
(continued)

NetLizard (maintained by Panamax, Inc., St. Petersburg, FL):

> http://www.netlizard.com/acqpol.html
> - Federal opportunities
> - Selected state opportunities
> - Selected international opportunities
> - Federal regulations and procedures
> - Statutes governing federal contracting and acquisitions
> - Court and other decisions affecting federal acquisitions and procurement

Acquisition Reform Network (ARNet):

> http://www.arnet.gov
> Run by the Office of Federal Procurement Policy, the Acquisition Reform Network is an enterprise created to foster and propagate measurable breakthrough improvements in the way that government obtains goods and services.

Small Business Resource Page:

> http://www.neonblue.com/buslinks.htm

Women Business Owners Corporation:

> *Federal Acquisition Reform:* http://www.wboc.org/pfar.html
> *US Government Resources:* http://www.wboc.org/usgovt.html
> *State and Local Government:* http://www.wboc.org/usmap.html
> The Women Business Owners Corporation, a national 501(c)(3) not-for-profit corporation, was established to increase competition for corporate and government contracts through implementation of a pioneering economic development strategy for women business owners.

Government Executive Magazine's Procurement Links:

> http://www.govexec.com/procure/proclink.htm
> Several dozen World Wide Web sites related to government procurement issues are now on-line. This site's reviews of select procurement sites, organized by functionality, can help maximize on-line time by minimizing the number of clicks it takes to reach the information needed. *GovExec.com* is government's business news daily and a premier Web site for federal managers and executives.

Service Corps of Retired Executives (SCORE):

> http://www.score.org/
> Formed in 1964, the SCORE Association is a nonprofit association comprised of 12,400 volunteer business counselors throughout the U.S. and its territories. There are 389 SCORE chapters in urban, suburban, and rural communities. SCORE members are trained to serve as counselors advisors and mentors to aspiring entrepreneurs and business owners. These services are offered at no fee, as a community service.

Table 6.1
(continued)

Federal Register:

> http://law.house.gov/7.htm

Electronic Commerce World (formerly *EDI World magazine*):

> http://www.ecomworld.com/
> *Electronic Commerce World*, a monthly magazine of electronic commerce helps integrate management and technology with comprehensive content on topics including electronic commerce implementation, financial EDI, electronic messaging, workflow automation, and imaging. The magazine provides a common ground for executives and middle-to-upper level managers to increase their understanding of electronic commerce. Tel. 1-800-336-4887 or (954) 925-5900.

Table 6.1
(continued)

6.13 Department of Defense Electronic Commerce Office

Established by the Deputy Under Secretary of Defense (Acquisition Reform) and led by the Director of DoD Electronic Commerce, the DoD EC Office manages the implementation EDI-based contracting systems at 244 installations within the DoD. These sites initiate 98% of DoD's small purchases. When completed, these EDI systems will enhance access by small businesses to small purchase RFQs. The EC Office is also responsible for facilitating the implementation of EC/EDI across all functional lines of the DoD.

6.14 DoD Electronic Commerce Information Center (ECIC)

This important center serves as the central repository of information on EC/EDI available to private-sector companies and government agencies. The Center was established to support and encourage the use of electronic commerce technologies within the DoD and other federal agencies. The Electronic Commerce Information Center assists potential DoD contractors with EDI and *Commercial and Government Entity* (CAGE) code registration.

A CAGE code is a five-character identifier for companies doing or wishing to do business with the federal government. The code is used to support a variety of automated systems throughout the government, including bidders' lists. In some cases, prime contractors may require their subcontractors to have a CAGE code also. The Defense Logistics Information Service (DLIS) in Battle Creek, Michigan, is the only authorized source of CAGE codes. A portion of the CAGE code application must be completed by an authorized Procurement Technical Assistance Center (PTAC) or Electronic Commerce Resource Center (ECRC). These DoD-funded centers provide assistance to small businesses and will assist your company in obtaining a CAGE code. PTACs are located nationwide; a complete listing can be found on the Web at http://www.fedmarket.com/tecassis.html. And ECRCs—with 12 regional locations across the United States—serve as a catalyst for a vast network of small- and medium-sized enterprises to adopt electronic commerce. You can learn more about ECRCs and the technical outreach and support they provide by visiting http://www.ecrc.ctc.com/about.htm.

6.15 DoD electronic commerce/electronic data interchange infrastructure

This infrastructure is a system of interconnected communications and computer systems that support the exchange of EDI transactions between the government and its private-sector trading partners.

6.16 Using value-added networks and value-added services

A VAN is generally a commercial entity (similar to a long-distance telephone company or a computer on-line service) that provides telecommunication services, electronic store and forward mailboxing, and other communications services for EDI transmissions. VANs are government-certified private companies that will "receive, share, sort, and electronically list all procurement opportunities routed through FACNET" [35]. In many ways, a VAN serves the same purpose as the U.S. Post Office, except that it is electronic. Mail is received by the VAN, sorted, stored, and routed between the government and businesses, as well as business to business. In addition, companies receive notification of awards, purchase orders, and submit invoices electronically through their VANs. The Federal Government sends all RFQs (also referred to as an "840") to the VAN, which then sorts through the thousands of daily RFQs to select only the type that your company has requested. The select RFQs are sent to your company electronically in EDI format. If your company elects to submit a bid, you send back a Response to Request for Quotation (an "843"). The VAN receives your 843 and routes it to the federal agency that issued the RFQ. The federal government will then award the contract via a PO (an "850"). Notices of award and invoices are also routed and delivered by the VAN.

VANs are necessary because it would be too expensive for the federal government to establish direct, point-to-point connections with all of its trading partners (for example, contracting firms such as your company) throughout the private sector (see Figure 6.3).[5] When selecting a VAN, your company may want to consider the following factors:

- Communication speeds and protocols supported by the VAN;
- Fixed or variable cost structure for the VAN's services;
- Data backup and recovery services provided by the VAN;
- Data security features that the VAN offers;
- Transmission status reports provided by the VAN;
- Transaction filtering (so that you receive only the government solicitation transactions (RFQs, for example) in which your company is interested, based upon your strategic plan);
- ANSI X12 standards compliance.

VANs as well as *value-added services* (VASs) are certified and tested by the DISA. VAN services can range from $60 to $1,200 per month, with an average being about $250. Most have a start-up fee—ranging from $150 to $300, and most require one-year contracts. Charges are assessed according to a company's trading partner profile and the number of ANSI X12 transactions per month.

The government envisions that many companies will use the services of VANs to access the FACNET system. A current listing of all DoD-certified VANs can be obtained by calling the DoDefense

---

5   An Electronic Commerce Processing Node (ECPN) is a collection of hardware and software systems that provides communications connectivity between VANs and government gateways to support EC. There are currently two ECPNs: one in Columbus, Ohio, and the other in Ogden, Utah.

Electronic Commerce Information Center at 1-800-EDI-3414 or sending a fax to (703) 696-0213. You may also connect to http:www.armyec.sra.com/knowbase/interim/main.html.

- Computer Network Corporation, Wilton, CT 06897, 1.800.828.0770; http://www.compnet.com
- GE Information Services, Inc. (GEIS), Rockville, MD 20850-1785, 1-800-742-4852; http://www.geis.com
- VANSAT, Oklahoma City, OK 73142, (405) 720-4745; 75362.2062@compuserve.com
- Technology Management Programs, Inc., Carlsbad, CA 92009, (760) 431-8133; http://www.rfqnet.com

A VAS may be a separate commercial entity (also known as an EDI service bureau) or a VAN that provides additional fee-based services ranging from translation and segregation of the data to complete turnkey business systems support for customers. VAS services may include EDI-to-facsimile (FAX) service as well as complete EDI-integrated business systems. Such VAS services are especially applicable to small government contractors who opt not to invest in EDI-related computer hardware and software. A wide variety of such services are available and are advertised in publications such as the *EDI Yellow Pages* (Phillips Business Information, Inc., 1-800-777-5006; http:www.biz-lib.com/ZPHEYP.html).

6.17 Electronic Bid Sets (EBS)

Whereas FACNET is an electronic two-way transaction system that posts and receives bids for procurement, the Electronic Bid Set (EBS) system is designed to expand gradually as information technology (IT) improves and changes. Currently, the EBS system disseminates information (one-way transaction) to contractors in electronic format. It does so officially via CD-ROM, and unofficially as of 1999 on the Internet. The EBS system is designed for service contracts in excess of $500K. FACNET is used for procurement actions under the simplified acquisition procedures threshold, which, for interim-FACNET-certified sites, is $100K. EBS is designed for service-type contracts in the Construction, A-E, Environmental, and Maintenance arena where there are significant volumes of data in the form of plans and specifications. Currently, EBS is targeted for service contracts greater than $500K and primarily in the construction arena. The USACE has been very proactive in developing this EBS concept.

FACNET receives bids electronically; however, legal issues have arisen with regard to the date and time a bid is received electronically as well as technical and managerial challenges. For large EBS projects, the official plans and specifications will be mailed by the contracting agency. The only difference is that, instead of mailing paper, the agency will mail electronic versions. Although plans and specifications will be placed on the Internet for convenience, due to security issues, the accuracy of this information will not be guaranteed. The official document will be mailed until secure electronic transactions are developed. Unlike FACNET, EBS does not require any subscription service because the official contract document is mailed to contractors on CD-ROM. However, because the information is available over the Internet, it is recommended that contractors subscribe to local Internet Service Providers (ISPs).

6.18 Electronic commerce outside of DoD

A specific "test case" of electronic commerce in the civilian agencies is the *Federal Aviation Administration*'s (FAA) implementation of a new acquisition management system in April 1996. Set in motion by the Department of Transportation's 1995 appropriations bill, this initiative was designed to move the FAA's procurement practices closer to the commercial market and to exempt the FAA from a broad array of federal procurement laws [36]. The new FAA procurement rules are included as part of the FAA's Acquisition Management System manual, available at http://www.faa.gov/asu/asu100/acqreform/acquis.htm (see Appendix A for the FAA web page and its contents). The FAA plans to use the World Wide Web in its procurement activities, and the National Science Foundation (NSF) has embarked on a project called "Fast Lane" that uses the WWW to build grant proposals online [37].

Given the trajectory of FASA (1994) and the Federal Acquisition Reform Act of 1996, the trend in federal acquisition is clearly toward electronic commerce, electronic data and information interchange, procuring commercial or *nondevelopmental items* (NDI) (that is, items previously developed for government rather than commercial use), and streamlining the complex procedural framework of the FAR and associated regulations.

Now that we have an understanding of RFPs and how they originate in the federal acquisition process, we will examine the contracting community's proposal response lifecycle and its critical components.

END NOTES

1. *Writing Winning Proposals*, Shipley Associates, 1988, p. 3-3. The phenomenon of increased competition is also noted by Cibinic, John, and Ralph C. Nash, *Formation of Government Contracts*, second edition, Washington, D.C.: The George Washington University, 1986, p. 522.

2. The National Archives has published the *CFR* annually since 1938. This compilation of executive orders, proclamations, and rules and regulations for departments and agencies does for administrative law what the *United States Code* (USC) does for statute law. Material for the *CFR* is drawn from the calendar year entries in the *Federal Register*, a daily publication of Executive Branch documents and notices of public applicability and legal effect. (*History at NASA* (NASA HHR-50), Washington, D.C.: NASA Headquarters, 1986, p. 11.)

3. Per FAR 2.101, "supplies" means all property except land or interest in land.

4. Per FAR 37.101, "services contract" means a contract that directly engages the time and effort of a contractor whose primary purpose is to perform an identifiable task rather than to furnish an end item of supply.

5. The Comptroller General of the United States and the *General Accounting Office* (GAO) that he heads "have been given a prominent role in the oversight of Government procurement and procurement related functions." Cibinic and Nash, *Formation of Government Contracts*, p. 53.

6. *Introduction to the Federal Acquisition Regulation Training Course*, Vienna, VA: Management Concepts, Inc., pp. 1-3.

7. *Writing Winning Proposals*, Shipley Associates, 1988, pp. 3-11.

8. Standards assist the government evaluators in determining how well a proposal meets the requirements. Several standards might be developed for one evaluation factor. Depending upon the evaluation factor, standards may be qualitative or quantitative. Standards are necessary because evaluators are not supposed to compare proposals. They must, therefore, have some standard against which to determine

if a proposal satisfies the RFP. See *Writing Winning Proposals*, Shipley Associates, 1988, pp. 3–23.

9. Cibinic, J., and R. Nash, Formation of Government Contracts, pp. 644–645.

10. McVay, Barry L., *Proposals That Win Federal Contracts*, Woodbridge, VA: Panoptic Enterprises, 1989, p. 11.

11. Cibinic, J., and R. Nash, *Formation of Government Contracts*, p. 311.

12. Holtz, Herman, and Terry Schmidt, *The Winning Proposal: How to Write It*, New York: McGraw-Hill, 1981.

13. "Federal Acquisition Streamlining Act Enacted," *Business Credit* 97(3) (March 1995), p. 6. See also "Annual Report on the State of Small Business," The White House to the Congress of the United States, 6 June 1996.

14. See "Reinventing the Business of Government: An Interview with Change Catalyst David Osborne," *Harvard Business Review*, May/June 1994.

    Mr. Osborne supported Vice President Al Gore's *National Performance Review* (NPR), which released a report in September 1993 that recommended sweeping changes in federal policies and procedures. NPR and the concept of "service to the citizen" involve putting people first. See Kniseley, Sina Fusco, "Point-and-Click Government," *Washington Technology*, 11 July 1996, pp. 44–45. See also Internet: http://www4.ai.mit.edu/npr/documents/text/npr/ovp/ eop/gov/us/1993/9/6/290.html for the National Performance Review.

    Much of the NPR effort to reinvent government has been pointed specifically at helping small businesses. The U.S. Business Advisor, which provides Internet access to information on all federal agencies, and the U.S. General Store for Small Business, which offers business owners one location for dealing with the federal government, illustrate how the reinvention of government can benefit small businesses.

15. Dreifus, Claudia, "Present Shock," *The New York Times Magazine*, 11 June 1995, p. 48.

16. Toffler, Alvin, and Heidi Toffler, *Creating a New Civilization: The Politics of the Third Wave*, Atlanta, GA: Turner Publishing, 1995 and 1994, pp. 27, 31–33, and 43.

17. Harrfeld, Heather, "Reinventing for Results," *Federal Computer Week*, 21 September 1998.

18. "President Signs Law Establishing First Performance-Based Organization," 7 October 1998; http://www.npr.gov/library/announc/100798.html.

19. "Congress Says Transportation Has Best Strategic Performance Plans," *Reinvention Express*, 26 June 1998, Vol. 4, No. 9; http://www.npr.gov.

20. The historical aspects of the NPR initiative were taken from "A Brief History of the National Performance Review," February 1997; http://npr.gov/lib...apers/bkgrd/brief.html.

21. Brockmeier, Dave, "Help Shape Federal Acquisition Regulations," *Business Credit*, 97(3) (March 1995), p. 40. See also the Information Packet on the DoD EC/EDI Program from the Office of the Under Secretary of Defense, Delores Smith, Director, DoD Electronic Commerce. This packet is obtained by FAXing your company's name and address to (703) 696-0213.

22. Ibid., p. 40.

23. Ibid., p. 42.

24. The Internet, "the global agglomeration of networks large and small that have come very close to realizing Marshall McLuhan's view of the 'global village,' has become a standard medium for international electronic communications. Platform independence, fast data transport, and a graphical interface—the Web—are just a few of the features that have helped to make the Internet both wildly popular and incredibly useful." See Tittel, Ed and James Michael Stewart, "The Intranet Means Business," *NetGuide Magazine*, July 1996, p. 121.

25. Slaughter, Jeff, "New Way of Doing Business with Uncle Sam: Electronic Commerce/Electronic Data Interchange," *Mississippi Business Journal*, Vol. 18, Feb. 1996, p. 10.

26. Kelleher, Kevin, "Feds, State Go On Line with Contracts," *San Francisco Business Times*, Vol. 9, April 1995, p. 3.

27. Ibid., p. 3.

28. Behr, Peter, "Just Say Know," *Washington Post*, 1 April 1996.

29. Behr, quoting Tom Hewitt of Federal Sources, Inc., in McLean, Virginia.

30. Donnelly, John, "Preston Touts Acquisition Reform Results," *Defense Week*, 7 August 1995.

31. "Acquisition Reform: Obstacles to Implementing the Federal Acquisition Computer Network," Letter Report, GAO/NSIAD-97-26, 3 January 1997.

32. Warburton, Charles, "Electronic Commerce," *The Military Engineer*, October–November 1997, p. 63.

33. Power, Kevin, "Agencies Opt Out of FACNET," *Government Computer News*, 10 February 1997.

34. McCarthy, Shawn P., "OMB Documents Pave the Way to Explore Electronic Commerce," *Government Computer News*, 23 March 1998.

35. Brockmeier, "Help Shape Federal Acquisition Regulations," p. 42.

36. See Yukins, Christopher R., "Relaxed Rules Won't Fix the FAA's Problems," *Washington Technology*, 11 April 1996, pp. 8, 12.

37. Graham, Charles E., "Electronic Proposal Submission Basic Concepts," *NCURA Newsletter*, Feb./March 1996.

Chapter 7

The proposal life cycle

You can't close the sale when you have not made the sale.

Let's now take a closer look at what proposals really are; how they fit into a small business's total marketing life cycle; and the important planning, decisions, "products," organization, and reviews required to prepare a successful proposal.

7.1 What is a proposal in the competitive federal and commercial marketplace?

Technically, a *proposal* is an offer prepared by a contractor (such as your company) to perform a service, supply a product, or a combination of both in response to an RFP document or *Commerce Business Daily* SF254/255 synopsis issued by a procuring agency. Proposals are legal documents, often incorporated by reference into the final contract. That means that they must be factual and accurate—including information provided in résumés and project summaries.

A proposal is designed to sell both technical and managerial capabilities of a company to accomplish all required activities on time and at a reasonable cost. Your company's proposal document(s) is scored, literally,[1] by the client's evaluators against formalized, specific standards. A proposal is the written, electronic, or oral result of informed thinking, knowledge assimilation, and marketing processes, supported in turn by hard and directed work and buoyed by a positive collective attitude within your company's proposal team. A proposal is, first and foremost, a sales document. To be sure, it includes a host of technical, programmatic, institutional, pricing, and certification information, but it must remain sales oriented.

A proposal involves *marketing* your company to a government, private-sector, or international client. It is a closing *sales presentation*, supported concretely by traceable and auditable credentials and tailored to persuade your client to select your company for the award because you are best-qualified to achieve the results your client wants to achieve [1]. In the words of a senior U.S. Department of Transportation official, write your proposal from the government COTR's perspective. Help him or her to feel comfortable that your company will meet schedules, budgets, and other performance parameters. Demonstrate that you understand and can function well in the COTR's world. Make the COTR want to link his professional fate with your company. A proposal is, necessarily, *continued* dialogue with your client. Proposals are, in fact, the primary vehicle for winning new federal government business in the competitive arena.

A proposal is a closing sales presentation, presented persuasively and supported concretely by traceable and auditable credentials.

To reiterate, a proposal (as well as a bid, that is, a response to an IFB) is an offer or response [2]. A proposal is not necessarily a contractual document. But under specific circumstances, an offer *can* be made into a contract by the government. The client uses your proposal and those of your competitors as the primary sources of information upon which to base their selection of a winning contractor. Government evaluators are in no way obligated or encouraged to review publicly available material about your company that is not included within your proposal. That means that

---

1   Scoring can take the form of numeric scores, color-coded scores, pluses and minuses, or qualitative (adjectival) narrative comments.

you have to ensure that all relevant and salient materials are included along with your proposal within the parameters of any page or font limitations.

To continue, a proposal is a package of carefully orchestrated and documented arguments. Nothing should be put into your proposal by accident! Each section of a proposal should present carefully constructed discussion and meaningful evidence to convince your client that you should be awarded the contract because of the superiority of what you are proposing. To do this, you have to support the following messages in writing:

- You understand your client's project requirements, critical issues, and success criteria.
- Your innovative, solutions-oriented approach satisfies all requirements (be careful not to imply that your approach *exceeds* requirements; this can result in the client thinking that you are proposing and charging for more than they are asking for).
- Your approach offers tangible benefits to the client.
- Your approach minimizes schedule, cost, and technical risk.
- You are better (that is, more reliable, more experienced, and less expensive) than your competitors.

Why us? And why not our competitors? These are the simple yet profound questions that internationally known and respected proposal consultant Hyman Silver[2] insists that proposals must answer effectively. Through the vehicle of your proposal, your company must *finish* the process of convincing your client to select you over your competition.

Proposals are important deliverable products. One carelessly written and poorly presented proposal can damage your company's reputation with your client or potential client. Experience suggests that clients tend to remember contractors that have submitted inferior proposal responses. I have heard many senior managers say that they want to submit a proposal to "introduce the company to the client." If this is not a waste of B&P funds, nothing is. Introduce your company through marketing calls, "leave-behind" brochures, or carefully tailored letters. But do not expend the human, time, and financial resources to respond to a solicitation document unless you intend to win.

---

2   This phrase was coined in 1975 by Hyman Silver. Mr. Silver, President of H. Silver and Associates (HSA) based in Los Angeles, California, held senior engineer and marketing positions with McDonnell Douglas and Rockwell International before founding HSA.

7.2 Where does the proposal fit into the total marketing life cycle?

The proposal life cycle (in particular, for procurement efforts on which your company is the prime contractor) begins well before the actual writing and production of a response to an RFP. Figure 7.1 presents an overview of the entire marketing and proposal life cycle.

There are a number of salient elements in the overall contractor marketing/proposal process. The proposal process may include additional "color" team reviews. Please note that the concepts and processes introduced in the following items will be expanded in subsequent chapters of this book.

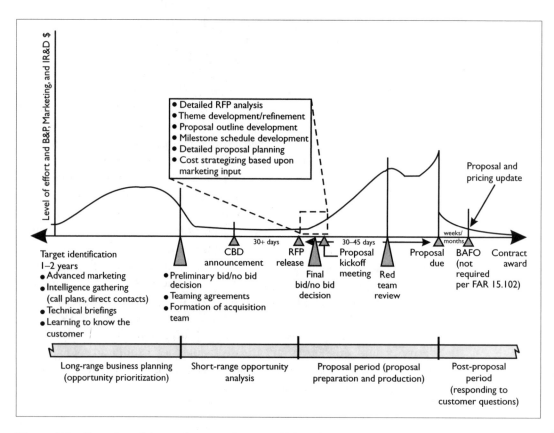

Figure 7.1 Overview of the marketing and proposal life cycle.

Long-term, preliminary marketing intelligence collection/direct client contact.

- Presell the client on your company's qualifications and commitment to complete the job successfully. Create "customer sentiment"[3] favorable to your company through technical briefings and discussions at various levels of the client's organization. Learn about such concerns as your client's technical requirements, financial concerns, scheduling issues, and decision-making processes. Verify that the project is a real, funded target.

For companies that provide support services, well-managed human talent and applicable contractual expertise are precisely what they are selling.

Formation of your company's acquisition team for a particular "qualified" procurement opportunity.

- Assemble a small team of management, marketing, and technical staff that will meet on a regular basis to develop a capture plan and share information gleaned from client contacts conducted in accordance with the call plan. Themes, or sales messages, for your proposal should begin to emerge. Identify your company's competitive strengths and weaknesses. Identify staff and relevant project experience to be used in your upcoming proposal. For companies that provide support services, well-managed human talent and applicable contractual expertise are precisely what they are selling.

- Résumés and project descriptions (project or contract summaries) can be tailored and fine-tuned well in advance of the release of the RFP. Take the time to gather the kinds of detailed educational, certification, technical, and programmatic experience information that your clients normally requires in their RFPs. Because staffing and past performance are critical evaluation factors in many RFPs, and because collecting specific information on staff and projects can require so much time, it is highly advisable to get started early.

---

3   Hy Silver coined the phrase.

Government's announcement of the procurement in the Commerce Business Daily (CBD) (governed by FAR 5).

Your company must provide a *written* request to the appropriate government procuring agency for each particular RFP. Telephone or facsimile requests are generally not accepted.

- The time to market your client for a particular project is drawing to a close. Once the RFP has "hit the street," most agencies will not allow any marketing relevant to the project under competitive bid.

Release of the RFP by the specific government procuring agency.

- If possible, arrange to physically pick up the RFP from the procuring agency. This will save several days delay in mailing time. Increasingly, RFPs are available for downloading from client Web sites.

Government preproposal (bidders) conference.

- Plan to attend the preproposal conference. Prepare questions for the client and submit them in writing in advance of the conference.
- Pay attention to client requirements that may be mentioned at the conference but not written in the RFP. Some conferences might include facility or site visits. Check to see if photography is permitted and, if so, take relevant photos or video.

Final internal bid/no bid decision.

- All such decisions should include (at a minimum) a member of senior management, the business development or advanced planning staff, the appropriate division manager, and the proposal manager.
- Good marketing intelligence will help to facilitate an appropriate decision.
- *What is the probability of winning?*
- Favorable financial return is an important consideration.
- Does your company have the people and facilities necessary to prepare a superior proposal and perform the project successfully?
- Does the project have potential strategic business advantages to your company?
- Teaming arrangements must be considered.

- Is your company the incumbent; if not, can you unseat the incumbent realistically?

Refer to Figure 7.2 for a representative bid/no bid decision matrix. Your company might use a total score of 75 to pursue a project.

Pre-kickoff planning meeting (conducted by the proposal manager).

It is often helpful for your company's project managers (PMs) or program managers—the people who most likely will be called upon to serve as *proposal managers*—to envision the upfront proposal planning process in terms of deliverable "products." These are very much analogous to the types of products that are integral to every program, project, task, and delivery order in your company's contractual experience base. Just as every PM develops and adjusts staffing levels, generates project milestone schedules in accordance with the client's requirements, and tracks project budgets to ensure cost control, so too must your proposal manager work with other staff in your company to develop a series of discrete "products" early in the proposal response life cycle, as shown in Figure 7.3. Particularly in small companies, there are insufficient overhead resources (nonbillable staff) available to maintain a proposal library, provide proposal coordination support, assist with résumé and project summary development, and solidify teaming agreements. Those critical tasks must be accomplished by the proposal manager, and in a very limited timeframe.

- Develop a detailed outline for each proposal volume, using RFP Sections L, M, C, and H, as well as the *Contract Data Requirements List* (CDRL), *Data Item Descriptions* (DIDs), and RFP attachments as appropriate. Take the time to assign responsible parties to each section. Allocate pages to each section in accordance with the relative weighting of Section M evaluation criteria. For example, if past performance will count for 20% of the total score and there is a 50-page limit to the proposal, plan on approximately 10 pages for the past performance section.
- Develop a workable proposal response milestone schedule that includes adequate writing, review, recovery, production, QC, and shipping time.
- Develop and modify existing proposal themes. Themes are conclusive reasons why the client should select your company over and above your competitors.
- Develop storyboards/scenarios/scribblesheets.

Client: _____
Project: _____
Prepared by: _____
Approved by: _____

GO: _____
NO GO: _____

Date: _____

	10 9 8 7	6 5 4 3	2 1 0	Ranking	Comments
Strategic Relevance: How well does project fit company marketing strategy	Fits well with marketing strategy	Fits somewhat into the marketing strategy	Does not fit into marketing strategy		
Relationship with the client	Close and excellent relationship with no problem projects	Fair/good relationship with some project problems	Strained relationship; problem project; selection questionable		
Knowledge of the project	Strategic project/prior knowledge and marketing of project	Know about project/no prior marketing	Know nothing about project		
Company staffing capabilities	Staff can perform all project requirements	Staff can perform most of the project requirements	Minimal or no staff capabilities		
Project competition	Competition manageable	Face formidable competition	Project "wired" for competition		
Company qualifications relative to competition	Technically superior; can demonstrate experience on similar contracts	Equivalent to competition; may have slight edge	Inferior to the competition		
Potential for future work	Project will lead to future work	Possible future work	One-time project; no future work		
How profitable is the project	Good profit potential	Moderate profit	Little profit		
Project schedule	More than adequate	Adequate	Not adequate		
Time, B&P costs, and resources to compete effectively for the project	Favorable	Reasonable	Excessive		

Figure 7.2 Bid/no bid decision matrix.

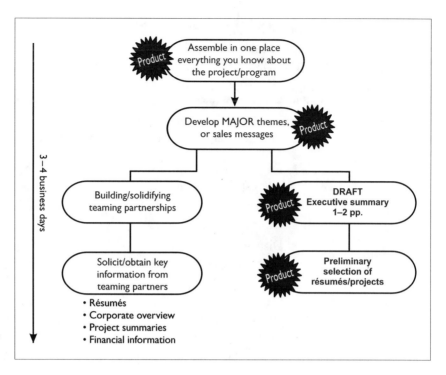

Figure 7.3
Critical pre-kickoff
proposal products.

- Select résumés and project summaries for the proposal.
- Generate a draft executive summary based upon your marketing intelligence and the evaluation factors found in Section M of the RFP. This will help to guide your writers to produce consistent, well-integrated narrative and graphics. Include graphics in your executive summary if at all possible. A full-color foldout page (11 × 17 inches) that captures the technical and programmatic essence of your proposal is one high-impact way to sell the benefits of doing business with your company early in the proposal.
- Develop a preliminary design concept for the proposal volume covers.
- Identify critical supplies so that they can be ordered in advance (such as three-ring notebooks, special colored paper, customized tabbed divider pages, and specialized software[4]).

---

4   Certain RFPs require that cost and technical volumes be submitted in specific software applications (e.g., Microsoft Word, Excel, WordPerfect, Lotus 1-2-3) and versions of those applications that your company may have to purchase.

Proposal kickoff meeting (with subcontractor(s), if any) directed by the proposal manager.

(to be attended by senior management, appropriate marketing staff, the proposal manager, the proposal coordinator, all key technical staff who will be engaged in writing major sections of the proposal, key reviewers, a representative from contracts, and if necessary, a representative from human resources and a representative from publications).

- Emphasize team spirit, a winning attitude, and the support and interest of senior management.
- Distribute detailed technical, management, and cost volume outlines.
- Distribute DRAFT executive summary. Having a draft of the executive summary at this stage will help everyone involved in the proposal-writing process to better understand what the client expects. It will also help to establish the major strategic themes your company wants to integrate throughout the proposal.
- Share critical marketing intelligence about your client, and discuss key win strategies.
- Technical and management writing assignments are made, along with page limits/allocations (in accordance with Sections M and L of the RFP) and writing conventions (such as capitalization, hyphenation, and punctuation). Themes, critical issues of the project, and success criteria are explained to the writers. Clear responsibilities for action items will be assigned.
- Stipulate electronic file formats, software application and version, and configuration control procedures.
- Assignments are made to contracts and human resources staff.
- Items that require the approval of senior management should be presented and finalized.
- Address the criticality of proposal security for both electronic and hardcopy versions of the resulting proposal documents.

Timely submittal of first draft technical and management input to proposal manager (text and graphics). Graphics should be prepared and produced in close association with the text.

- Ongoing interaction and feedback between the proposal manager and the proposal writers are critical for developing a viable draft document quickly and efficiently.

Edit/review of technical and management input.

Submittal of technical and management input to publication group.

- Ensure that electronic files are compatible.
- The Internet can be used for rapid file transfer from remote sites or teaming partners in other geographic locations.

Submittal of cost and staffing information to Contracts.

- It is critical to conceptualize and develop the cost volume in close association with the technical and management volumes (see Figure 7.4). Staffing selections, technical solutions, facilities and computer resources proposed, and your proposed management plan and organizational structure can all have profound cost impacts. Proposal costing is sophisticated business, not mere "numbers crunching." The key is to have financial staff with direct program or project experience who can interact appropriately and effectively with the technical proposal team.

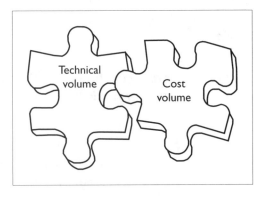

Figure 7.4
Develop the cost volume in close association with the technical volume.

Preparation of complete first technical/management draft/one master copy only.

Proof/review of first draft.

- Entire document frozen 24 hours prior to red team review.

Red Team Review of technical, management, and cost volumes (one to two days only).

Integration of all Red Team review comments (one to two days) by proposal manager, selected technical staff, and the proposal coordinator (if needed).

Quality check of format and content by publication group.

Preparation of second complete technical/management draft.

Proof/review of draft.

Preparation of revised cost volume.

Gold Team final review of technical, management, and cost volumes.

- Entire document frozen 72 hours prior to final delivery to client.

Final production, integration of text and graphics, quality check, photoreproduction, collation, assembly, and preparation for delivery (three days).

Signatures on cover letter and cost proposal forms (it is suggested that you purchase a stamp that reads "ORIGINAL").

Final delivery.

- Late proposals most likely will not be accepted by the government (see FAR 14.304). Even 1-min past due is still late! Request, obtain, and archive a time- and date-stamped receipt.

Archive hardcopy/electronic copy of proposal text and graphics in your corporate library.

- These materials should be stored along with all of the marketing intelligence documents, *Freedom of Information Act* (FOIA) requests, competitor information, the original RFP and amendments, and other materials relevant to this procurement. Being able to easily retrieve this material, even months or years later, is critical if you are required to respond to *Clarifications and Deficiencies* (CRs & DRs), submit a BAFO, or deliver an oral presentation.

Oral defense/BAFO/facilities inspection.

- Certain agencies require that BAFO submittals or CRs and DRs be submitted as "change pages" to your original proposal submittal. Being able to retrieve your original submittal easily is a vital time saver.
- Take time to strategize and develop a powerful, well-rehearsed presentation.

Attend government debriefing when your company is an unsuccessful offeror and also when you win (see FAR 15.1003). Learn why you won.

Develop "lessons learned" and discuss with key proposal staff. Share lessons learned with other proposal managers.

7.3 Bid/no bid decision-making process

Being very selective with the procurements your company decides to pursue is a first step toward long-term proposal and general business development success. Careful selection of bid opportunities improves your proposal win ratio and maximizes the return on general and administrative (G&A) dollars your company has invested in the proposal effort (see Figure 7.5). Identifying sufficient high-dollar, high-probability opportunities for which your company is qualified is a necessary step prior to selection of particular targets. The essence of the bid/no bid downselection process relates to four primary factors.

- *The probability of winning.* Does the client know your company and does your company know the client? Is there an incumbent contractor? Is that contractor performing satisfactorily? What is the nature of the relationship between the incumbent and the client? Does marketing intelligence indicate that your company is in a favorable competitive position?
- *The reality of the procurement.* Will it receive funding and will the competition be genuine, or does the government really intend to retain the incumbent?

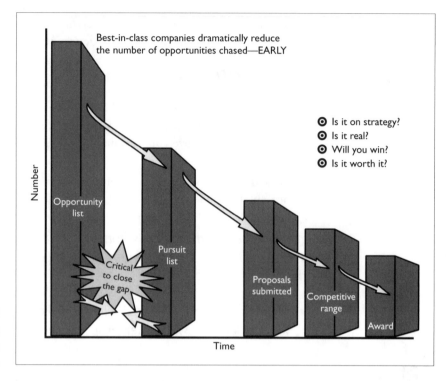

Figure 7.5
Downselect
opportunities
early.

- *The desirability of obtaining the contract.* What are the profit expectations? Will the contract result in a strategic business advantage for your company? Will the procurement support your company's stated objectives as presented in your Strategic Plan and Mission Statement? What are the inherent risks and liabilities? What is the add-on or follow-on potential? Does the procurement require any up-front capital expenditure or cost sharing by your company? Will new staff have to be hired?

- *The necessary facilities, resources, and human talent to perform the work on schedule and within budget, as well as to prepare a compliant proposal.* Are the very best staff available to write the proposal documents? Are there other key competing proposal priorities that will dilute the effectiveness of the proposal operations group? Is there secure floor and wall space available in your facilities to house a proposal team? Are there sufficient computers and printers that will be able to be dedicated to the proposal publication effort?

The bid/no bid process must balance the cost of responding; the availability of human, material, and other resources; and the magnitude of the business opportunity. The availability of in-house or consultant staff, facilities, and equipment resources to provide effective proposal support cannot alone *drive* the decision-making process but certainly must be considered. If your company cannot commit the best resources to prepare the proposal, then you should strongly reconsider any decision to bid the procurement. For the small business, the vice president of business development must perform this downselection process, working with a multidisciplinary team. The company's owners should get involved as well.

The major objective of the bid/no bid process is to arrive at a consensus within your company to pursue or not pursue a given business opportunity. It is advisable to conduct both a preliminary (pre-RFP release) and final (post-RFP release) bid/no bid decision. This applies to small as well as large companies.

Poor bid/no bid decision making is arguably the most common cause of business development failure in the federal contracting marketplace. Merely because your company has related technical expertise does not mean that you have a high probability of winning the business or that the business fits strategically with your goals and direction. The "shotgun" approach to business development and proposal development will be unsuccessful in the long term.

7.4 Planning and organizing

Once you receive an RFP, your acquisition team for that procurement should conduct an audit of the document, carefully mapping and accounting for all requirements, including the CDRL and DIDs. Review the staffing level requirements and position descriptions in the RFP. Formulate appropriate questions to be submitted in writing to the government for clarification. Be prudent in asking questions because your competitors will be given copies of those questions along with the government's response.

Various automated tools are available to support the RFP analysis and proposal development processes. These software products include Proposal Master and RFP Master (Sant Corporation), PME (Power NET, Inc.), RESTRIEVE (Applied Solutions, Inc.), and Proposal Strategist (Wiseware).

7.4.1 Draft executive summary

Prior to your company's formal kickoff meeting for each procurement effort, an individual appointed by the proposal manager should prepare a one- to two-page executive summary that introduces high-level themes, demonstrates your company's understanding of the importance of the client's project or program, indicates that your company has the credentials and talent to do the project, provides an overview of your company's approach, and highlights key project staff.

Your executive summary should applaud your client and your client's achievements. In addition, it should be structured to address in summary form the evaluation factors found in Section M of the RFP. Help your client understand exactly how your company will enhance their achievements, funding, congressional notoriety, technical excellences, and so forth.

7.4.2 Theme development

A theme is a substantiated sales message, a point of emphasis, an advantage, a unique or superior benefit, or a supported claim. Themes reappear, *woven inextricably* throughout the proposal volumes and serve to focus and unify the entire presentation. A good theme is direct, addresses a program issue or customer concern, and can be supported by concrete evidence. Your company's strengths and the competition's weaknesses can be the bases for themes. Themes are highly sensitive material and should always be marked as "Competition-Sensitive Materials" when distributed to staff throughout your company for review and comment. It is suggested that your company purchases stamps that read "Competition-Sensitive."

There are several different types of themes: (1) common themes and (2) unique themes, or *discriminators*. "Common themes," although obvious and unprofound, are still important to utilize in your proposal. "High performance, low risk," is an example of a common theme. "Unique themes" or "discriminators" are premised upon particular advantages your company can offer and upon distinct disadvantages of your competition. Discriminators are vital, credible points that are meaningful to your client. You need to do your marketing homework to identify these discriminators and then locate them at appropriate points in the proposal documents. Perhaps your company is the only known proposer that has developed personal computer–based air traffic controller instructional materials for a particular office of the FAA. A discriminator that you would certainly include in your proposal is a statement that highlights this unique contractual experience. It should appear, as appropriate, in section headings, the narrative, graphics, and captions. This message can also be conveyed via photographs or your proposal volume cover design. You can even reinforce this important unique theme in your compliance matrix.

In small proposals (less than 50 pages), your company may wish to develop 8 to 10 major themes and 15 to 20 minor (or subsection) themes. In larger proposal efforts, there may be 25 or more major themes and several hundred minor themes.

The following are select examples of themes. The particular procurement situation dictates whether these may be common or unique, major or minor.

- Lessons learned from our phase-ins with ONR, NAVAIR, and ARDEC will be applied to ensure a smooth transition from Day One of this important contract.

- Our real-time supervision and automated performance assessment methods have proven to be responsive and timely for other USACE clients.

- Our past performance with the U.S. Navy is the credibility base for this procurement.

- Our innovative technical approach based upon _____ allows for software and hardware "technology insertions" in the future.

- We offer established relationships with world-class computer science departments at universities to ensure an ongoing, accessible base of talent.

- We will ensure continuity of service through industry-trendsetting employee retention rates made possible by superlative, yet cost-competitive fringe benefits.

- We will use metrics to quantify and monitor customer perceptions of our level of technical proficiency and interpersonal skill.

- As extensions of the Air Combat Command's staff in Europe, much of our energy will be dedicated to developing and supporting solutions-based decision making consistent with the external and internal drivers that impact day-to-day operational activities.

- We will address dispute resolution proactively and positively in the context of partners working toward common goals.

- We will conduct business at the operational level in strict accordance with FIP, commercial leasing, and software licensing procedures and protocols.

- The ABC Team: Superlative Service Built upon Knowledge, Common Sense, and That Intangible Personal Touch.

- We will prove to be highly capable representatives as we honor the daily and "big picture" needs of an agency whose work and mission is critical to the economic health of America within the global marketplace.

- Our management and procurement structures have been specifically designed to support rapid schedules, immediate mobilization, and midcourse changes.

- Our cost-effective approach to environmental solutions meshes closely with the USACE's nationwide initiative to improve business practices and to produce quality results.

- Our well-established advocacy, mentoring, and practice of process control and checking also links with the Navy's emphasis on process improvements directed toward revolutionizing effectiveness, being more responsive to customer needs, and being more accountable to taxpayers.

- Our networked environment, business culture of technical resource sharing, proactive administrative support, and ongoing project communication channels will ensure that staff and resources are on the job when and where they are needed to support the U.S. Postal Service.

- We are able to propose proven leadership that allows program continuity while preserving program knowledge.

- ABC Company has crafted a very effective, successful, and proven recruitment process. We have cultivated a company-wide business culture that attracts and retains highly talented professionals.

- This "single-vendor" management approach will eliminate program accountability complexities, reduce costs, and minimize overall management risks and personnel issues/concerns to the Department of Health and Human Services.

- Our demonstrated record of careful cost controls will result in best-value support to NOAA throughout the life of the contract.

- In-house IR&D on interactive software is directly applicable to the technical solution for this important NASA/MSFC project.

- The total Washington, D.C.-based data management resources of our company will be available for application to specific needs of this important EPA program.

- COTS hardware that we have proven in similar situations will minimize start-up time requirements, development costs, and reduce the phase-in period.

- Dedicated, hand-picked Software Engineer offers 18 years of documented success in developing and managing automated systems for the DoD.

- All our proposed professional staff are currently working on the existing contract and hold active Secret clearances.

- Proximity of our company's headquarters to Fort Bliss will ensure effective, real-time communications.

- We are a progressive minority-owned firm with an established $3M line of credit.

- Our company staffs and maintains an office within 5 miles of the work site.

- Our company was the incumbent on the last modeling and simulation project, which successfully formed the basis for this one.

- Our company has served the *Air Force Center for Environmental Excellence* (AFCEE) successfully for the past 12 years. Our legacy of field-tested performance...

- State-of-the-art technology meets the NRC's current and projected requirements; that is, no new technology is needed.

- An excellent team of well-qualified, well-integrated subcontractors reduces program risk.

- Our company's organizational structure has a one-to-one correlation with the Air Force's project organization for ease of interface and communication.

- Our QA/QC Program, as already implemented at McClellan AFB, will ensure rapid problem isolation and resolution.

- Our integrated team approach will focus upon the formulation of small, flexible teams to evaluate emerging technologies.

Your company's Acquisition Team members should convey appropriate themes to your client during the client contact/marketing phases of the proposal life cycle. Then, when those themes reappear in the proposal, the client will tend to view them as *natural outcomes* of previous conversations and meetings rather than unvarnished marketeering tactics. And once again, themes should be woven into the proposal body, carefully integrated into headings, text, captions, figures and tables, and even compliance matrices (under a "Comments" section).

7.4.3 Bullet drafts/storyboards

Proposals are choreographed exercises in knowledge and information management, assembly, and packaging.
Storyboards, bullet drafts, scenarios, mockups, and scribblesheets[5] are proposal tools designed to break the proposal planning, control, and tracking process into "bite-sized" pieces. Proposal choreography provides the proposal manager with a mechanism designed to develop a *coherent,*

consistent document from multiple authors with varying degrees of writing skill in a limited timeframe. Storyboards help to ensure that all RFP requirements are addressed. They also assist in linking text with graphics and help technical authors overcome any initial "writer's block." Many times these tools prove vital for coordinating subcontractor input. Choreographic techniques are most effective and appropriate for the Technical Volume of a proposal.

Select proposal choreographic guidelines include:

- Identify the scope and context of the section or subsection to be choreographed.

- Identify appropriate sections of the RFP that correspond to that particular proposal section or subsection.

- Develop an introductory sentence that sets the tone and provides the emphasis for the narrative of the section or subsection.

- Prepare major and minor theme statements for the section or subsection.

- Prepare five to eight statements that clarify and direct your writing approach for the section or subsection. These short statements are to include the major and minor points to address in the narrative. The bullet statements are to build upon and resonate with the thematic statements.

- Generate at least one graphic concept and compose a thematically meaningful caption.

- Prepare a concluding or summation sentence for the section or subsection.

The number of storyboards, bullet drafts, or scenarios to be developed for any given proposal depends upon the complexity of the RFP. Small procurement efforts, for example, might only require 10 to 15 storyboard sheets. A sample bullet draft worksheet is shown in Figure 7.6.

---

5   Proposal "choreography," that is, the manner by which a proposal response is planned, controlled, and tracked section-by-section, can be referred to by a variety of names: storyboards (a term borrowed from Hollywood moviemaking), bullet drafts, scenarios (the method of Hy Silver and Associates, Inc.), mockups (a process used by Shipley Associates), and scribblesheets (a method employed by TRW, Inc.). *Choreography* will be used herein to refer interchangeably to storyboards, bullet drafts, scenarios, mockups, and scribblesheets.

Propietary
Bullet Draft Worksheet

Proposal volume:	Subsection number:	Writer:	
Subsection author:		Page limit:	Due date:
RFP requirement:			
Discriminator:			
Themes:			
Features:		**Benefits:**	

Introductory sentence (Verbatim):

Major points (Bullet format):

Graphic:

Concluding sentence (Verbatim):

Action Caption:

Approved by: _____   Date: _____

Figure 7.6
Sample bullet draft
worksheet.

7.5 Kickoff meeting

The formal kickoff meeting is your proposal manager's best opportunity to inspire and imbue his or her team with a winning attitude. The proposal manager can do this most effectively if he can demonstrate that he has planned carefully, organized a workable proposal response schedule, and made the best use of company resources. The attendance of your company's senior management during the kickoff meeting is highly desirable in order to ensure that the importance of the procurement to your firm's future is communicated.

7.6 Writing

The goal of each proposal author is to obtain the maximum score for each subsection. This is accomplished most effectively when writers follow the storyboards or bullet drafts carefully. Proposal writing is not recording as much as one knows about a given technical or programmatic topic. Too often, technical staff who are assigned to write proposal sections merely provide a voluminous quantity of material that does not address the specific elements contained within the RFP, nor does it mesh effectively with the narrative and graphics prepared by the other authors. Quantity alone is totally inadequate to yield a winning response. Crisp, distilled writing complemented by appropriate graphics and photographs is a major step in the right direction.

One goal of pre-proposal intelligence gathering is to have assembled a technical library of documents relevant to the procurement at hand. Your proposal writers should consult this resource material in advance of the release of the RFP. In addition, the government often provides contractors with access to a technical library of project documents once the RFP is released. These resources can provide the basis for the detailed, client- and program-specific narrative that is a requirement for successful proposals. Papers and technical documents will help illuminate the types of technical issues of interest and concern to your client. You will want to demonstrate an understanding of these issues.

Writers should also keep in mind that in the case of multiyear projects, they must prepare a response that factors in evolutionary program growth and change as well as technology insertion. Technology insertion refers to the addition of advanced technological "modules" or "elements" into an existing infrastructure.

7.7 Major contractor review cycles

All proposals your company produces should be subject to careful internal review. Senior management endorsement of and participation in the review cycles is strongly encouraged. The purpose of the "color" team reviews is to assess the responsiveness of the proposal to the given RFP; to ensure that your company's themes are apparent throughout the entire proposal; to identify statements that could be misrepresentative of your company or of the potential client; and to provide concrete suggestions for improvement in such areas as structure, strategy, technical approach, risk management, and pricing. Reviews are not exercises in fault-finding but rather are the objective critique of vital company documents.

7.7.1 **Blue Team/Pink Team**

The Blue or Pink Team evaluations are generally early course correction reviews. The fundamental technical architecture and programmatic direction of your company's proposal should be evaluated at this stage, deficiencies identified and articulated, and corrective action offered. Avoid focusing on formatting, aesthetic, editorial, or page-count aspects of the proposal documents at this early stage.

If possible, electronic configuration control of the proposal documents should remain in the hands of the proposal technical writing team up through the response to and integration of Pink or Blue Team comments. An interactive meeting should take place between the reviewers and the proposal writing team once the review is complete in order to clearly communicate the strengths and weaknesses of the proposal effectively. The focus is on substantive corrective solutions. Comments such as "Beef up this section" or "Rewrite this part" are not helpful. What are beneficial to the proposal writers are comments such as, "Enhance Section 4.6 by discussing specific automated cost and schedule control applications, particularly as they are compatible with…" or "Augment your discussion of technical and contractual interfaces with the Navy by including a graphic that clearly indicates points of contact…." Reviewers should attempt to provide specific guidance, direction, and solutions, as well as encouragement.

7.7.2 **Red Team**

The Red Team functions as an internal SEB, critiquing the proposal for compliance with RFP instructions and evaluation criteria (as found in Sections L and M, respectively, of the RFP). The Red Team looks for consistency and continuity among sections and volumes (intervolume compatibility), inclusion of sales messages and win strategies, presence and substantiation of themes, clarity of text and artwork, and overall believability and persuasiveness. The proposal manager will determine how to respond to the Red Team comments, particularly when there is disagreement among reviewers. Figure 7.7 presents a sample proposal reviewer's comment sheet.

The Red Team serves as a recommending group. Logistically, the Red Team reviewers should be fully dedicated to the effort, preferably co-located and isolated from the day-to-day demands of business during the course of their review. Red Team reviewers should include a number of people with appropriate technical competence and persons with high degrees of marketing and management competence. No reviewer should have participated in the proposal writing effort. Red Teaming is not merely

Red Team Evaluation Report

Proposal Title _____   Reviewer _____
Volume No./Title _____   Date _____
Section or Block No./Title _____

Criteria	Evaluation				
	Good	Average	Inadequate	Not Applicable	Cannot Judge
1. Overall impact (scoring potential)	❑	❑	❑	❑	❑
2. Readability/understandability	❑	❑	❑	❑	❑
3. Responsiveness to RFP requirements	❑	❑	❑	❑	❑
4. Responsiveness to evaluation criteria	❑	❑	❑	❑	❑
5. Theme inclusion	❑	❑	❑	❑	❑
6. Presentation format (including graphics)	❑	❑	❑	❑	❑
7. Technical content quality	❑	❑	❑	❑	❑
8. Believability/persuasiveness	❑	❑	❑	❑	❑

Specific suggestions for improvement:

Figure 7.7 Sample proposal reviewer's comment sheet.

an editorial exercise. Make available a copy of the full RFP to Red Team members, along with a full kickoff meeting package (proposal directive) that includes your major and minor themes and storyboards or bullet drafts. The proposal manager should provide written instructions to the Red Team reviewers explaining what the expectations of the review process are and asking that comments be detailed and explicit so that the Proposal Team can respond effectively. The Red Team proposal draft should include a table of contents, compliance matrix, and list of acronyms.

The Red Team review is designed to assess:

- Overall impact and competitiveness of your proposal;
- Readability and understandability;

- Compliance with RFP requirements—both overt and hidden;
- Inclusion of major and minor themes;
- Effectiveness of the presentation, including graphics and photographs;
- Quality of technical and programmatic content;
- Believability and persuasiveness.

The recommendations and comments that emerge from the Red Team review should be presented to the proposal writing team in a briefing and weighed and integrated (as appropriate) into the proposal volumes under the direction of the proposal manager. Specific people should be assigned to completing all action items identified as essential.

Red Team reviews should be scheduled early enough in the proposal response cycle to allow sufficient time for the proposal manager and proposal writers to implement an adequate recovery. Typically, on a 30-day response cycle, this is 15 days prior to the final production and delivery of the proposal documents.

7.7.3 Gold Team

In some companies, the Gold Team is a final review of the proposal volumes by members of senior management. Other firms use the Gold Team as a final proofreading and quality assurance stage. Figure references, pagination, and table of contents should be verified at this point. Participants include the proposal manager and senior publications staff.

7.7.4 Black Team

Black Teaming is a low-profile, highly confidential activity designed to assess the strengths, weaknesses, and current business activities of the competition. A document known as a "Competitor Assessment" might be generated from Black Teaming. Black Teams generally report to a member of senior management. Black Teams may include an outside legal firm or marketing intelligence consulting group who can investigate the competition without involving your company directly.

7.8 Preparing for orals and best and final offer

Proposal team activity, therefore, does not stop following submittal of the volumes to the client. Once a company's proposal is determined to be within the competitive range based upon the evaluation factors, the client may elect to have the company deliver an oral presentation to the evaluation panel and/or submit a BAFO. The BAFO is, in effect, the opportunity to enhance your proposal. Planning and rehearsing for orals and generating an effective presentation are very important. Many times, BAFOs involve providing the client organization with a revised cost volume that reflects the lowest possible cost for which your company can successfully perform the statement of work.

Increasingly important in the federal procurement process, oral presentations provide an excellent opportunity to reaffirm your company's distinctive capabilities, innovative technical solutions, and risk-aware program management approach. Capabilities include qualified personnel (proposed project manager and key staff), phase-in experience, cost-effectiveness of your approach, and records of outstanding contractual experience, among others. The oral presentation cannot exceed the government-assigned time limit, which can range from 15 minutes to several hours. Multiple rehearsals or dry-runs are crucial to acquaint your company's staff with public speaking, tight time limitations for presentation, and smooth computer/audiovisual equipment operation. Some government agencies and private-sector clients are now requiring multimedia presentations, replete with color, sound, and highly professional imagery. Clients want to meet and interact with the proposed project team face to face.

7.9 Debriefings (refer to FAR 15.1003)

Contractors that have submitted proposals to the U.S. government for a particular procurement are legally entitled to debriefings. Debriefings may take the form of a face-to-face meeting, teleconference, or letter and are generally attended by representatives of the government's contracting office, project technical staff, and financial office. Request a debriefing in writing through the client's contracts office. Lessons and information that your company can learn from a debriefing experience include:

- Breakdown of your technical score by proposal section (this is provided in very general terms);
- Process by which the scores were derived by the government evaluators.
- Strongest and weakest points of your proposal;
- How best to proceed with the next proposal to the same procuring agency; What specifically is important to that agency, for example,

a reluctance to accept complex teaming arrangements, close and effective control of subcontractors and consultants that may be proposed, an emphasis upon a particular project management configuration, and incumbent capture strategy;

- Advisability of making further offers to the particular client.

Do not expect to have your proposal compared to your competitors' proposals during a formal debriefing process. Competitor proposal performance and cost strategies are strictly off limits. You may, of course, request a copy of the winning proposal volumes under the FOIA. Company-sensitive trade secrets and commercial or financial information contained in the competitor's proposal will not be released indiscriminately by the government. Instead, the proposal documents will undergo review both by the competitor (preliminary) and the government to ensure that only information in accordance with the FOIA is released.

We are now ready to take a closer look at the major parts of a proposal submitted in response to a formal RFP.

END NOTES

1. Holtz, Herman, and Terry Schmidt, *The Winning Proposal: How to Write It*, New York: McGraw-Hill, 1981.

2. The federal IFB, or "sealed bid" procedure, should result in a firm fixed price contract. This procedure is especially well-suited for government purchases of COTS items. Bids are opened publicly and the award is made immediately to the lowest priced, responsible, responsive bidder. No discussions are allowed, and no exceptions may be taken by the contractor.

   The Federal Government has been moving in the direction of procuring dual-use, COTS, and NDI. An example of this procurement trajectory is in the purchase of *intelligence and electronic warfare* (IEW) equipment for the U.S. Army. (See "DOD's Airborne Recon Plan Pushes Sensors, Image Recognition," *Sensor Business News*, 8 May 1996.) In FY1997, DoD seeks a $250M *Dual-Use Applications Program* (DUAP). With COTS/dual-use embedded in DoD acquisition, MIL SPECS will no longer provide giant defense firms with an artificial edge over agile new defense entrants into the marketplace. (See "DOD Acquisition Chief Pushes COTS/Dual-Use, Wants More Suppliers," *Technology Transfer Week*, 30 April 1996.)

The concept of "responsiveness" was developed in sealed bidding, wherein a Contracting Officer is prohibited from considering any bid that deviates from the IFB. This prohibition does not apply to negotiated procurements. (See Cibinic, John, and Ralph C. Nash, *Formation of Government Contracts*, second edition, Washington, D.C.: The George Washington University, pp. 523–524.)

An RFP, which results in "competitive proposals" for competitive acquisitions under Public Law 98-369 (Competition in Contracting Act of 1984), is a more flexible approach used in sole source or competitive situations. Any contract type from fixed price to cost reimbursement can be used. The government opens and reviews proposals privately. The number and identity of offerors are not revealed (immediately), and award may be based upon considerations other than price. Considerable discussion between the government and the offeror may or may not take place. The offeror may take exception to the solicitation terms or may submit alternate proposals. And modifications to proposals by the offerors are permitted. Upon acceptance, the proposal or offer becomes a contract in some government procurements wherein the technical and contractual requirements are tightly specified.

Chapter 8

Major proposal components

Overview

Your company's response to a competitive RFP will generally consist of a set of volumes: technical, management, and cost. These volumes may be called different names in different RFPs; "technical approach," "business management," and "cost/price" are several variations. In addition, there will probably be a transmittal or "cover" letter; an executive summary, which is often part of the technical volume but may be a standalone volume; and various supporting volumes such as quality assurance plan, configuration management plan, and health and safety plan. Sometimes, the technical and management volumes are contained in one single document. A U.S. government RFP may require that SF 33s, "Solicitation Offer and Award," and SF 1411s, "Contract Pricing Cover Sheet," be submitted as separate "volumes." Technical volume page counts for federal proposals can range from 10 to 15 sheets to hundreds or even thousands of pages. And certain major aerospace and defense proposals can contain more than

70 individual volumes, considering executive summary, technical volume, management volume, cost volume, make or buy policy, quality assurance plan, recruitment plan, health and safety plan, subcontractor plan, and so forth. The trend, however, even with billion-dollar procurements, is for proposals on the order of 100 to 150 pages to be required. Focused, highly refined text and graphics are required to convey your messages and meet all of the RFP requirements in a concise manner.

8.2 Transmittal letter

The transmittal or cover letter should do more than simply transmit your proposal documents to the government or private-sector client. The letter should be viewed as one more opportunity to call your client's attention to the advantages to be gained from doing business with your company. This letter should be prepared on your company's letterhead with the appropriate office location address and be signed by your company's president or executive vice president. It should emphasize your president's personal commitment to the successful completion of the particular project or timely delivery of the specified product or service. It is advisable to attach a copy of the transmittal letter with *each* volume of the proposal because the volumes are often distributed to a committee of client evaluators.

Since the transmittal letter requires the signature of a member of senior management, the proposal manager should plan to have the letter completed for signature several days in advance of the proposal due date. The fluctuating schedule of business executives could result in their being out of the office at the very time their signature is required on the letter and cost volume forms.

Definitely avoid the following hackneyed verbiage in your transmittal letters:

We are pleased to submit this proposal in response to Solicitation No. 123-99-R-0009…

Be creative and inventive, yet to the point and in accord with any specific RFP instructions. Briefly convey your company's enthusiasm to contribute to your client's technical, programmatic, and budgetary success on this important project. Consider the following example.

When DOT partners with our company, you are harnessing the talents and focused energies of businesspeople who are also information technology specialists. Four years of outstanding IT service provide the technical matrix in which fiscally prudent and market-driven business decisions are made every day. We very much appreciate that you and other DOT staff members have invested your time and effort in reviewing our submittal. From the vantage of our knowledge and experience base, we certainly see the challenges—and the genuine mutual successes—that are possible under this important program...

8.3 Technical volume

From a client evaluation perspective, the technical volume of your proposal is frequently weighted most heavily. This means that this volume is absolutely crucial to the success of any proposal effort. Sometimes, however, contractor technical staff and senior management place *too much* attention on responding to this volume, to the detriment of the remainder of the effort. Proposals are not *only* technical approaches and solutions, though to be sure, these items are critical. As noted previously, proposals are fundamentally sales documents.

The technical volume is the volume most often "storyboarded," that is, decomposed into logical subparts and "choreographed" to encourage written responses from the proposal team that are consistent thematically and compliant with the RFP requirements (see Figure 8.1). From a "big picture" vantage, the technical volume presents:

- What your company is selling to your client;
- Why you are selling it;
- How your company will perform its function(s);
- How it is better than what the competition has to offer.

8.3.1 Front Cover

Your company's image,[1] or collective identity, should be conveyed in the cover design of your proposal documents, provided that the RFP does not stipulate otherwise. The cover should attempt to: (1) attract positive attention with an appropriate high-impact presentation, (2) identify the RFP and the offeror (your company) in words and images, and (3) convey the most significant features or themes of your proposal. Almost every RFP states somewhere in the preparation instructions that attractive, colorful

---

[1] Company image should be conveyed uniformly throughout all of your externally distributed materials, including business cards, letterhead, envelopes, marketing brochures, annual reports, marketing fliers, advertising, trade show booth, and signage.

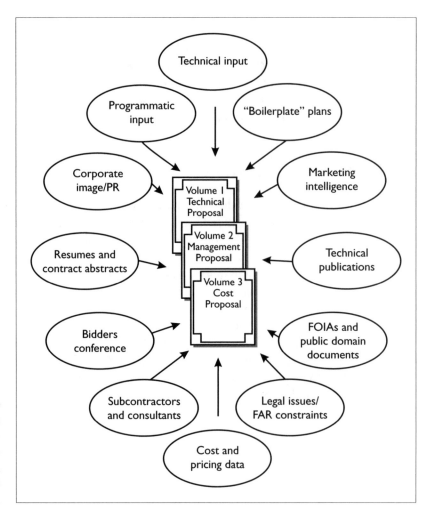

Figure 8.1
Proposals are choreographed exercises in information management, assembly, and packaging.

covers, diagrams, and the like are uncalled for and, if used, demonstrate that the offeror is unresponsive and not very cost-conscious. Real-world surveys indicate, however, that attractive covers are noted positively by government evaluators. They are, after all, human beings who respond more favorably to attractive images than bland cover stock. In addition, some clients will assess your proposal as being indicative of the caliber of deliverable your company would produce during the contract life cycle.

Your company's logo (and those of your teaming partners, if appropriate), attractive photographs and/or artwork, and thematically meaningful words can be combined to create a stunning cover design using Corel-DRAW! and Photoshop, for example. Numerous companies offer stock photographs as electronic images (e.g., .WMF, .TIF, JPEG formats) that are legally usable in your proposals from a copyright and intellectual property standpoint. To avoid unwanted and potentially costly litigation, never scan and use photographs or artwork from published books or magazines without the written consent of the photographer, artist, and/or publisher. Modifying a photograph is insufficient to reproduce it without permission. The Corel Gallery of images at http://www.corel.com/galleryonemillion/ index.htm offers a huge selection of clipart (noncopyrighted line art) and photos that can be purchased and used for your proposal. Other suppliers include Image Club Graphics at http://www.imageclub.com/digitalvision. You may require the support of an outside printing vendor to generate multiple copies of two- or four-color covers. And if you elect to use three-ring notebooks to package your proposal volumes, be sure to have "spines" designed and produced, too. No back cover design is usually necessary, although some companies elect to prepare them.

8.3.2 Disclosure statement on the title page

The disclosure statement, which should appear on the title page of each proposal volume, is designed to protect your company's proprietary information from illegal disclosure (see Figure 8.2). It serves notice that the content of your proposal documents is sensitive information. Reference to the disclosure statement should be stamped or printed onto each page of each volume of your proposal (for example, "Use or disclosure of data on this page is subject to the restriction on the title page of this document.") The disclosure statement will not, however, prevent the government from releasing information from your submitted proposal in accordance with the FOIA. Sometimes an RFP will stipulate the exact verbiage for the disclosure statement.

Figure 8.2
Sample disclosure statement for the title page.

> The information contained in this proposal shall not be disclosed outside the Government (or client organization, as appropriate) and shall not be duplicated, used or disclosed, in whole or in part, for any purpose other than to evaluate this proposal. If a contract is awarded to this offeror as a result of or in connection with the submission of the data herein, the Government shall have the right to duplicate, use, or disclose the data to the extent provided by the contract. This restriction does not limit the Government's right to use information contained in the data if it is obtained from another source without restriction. The data subject to this restriction comprise the whole of this proposal.

Similarly, for commercial proposals, the disclosure identifies the document as containing proprietary information not to be disclosed outside of the client organization, unless authorized by the proposer. The protection obtained in this manner is less than desired, but it at least puts the reader on notice.

8.3.3 Executive summary

The executive summary should be written so that even an uninformed reader will absorb and understand the principal ideas expressed in each of your proposal volumes. The executive summary, which in many proposals is only one to two pages in length, should be structured to:

- Introduce high-level themes that will reappear throughout the proposal volumes (technical, management, and cost).

- Enumerate your company's applicable credentials and applicable contractual track record.

- Demonstrate that your company understands the *importance* as well as the technical requirements of the project or product deliverable.

- Indicate that your company can perform the project successfully or provide the product or service on schedule and within budget.

- Provide an overview of your technical and management approaches for this project in accordance with Section M of the RFP.

- Highlight your key project staff for service contracts in particular (by name if your client knows them).

- Focus on the primary reasons why the client should select your company as its partner on this important contract. How will the client's selection of your firm benefit them from a technical, programmatic, risk mitigation, and cost perspective? What specifically makes your company the "best value" for this client?

- Demonstrate your local, established presence in the applicable geographic area that ensures your preparedness to meet client requirements in a timely manner and comprehend the local political and regulatory decision-making environment.

- Convey your firm's successful consulting work and your seasoned ability to manage multiple tasks simultaneously.

- Highlight your successful performance history with the client. Provide one or two *success stories* as well as key *lessons learned*.

- Point to your solid knowledge base in applicable technical areas that will provide *value-added* technical support and direct benefit to the client.

- Outline your highly effective management strategies that ensure schedule adherence, cost control, and superior-quality deliverables. Introduce your project management leadership to the client, and demonstrate their pivotal programmatic, technical, and teamwork strengths.
- Bring to life your company's corporate commitment to partnership with the client to support its mission.

The benefits of your company must be sold right here.

Use photographs of related projects, equipment, and systems/installations to help authenticate your contractual experience and technical approach. A busy executive should be able to scan your executive summary in about five minutes and understand at least the high points of what it is you are selling and why you are the company to be selected.

If at all possible and in accordance with RFP instructions, develop a full-page (8.5-in × 11-in) or foldout page (11-in × 17-in) graphic as part of the executive summary that conveys the primary benefits to the client of your company's approach. Tell your proposal story in one graphic image through the judicious use of photos, line art, clip art, and carefully selected words. If prepared well, this can be a powerful tool for capturing reviewer attention and interest in the remainder of your proposal.

8.3.4 Building a compliance (cross-reference) matrix

Compliance, cross-reference, or traceability matrices are all terms for tables that clearly *map* the RFP requirements (from Sections C, L, M, and H as well as the CDRL, DIDs, and RFP attachments as appropriate) to the specific sections within your proposal in which your company responded to a given RFP requirement. These matrices often appear in the front matter of your proposal volumes but may be placed at the very end of each proposal volume as an 11-in by 17-in *foldout page* that will allow the government evaluators to be able to refer constantly and easily to the matrix. Tables 8.1(a–c) illustrate sample compliance matrices. Table 8.1(a) is a common, straightforward approach. Its format was dictated by the RFP. Table 8.1(b) includes reference to both text sections/paragraphs and figures contained in the proposal response. And Table 8.1(c) presents a detailed mapping of elements from Sections C, L, and M of the RFP to proposal response sections and subsections. Note that in certain cases, the response to specific RFP elements is found in more than one proposal section or subsection. The number of columns in a compliance matrix varies depending upon whether you elect to delineate your responses into "text" sections and "graphics." You might also consider adding a "COMMENTS" column in which important themes can be introduced, as shown in Table 8.1(c).

8.3.5 Narrative body of the technical volume

Translate your company's features into benefits to your client.
If Section L of the RFP allows, *each section* (and major subsection) should be written to include an introduction, body, and summation. The "body," or main narrative portion, might include a discussion of: (1) your company's understanding of the client's requirements, including how they might evolve over the lifecycle of the contract; (2) critical issues associated with the requirements or task performance; (3) your company's innovative approach to addressing the requirements, including lessons learned and success stories from similar contractual experience; and (4) success criteria for the task or effort. Strive to prepare a solution-oriented proposal—one that offers the least technical, schedule, cost, and political risks to your client. And weave your themes throughout the narrative sections of your

Table 8.1(a) Sample Compliance Matrices: (a) Common, Straight-forward approach; (b) With Reference to Text Sections Paragraphs and Figures Contained in the Proposal Response; and (c) Detailed Mapping of Elements from Section C, L, and M, of the RFP to Proposal Response Sections and Subsections

SOW element	Proposal section
1	2.1.1.a.
2	2.1.1.b.
3	2.1.1.c.
4	2.1.1.d.
5	2.1.1.e.
6	2.1.1.f.
7	2.1.1.g.
8	2.1.1.h.
9	2.1.1.i.
10	2.1.1.j.
11	2.1.1.k.
12	2.1.1.l.
13	2.1.1.m.
14	2.1.1.n.

RFP Number		Proposal	
		Part	**Part**
	Section/paragraph	**Figure**	**Section/paragraph**
L.2.A.5	Separation of technical and business proposals		Part I, Part II
L.2.B.	Technical proposal instructions	Part I	Part I
L.2.B.1.a(1)	Objectives		I,2 all
L.2.B.1.a(2)	Approach		I,3 all
L.2.B.1.a(3)	Methods		I,5 all
L.2.B.1.b(1)	Experience	6-1, 6-3	I, 6.4; Appendix C
L.2.B.1.b(1)(a)	Pertinent contracts	6-1	I,1 all; I,6 all and I, Foreword supporting materials (attached)
L.2.B.1.b(1)(b)	Principal investigator	6-4	I, Foreword; I,3 all; I,4 all
L.2.B.1.b(2)	Personnel	6-3	I,6.4; I, Appendix C
L.2.B.4	Other considerations		I, Foreword; I,3 all; I,4 all
L.2.B.5	Special instructions		I,5
L.2.C.1	Cost and pricing data	Part II	Part II
L.2.C.1.a(1)	Materials		II, SF 1411, Attachment
L.2.C.1.a(2)	Direct labor		II, SF 1411, Attachment
L.2.C.1.a(3)	Indirect cost		II, SF 1411, Attachment
L.2.C.1.a(4)	Special equipment		II, SF 1411, Attachment
L.2.C.1.a(5)	Travel		II, SF 1411, Attachment
L.2.C.1.a(6)	Other costs		II, SF 1411, Attachment
L.2.C.1.b(1)	Judgmental factors		II, SF 1411, Attachment
L.2.C.1.b(2)	Contingencies		None
L.2.C.2	Total plan compensation		II 4, all

Table 8.1 (b) (continued)

RFP Element		
Attachment A (SOW)	Proposal Section	Comments
SOW 1.0	B.1.1	The Team has the specific experience, capabilities, and knowledge to support all of the requirements of the SOW.
SOW 2.0	B.1.2	Our staff, including our proposed senior scientist, have years of prior involvement with this project. We understand the mission objectives as well as the system requirements.
SOW 3.0	B.1.3	Our knowledge coupled with our extensive experience supporting the scientific projects provides us with an understanding of the critical project elements.
SOW 4.0	B.1.4	The Team has taken an integrated approach to all functions of the development effort.

Table 8.1(c)
(continued)

proposal. Avoid the phraseology: "Our major theme for this section is...."
That is not effective proposal writing. I worked with one enthusiastic proposal manager who, upon learning about themes in my proposal training seminar, included "Major Themes" as a heading in his proposal. We had some restructuring to do!

Table 8.2 is a sample structure for a section of a technical volume in a task-order procurement. In responding to a particular RFP, you should structure your outlines and proposal volumes in accordance with the numbering scheme and conceptual scenario provided in Section L. And use the exact verbiage from Section C as outline headings and subheadings, if possible. When the client asks for "ABC," give them "ABC" and not "BCA" or "XYZ." In other words, focus on and respond directly to the client's requirements. Translate your company's distinctive capabilities and extensive expertise into benefits to your client. This can be done in the narrative, in "features/benefits" boxes (as shown in Table 8.3), and in figure captions and table legends.

Body	3.3	Technical task #3
	3.3.1	Introduction
	3.3.2	Understanding the EPA's requirements for Task #3 (include how requirements may evolve over the contract lifecycle)
	3.3.3	Critical issues associated with task performance
	3.3.4	Your company's (or team's) innovative technical approach
	3.3.5	Success criteria for Task #3
	3.3.6	Summation

Table 8.2
Sample Technical
Volume Outline

Features	Benefits
Single-vendor solution, with a lean organizational structure	Straightforward lines of authority and communication with the IRS, resulting in more effective, efficient, and correct service and problem resolution
Proven approach to transitioning incumbent personnel	Retention of institutional IRS knowledge and smooth contract startup within 30 days of award
Project structure tailored specifically to the IRS's unique requirements	Responsiveness, flexibility, and value of service
Project management team highly experienced and successful in the management of eight similar task order contracts	Proactive approach to meeting IRS needs due to complete familiarity with the contract process; the process functions as a mechanism to facilitate service, rather than a barrier
Three years of experience in facilities management	Effective maintenance of IRS baseline, with associated cost savings estimated to be $100K over the life of the contract
Four years of experience in prototyping and implementing advanced application technologies	Capability to plan and execute the IRS's migration to newer technologies

Table 8.3
Features/Benefits
Box

8.4 Management volume

The management volume of a proposal is many times the most over-looked and poorly constructed document in the entire proposal package. Too often, the management volume sections are viewed to be mundane "boilerplate," in which each government agency is interchangeable. The management volume is a project- and agency-specific narrative that clearly demonstrates how your company will manage staff, maintain reasonable spans of authority, control schedules and costs, assess technical perform-ance, address technical issues, and access resources companywide as nec-essary. In general, your management volume must describe and illustrate clearly:

- How your company proposes to organize and marshal its resources to perform on the project or deliver the products or services;

- The roles and span of authority of your proposed project manager;

- How your company's project staff will communicate and interface with the client on an ongoing basis over the life of the contract;

- Why you will manage the effort more efficiently and effectively than your competition;

- Your company's identity (through staff résumés and related and past contractual experience);

- That you can accomplish the project successfully, within schedule constraints and under or within budgetary margins;

- The rationale for forming your particular teaming arrangement, if there are subcontractor(s) involved;

- That you have successfully managed projects of similar size, scope, complexity, and contract type, along with relevant examples of such projects.

A suggested outline for the management approach is presented below.

- Introduction
- Project Organization (include *functional* project organizational chart populated with staff from your firm as well as with subcon-tractor staff)
- Project Leadership
  - Project manager's role and span of authority
  - Accessing corporate resources (discuss your company's commitment of resources to project success, and the specific

mechanisms that you employ to make this happen at the level of day-to-day business)

- Project initiation and work order kickoff strategies

- Ongoing Communication with the Client Organization
  - Client/company technical and contractual interfaces
  - Reporting schedule

- Project Execution
  - Effective schedule control approaches
  - Capacity to meet fluctuating workloads and accelerated schedules
  - Automated scheduling tools
  - Appropriate task prioritization
  - Demonstrated history of meeting critical project deadlines
  - Proposed schedule of project by task
  - Work planning and monitoring
  - Response to evolving project requirements
  - Cost control approaches
  - Efficient invoicing mechanisms
  - Automated cost tracking/control tools (e.g., Primavera)

- Subcontractor Management
  - Rationale for teaming
  - Project manager's approval of subcontractor staffing
  - Ongoing communication with project team
  - Invoice verification
  - Integration of subcontractor staff into task activities
  - Work assignments
  - Issue resolution

- Quality Assurance/Quality Control

You must demonstrate convincingly how your company will perform the work, manage and control the work, manage and control *changes* to it, and be responsive to the client's needs. You should demonstrate a staffing and facilities capability sufficient to handle a project of the size proposed. Depending upon the requirements set forth in Section L of the RFP, the management volume might include the following sections and subsections.

Introduction.

An introduction should highlight the key aspects of your company's management approach. These might include a project organization that accommodates evolution and growth within the contract; assignment of seasoned and qualified personnel who understand the specific needs and goals of their particular areas of support; clear definition of roles, responsibilities, and lines of communication; and the support of senior management for the contract.

Executive summary.

An executive summary is similar to that used in the technical volume.

Project organization.

Project organization addresses such items as project leadership, spans of authority of your proposed supervisory staff, and the clear lines of communication between your project manager and his or her technical and support staff. Emphasize your project manager's access to required company resources and this particular project's visibility and importance to your senior management. Demonstrate that your project organization can accommodate evolution and growth within the contract.

Corporate organization.

The corporate organization section presents an organizational chart in graphic form, highlighting how the specific project under consideration is linked managerially and structurally to the company as a whole. Demonstrate a streamlined managerial structure with direct access to senior-level support and resources company-wide.

Client/contractor interfaces.

A section on client/contractor interfaces presents your company's proposed working relationship with the client at technical, programmatic, and contractual levels. *Answer the question*: If I am the client, whom do I contact within your company for project status in terms of technical accomplishment, schedule, cost, and deliverables, and so forth. Demonstrate clearly in narrative and graphic form the lines of interaction between your company's technical, programmatic, and contractual staff and the client's counterparts.[2] Discuss the clearly defined reporting pathways within your company.

Management plan.

A management plan discusses your proposed project manager's authority, roles and responsibilities and his or her relevant project management experience. Describe how your project manager will interface with your company's senior management and key government staff to manage the project effectively and efficiently. Demonstrate through reference to "lessons learned" from other similar contractual experience that your management approach can accommodate evolving, fluctuating, and unforeseen workloads; multiple task assignments; "requirements creep";[3] and unanticipated employee turnover. Discuss work planning and monitoring, allocation of project resources, review of project work, and regular reporting to the client. The management plan may include contract phase-in and phase-out plans.

Cost and time controls.

In a section for cost and time controls, describe how your company plans to maintain control over the schedule and cost once the project begins. Illustrate specific and innovative control techniques proposed, emphasizing *automated* daily cost tracking and accounting, scheduling, and purchasing techniques (e.g., micropurchases), if appropriate. Explain what your company does to correct cost exceedances in cost reimbursable contracts. For the U.S. government, clearly indicate that your cost accounting and purchasing systems are approved by the *Defense Contract Audit Agency* (DCAA) and that your company complies with *Cost Accounting Standards* (CAS) and FAR Cost Principles. If a management information system is used to forecast resources and track task progress/performance, articulate how it will be employed on this particular contract. State that your firm has the ability to control its workload through prudent bid/no bid decisions.

Scheduling.

Demonstrate in graphic form the major schedule milestones and contract deliverables. Discuss automated scheduling tools that your company has employed successfully on other contracts of similar size, scope, complexity, and contract type.

---

2   It is very effective to develop a graphic that illustrates your company's contracts group interacting with the client's contracts office, and your proposed project manager (and task leaders or delivery order managers) interacting directly with, for example, the client's *contracting officer's technical representative* (COTR) and *assistant technical representative* (ATR).

3   "Requirements creep" refers to the gradual expansion of the scope and complexity of the tasks or project.

Staffing plan.

Describe staffing levels required in each labor category over the life of the contract. Discuss your company's successful recruitment policies, ongoing staff training programs, and internal programs to ensure employee motivation and retention (compensation plan).

Key and other résuméd personnel.

Present short biosketches that highlight experience relevant to the procurement.

Résumés.

Present résumés in the proper format as specified by the RFP, tailored to the specific requirements of the procurement. Map or link the individual's expertise and experience to the position description found in the RFP. Be certain that the government evaluators will understand precisely why the proposed staffperson was selected for the particular project.

Company capabilities and experience.

Describe your company's quality and reliability record on projects of similar scope, schedule, budget, and contract type.

Clients often use past performance as the single most important, reliable predictor of future performance.

Related experience.

Bring your company's relevant contractual experience to life. Focus on project successes such as cost savings (tell how many dollars), schedule adherence (tell how much time saved), repeat business from a client, value-added service, and the application of innovative technologies. Highlight lessons learned from previous and existing experience that can be applied quickly and transparently to the contract under consideration. Pay particular attention to preparing and tailoring the narrative of past and related experience because of the increasing importance of this area as an evaluation factor.

Subcontracting plan.

For major subcontractors, include a letter of commitment from the president of the subcontracting firm(s). This is in addition to a formal teaming agreement. Provide evidence of sufficient subcontractor control. Discuss your project manager's oversight of the subcontractor(s)' staffing and invoices, issuance of subcontractor work assignments, planned methods for evaluating subcontractor performance, clear definition of subcontractor tasks and responsibilities, communicating and reporting practices and procedures between your company and the subs, meeting times/schedules, and problem resolution. *Ensure that your company's interactions with your subcontractor team are transparent to your client.* Demonstrate that you will hold your subcontractors to the same rigorous quality standards for technical performance as you do your own company.

Company facilities and special resources.

Describe the mechanism in your company to make resources and facilities available as necessary to support the project successfully. Indicate any *government-furnished equipment* (GFE) or contractor furnished equipment (CFE) required. Describe how your facilities and equipment are applicable to the SOW, and how you will use them in meeting critical project requirements.

Risk management.

Based upon marketing intelligence about your client's concerns and goals, demonstrate that your company will keep the project on schedule, control costs, retain key technical staff, maintain more efficient staffing levels, respond to fluctuating workloads, and so forth. You must convince your client that your company will make them look good in the eyes of their superiors, including possibly the U.S. Congress. Demonstrate that you are fully capable of risk identification, analysis, and mitigation. Given the growing emphasis on performance-based contracting (PBC; defined at FAR 37.101), this is a critical element in the management proposal.

Configuration management.

Depending upon the technical scope of work, and particularly for software and hardware systems development, discuss how your company will address configuration control, design, development, and implementation.

Data/information management.

Depending upon the SOW, discuss how you will apply automated tools to assist in data and information collection, analysis, archival, and retrieval. Also describe your techniques for ensuring data integrity and controlling data access. Indicate if your *management information systems* (MISs) are compliant with government standards.

Work breakdown structure.

The *work breakdown structure* (WBS), which is most often included in the management volume, is an organizational tool that can be used to assist with project management structuring, cost estimates, and cost substantiation. (See Figure 8.3.) Sometimes referred to as a work element structure, the WBS helps to "interrelate the technical, management, and cost aspects of the work" [1]. Specifically, the WBS is developed to ensure that all required work elements are included in your company's technical approach and management plan without duplication. Lower level tasks and subtasks that can be more easily scheduled and estimated in terms of staffing are identified and arranged in hierarchical fashion to assist in the overall accumulation (or rollup) of resource requirements. During actual project performance, the WBS serves as a tool for cost control, along with a monthly "estimate-to-complete" analysis.

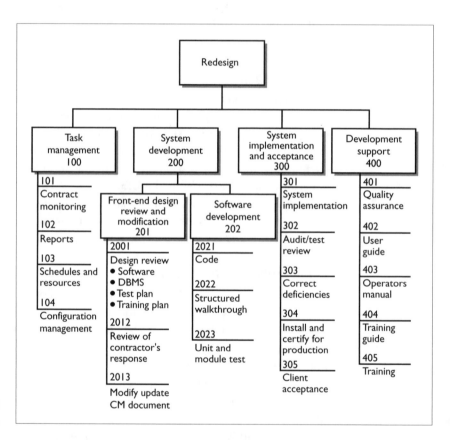

Figure 8.3
Sample WBS.

Technical performance measurement system.

Highlight any performance metrics that your company may employ to estimate, monitor, assess, and report project performance in an effort to identify potential problems before they devolve into major issues.

Deliverable production and quality assurance.

Demonstrate through reference to other successful project execution that delivering superior-quality deliverables which have undergone internal quality review prior to submittal to the client is a top priority to your company. Discuss formalized *quality assurance/quality control* (QA/QC) procedures and proactive process planning and review mechanisms that have been established within your company. For example, perhaps you have implemented a program that includes document design meetings and *senior technical review* (STR) for all contract deliverables. Describe this program and how it will benefit your client on the particular contract at hand. You might also discuss your deficiency tracking program and include a sample deficiency tracking form.

Summation.

Reinforce your primary management theme statements in a set of strong closing paragraphs.

8.5 Cost volume

Cost is always pivotal in competitive procurements. It is, therefore, important for your company to develop an early technical design and/or approach upon which cost strategies and preliminary "bottom line" prices ("bogeys") can be built. The costing effort will also be facilitated by generating a staffing and management plan as soon as possible after RFP release. Develop your cost volume in close association with the technical and management volumes. The technical solution and staffing plan are integrally linked with the pricing strategy.

The cost volume must contain resource estimates, costing methodology, and a rationale sufficient to allow the client to establish the fact that the product or service to be provided will be cost effective and that the cost estimates are realistic to perform the contract work successfully [1]. Resource estimates include direct costs (primarily labor[4]), *other direct costs* (ODCs) (for example, material, travel, computer costs, and consultants), and indirect costs (overhead and *general and administrative* (G&A)). Credibility is essential in deriving the optimal balance of performance and cost. Cost volumes also include terms and conditions and responses to representations and certifications (Section K of the RFP). Provide sufficient

---

4   Because of holidays, vacation, and sick leave, the total number of labor hours in one staff-year equals approximately 1,860.

detail to satisfy the client. Ensure that your marketing efforts have provided this kind of detailed guidance from the client.

There are automated tools available on the market to assist your company in the preparation of various costing estimates. MicroFusion for Windows, which is produced by Integrated Management Concepts (805-376-3306) in Newbury Park, California, is one such application. With MicroFusion, you can develop multiple rate structures, complex burden structures, and escalate program costs.

Cost proposals are analyzed by the government to assess realism and the probable cost to the government. If total or cost element ceilings are included, the evaluation will also assess the maximum probable cost based upon such ceilings. The client is interested in proposed methods to control costs that do not have a negative impact on contract performance. Cost and pricing data must be current, accurate, and complete.

The trend in cost volumes is to make that document more sales-oriented—with more narrative, themes, graphics, and attractive formatting. They too can contain a copy of the executive summary.

The cost volume has a higher level of confidentiality than the other proposal volumes. Fewer copies should be made, and it should be distributed for internal company review on a need-to-know basis. Clients often require that the cost volume be submitted separately from technical and management. Many companies produce their cost volumes within the contracts or procurement departments rather than through a publications group.

8.6 Government contract requirements

The federal government imposes standards, procedures, guarantees, and documentation requirements on companies with which it does business; and it exacts significant penalties for noncompliance [2]. The *Truth in Negotiations Act* (TINA) was enacted in 1962 to protect the government from unscrupulous DoD contractors who supported their cost proposals with erroneous information. Congress extended the Act in 1984 to include all government contracts. On all negotiated contracts and subcontracts in excess of $500,000, your company will be required to furnish certified cost or pricing data. The government will include a "Price Reduction for Defective Cost or Pricing Data" in any resulting contract.

Small businesses must also be aware and knowledgeable about federal CAS. Created by Congress in 1970, the CAS was designed to "promote uniformity and consistency in the way defense contractors measure, assign, and allocate contract costs" [3]. The applicability of the CAS was extended in 1988 to include all negotiated government contracts in excess of $500,000. The "contract clauses involving CAS also restrict changes in company accounting practices…. Cases exist where a company has made an accounting change only to learn later that it owed the government money" [4]. As an entrepreneur, you will be well served to implement and energize a well-organized, highly trained, and watchful contract administration group.

The proposal effort is not over when the technical, management, and cost sections are written. There still remains editing, proofreading, configuration management of multiple proposal iterations, integration of iterations and review comments, word processing/publishing, graphics generation, outside printing coordination, photoreproduction, collation, assembly, letters of intent to obtain, volume coordination and consistency cross-checking, and delivery.

END NOTES

1. Stewart, Rodney D., and Ann L. Stewart, *Proposal Preparation*, New York: John Wiley and Sons, 1984, p. 126.

2. Wall, Richard J., and Carolyn M. Jones, "Navigating the Rugged Terrain of Government Contracts," *Internal Auditor*, April 1995, p. 32. This section draws upon Mr. Wall's and Ms. Jones's article.

3. Ibid., p. 33.

4. Ibid., p. 33.

Chapter 9

Acquisition and proposal team activities

W HEN A COMPANY is tracking a particular marketing opportunity, advanced planning and strategizing are crucial to ultimate success. One mechanism that companies use in this upfront marketing planning is the concept of an acquisition team.

9.1 Formation and function of acquisition teams

The *acquisition team* is formed in advance of the release of the formal RFP by the client organization. Under the direction and guidance of an acquisition manager, team members (usually three to five people) participate in pre-proposal intelligence gathering (including FOIA requests), strategizing, briefing the potential client, developing strategic proposal themes, and generating bullet drafts or storyboards to use as planning and compliance-verification tools during the proposal process. This team is superseded by the *proposal team* when the RFP is released. There should be some overlap in staff between these two teams to ensure continuity of approach and direction as well as seamless exchange of project-specific marketing information.

- Crucial to proposal success is the transfer of relevant marketing information to appropriate members of the proposal team. And from there, that information must be interpreted and translated into the actual proposal documents—into narrative, theme statements, graphics, captions, and the compliance matrix. If your marketing intelligence does not find its way into the proposal, then it cannot be evaluated and point scored. Therefore, it is valueless.

9.2 Pre-kickoff activities

Time spent in effective planning and strategizing will be critical to the success of the overall proposal effort.

Once an RFP to which your company will be responding is released, the proposal manager and key members of the Proposal Team should dedicate at least one if not more days to detailed analysis of that RFP along with in-depth proposal planning. *Don't panic or become overly anxious*, because the time spent in effective planning and strategizing will be critical to the success of the overall proposal effort. Too many times, proposal drafts emerge from review cycles with the recommendation for fundamental restructuring of the overall outline, proposed staffing, related contractual experience, primary strategic themes, and so forth. Building concurrence, based upon the sound marketing intelligence you have collected, early in the proposal planning process will save innumerable hours later in the proposal lifecycle.

The goal of pre-kickoff planning is to ensure that the formal kickoff meeting is an efficient and effective forum of information exchange and assignment making. During the pre-kickoff planning phase of the proposal response lifecycle, you should do the following.

Perform a detailed RFP analysis.

- Deduce probable weighting of evaluation criteria from Section M of the RFP.

The goals of RFP analysis are to: (1) become familiar with the proposal structure and format (Section L), (2) understand each technical and programmatic requirement in detail, (3) understand the source evaluation criteria and client success criteria (Section M), (4) evaluate your company's pre-RFP assumptions about the procurement and the client, and (5) develop written questions for submission to the client.

Prepare a draft executive summary.

The draft executive summary will be invaluable in assisting writers to fully understand the scope of the procurement and your company's primary sales messages. The executive summary should be one to two pages in length.

Develop and hone your strategic win themes.

Prepare bullet drafts.

- Vital for subcontractors;
- Help technical authors over their initial "writer's block";
- Ensure that all RFP requirements are addressed in your outline;
- Foster the smooth integration of writing from various authors;
- Link text and graphics/photographs;
- Primarily associated with technical proposal volume.

Review the proposal debriefings files, related proposals, project descriptions, and résumés in your company's proposal data center or library.

Plan to get the right people to write proposal sections at the right time; obtain senior management's support to make them available.

Prepare the proposal directive or kickoff package.

Appendix B presents a modified example of an actual completed package. The outline for the proposal does not follow the standard A, B, C format; rather it was mapped directly to the numbering scheme in the client's RFP. In this particular case, technical and financial information was included in one volume.

When the kickoff package has been prepared, it is then appropriate to convene the formal kickoff meeting.

Another example of a technical volume outline is presented in Table 9.1. Again, this outline was built using the numbering scheme presented in the client's RFP. This type of outlining configuration effectively serves multiple purposes and is highly recommended. First, it assists in the generation of a cross-reference or compliance matrix for the proposal volume. Second, it links themes and critical issues with specific proposal sections and subsections so that your company's proposal writers know what to infuse into their narrative. Clearly defined points of contact for each section and subsection are noted, and the status of each can be tracked. A detailed proposal summary sheet template is provided in Table 9.1. This overview provides an excellent starting point for the proposal kickoff meeting discussion.

Technical Volume Outline Procurement Name						
Company outline number	RFP Section	Technical Item	Person Responsible	Page Allocation	Themes/ critical issues	Status
—	—	Executive summary	John Smith	2		Closed
1.	L.5.1	Mission suitability factors	Jane Jones	0.5 (Intro)		Open
A.	L.5.1.A	Key personnel	James Goodfellow	0.5 (Intro)		Open
A.1		Project factors	Sally Bates	1.5		Open
A.2		Deputy project manager	John Smith	1.5		Open
B.1	L.5.1.B.1	Overall understanding	James Goodfellow	0.5		Open
B.1.1	SOW 1	Data center operations	Sally Bates	3		Open

Table 9.1 Effective Outline Configuration Tracks Status and Ensures Compliance

9.3 Proposal kickoff meeting

The kickoff meeting for a proposal activity is chaired by your company's proposal manager. This is the opportunity to build upon the successful activities of the acquisition team.

Invite all members of the proposal team, including major subcontractors,[1] if appropriate.

- Principal writers/contributors;
- Representative of senior management to demonstrate corporate commitment to the proposal and the project;
- Acquisition manager (leader of the acquisition team);
- Representative from contracts department;
- Publication group manager;
- Human resources manager;
- Key reviewers (Red Team participants).

Distribute key materials (proposal directive or kickoff package).

- Full RFPs, not only Section C, the SOW (now is not the time to save money on photocopying costs);
- Proposal summary sheet (see Figure 9.1);
- Complete listing of the proposal team membership;
- Evaluation factors for award;
- Milestone schedules for each proposal volume;
- Proposal volume outlines and writing assignments, including page and word allocations (see Table 9.1);
- Thematic and writing guidelines;
- Draft executive summary;
- Résumé format requirements;
- Action items sheet;
- Telephone list.

---

1   You may want to hold a separate meeting with the subcontractors since they may not have exclusive teaming agreements with your company. You do not want anyone with divided loyalties gaining access to company-sensitive proposal information.

Proposal Summary Sheet

Proposal title:
Internal charge code:
Proposal manager:

1. Solicitation of RFP #:
2. Proposal due date and time:
3. Client name (point of contact):
   Telephone #:

4. Type of contract:
5. Period of performance:
6. Place of performance:
7. Contract start date:

8. Proposal copies required
   Technical:
   Management:
   Cost/price:
9. Page limitations:

10. Overview of scope of work:
11. Offer acceptance period:
12. Packing label address:

13. SIC code:

14. Subcontractors:
15. Key personnel:
16. Other résuméd personnel:

17. Project summaries your company will use:

18. Defense Priorities & Allocation (DPAS) rating:

Figure 9.1
Proposal summary sheet is part of the proposal directive package.

Emphasize senior management commitment to this important project. This is the reason for the member of senior management to be in attendance.

Build motivation. Emphasize your company's business development bonus policy.

Highlight the nature of the procurement, and what marketing intelligence your company has collected regarding the opportunity.

Review your competitors' as well as your own company's strengths and weaknesses.

Discuss the concrete reasons for bidding on this procurement.

Highlight the strategies for winning (technical, management, cost). Why your company and not your competition?

Discuss any major or unexpected changes in the RFP from what had been expected during your intelligence-gathering efforts (capture plan activities).

Discuss the role of your teaming partners and subcontractors.

Review the proposal outlines.

Review the proposal milestone schedules. Emphasize the importance of adhering to the schedule.

Make well-defined writing assignments; assign specific action items for tracking to appropriate staff.

Discuss writing guidelines (such as style, page limitations, and words per page).

Review your major sales messages (themes) for this proposal.

Discuss the résumés and related contractual experience that will be included with your proposal.

Obtain approval of the concept for the cover design of the proposal volumes.

Review the staffing plan and WBS.

Identify special needs (such as binders, tabs, photography, and colored paper).

Determine who will deliver the proposal to the client and by what means (for example, drive and air travel).

9.4 Post-kickoff activities

In the several days following the kickoff meeting, the proposal manager should spend sufficient time with members of the proposal writing team in order to:

Make certain they understand precisely what is expected from them during the proposal effort;

- Help the writers understand how their input relates to and fits with the entire proposal;
- Reinforce the need to incorporate appropriate proposal themes throughout their writing;
- Impress upon them the importance of the milestone schedule.

Keep in mind that all of these activities are designed and intended to assist your company and your staff to remain on track over the entire course of the proposal response lifecycle.

We will now turn our attention to the role of the proposal manager, the individual charged and empowered with the responsibility and authority to oversee and orchestrate the entire proposal and post-proposal response lifecycle.

Chapter 10

The role of the proposal manager

The job of leaders is to paint an actionable vision…
—Roger N. Nagel, *Industry Week*, July 7, 1997

10.1 Overview

The proposal manager who is appointed to plan, organize, schedule, manage, arbitrate, and execute a given procurement response for a small business is *the* pivotal individual in the entire proposal process. He or she must skillfully balance the business opportunity against severely limited B&P monies and scarce human and material resources. And if those elements were not enough, the entire proposal response cycle unfolds in a matter of a few *weeks*—indeed, most smaller procurements have only 15- or 45-day response fuses. And certain state and local RFPs have response times measured in several days!

Your company needs several people (in very small firms, two people may be sufficient) to be well trained in the art and science of proposal management. Candidate staff need to possess (and ultimately enhance) the ability to plan and organize their own activities as well as those of a wide variety of other staff. Proposal managers should be effective leaders and positive motivators; exercise tact and diplomacy; display drive and enthusiasm; and practice thoroughness, organization, and attention to detail. Most importantly, they must take full *ownership* of the proposal development process.

Specific knowledge required for proposal manager candidates includes: (1) a broad understanding of your company including its history, products, services, personnel, facilities, technologies, and costing policies; (2) familiarity with the particular client's organization, technical requirements, key staff, acquisition policies, and special methods of evaluating proposals; (3) broad exposure to the operations of the particular marketplace, such as the federal regulatory and statutory structure; (4) awareness of the potential competitors for a given procurement; (5) knowledge of public relations and marketing techniques; and (6) appreciation of documentation and computer-based publication methods so that the proposal is a *well-presented sales document*.

The proposal manager is charged with complete responsibility and authority for the technical, management, and cost volumes of a proposal.

In many small companies, the proposal manager for a given procurement is drawn from the technical ranks because of the individual's knowledge of the project. Often, the proposal manager will also be proposed formally as the project or program manager. Because of the lack of depth of human resources in small firms, the proposal manager will likely have to provide the client on an existing contract with 40 hours a week of technical and programmatic support *in addition to* his proposal management responsibilities and obligations. Some larger companies have the comparative luxury of a full-time proposal management "pool," which can be drawn upon for proposal support as required. The proposal manager will most likely interact directly with the proposal coordinator or specialist within the business development group. The proposal coordinator or specialist can augment and complement the skills, knowledge, and time resources of the proposal manager. Small companies should avoid the very natural tendency of allowing the proposal coordinator or specialist, because that individual is probably a *full-time* "overhead" person, to assume too many of the proposal manager's functions and obligations. Both individuals should work together closely, but tasks should not merely be transferred from one person to the other. Proposal coordinators often participate in more than one proposal effort at a time. In fact, there are instances in small firms of 1 person coordinating 10 or more proposals simultaneously.

10.2 Generalized job description

The job description of the proposal manager is extremely open-ended, but in general he or she must provide technical, programmatic, cost, and organizational direction for the entire proposal effort, extended to include the following:

Recruit, organize, and direct the proposal team.

Arrange for necessary human, material, and computer resources. Requisition or obtain dedicated, secure floor and wall space to prepare the proposal.

Getting enough of the right resources for a proposal at the right time can be challenging in both small and large businesses. This is one of several points at which senior management's direct involvement is critical. Give every proposal that your company designs and prepares the resources necessary to win, or don't bid it.

Attend preproposal (bidders) conference (see FAR 15.409).

- Identify competitors and potential teaming partners;
- Obtain the client's official definition of evaluation criteria;
- Note any last-minute changes to the RFP;
- Listen for hidden messages from the government, such as "We are happy with the incumbent";
- Obtain details about the contract job site.

Ensure adequate planning of the proposal effort.

Time spent on planning and strategizing early in the proposal response cycle before the formal kickoff meeting is crucial to the overall success of the proposal team. Participate in RFP analysis, schedule and outline development, page allocation limits, writing assignments, and bullet draft preparation and review. Talk often with key technical writers, reviewers, and your publication staff to ensure that people understand your expectations and your role and responsibilities as well as theirs.

Prepare and then negotiate B&P budget estimates with senior management, and monitor and control B&P and ODC expenditures.

Preparing B&P projections results in an up-front commitment to make select key technical and programmatic staff available to prepare the best possible proposal.

It is highly advisable that your company develop and implement a method for projecting and tracking bid- and proposal-related costs. Direct labor for marketing a potential business target as well as for the technical writing, cost development, proposal development, and production of the proposal and BAFO should be estimated in detail for the entire proposal lifecycle. In addition, ODCs, including travel, classified advertising for new hires, subcontractor costs, temporary staff, and proposal consultants should be estimated as well. These estimates should be completed by the assigned proposal manager in advance of a proposal tracking number (for time-sheets) ever being assigned and approved by a member of senior management. The approved forms (hardcopy or electronic) are then used as an extremely effective tool to manage the cost of the entire proposal process in a manner analogous to managing a project. B&P projections can then be approved or disapproved by senior management based upon what the company had allocated for B&P expenditures for a given fiscal year.

The benefits of this structured, planned approach include that actual costs for proposal development, including marketing and post-proposal activities, are more accurately projected as real data are collected and that new estimates are calibrated against these real data. Proposal managers should be held accountable for any significant proposal-related cost over-runs. Another benefit is that there is an up-front commitment to make select key technical and programmatic staff available to prepare the best possible proposal.

Due to staffing constraints, small companies may need to bring temporary publication staff on board for a portion of the proposal response cycle. In addition, outside professional proposal consultants with technical or proposal development specialties and expertise may be required to augment your company's internal resources. And there are companies throughout the country that offer end-to-end proposal development services.[1] They will write, review, and produce your proposal for a negotiated fee.

---

1   Full-service consulting firms are located coast to coast, including OPTYM Professional Services, Inc., in McLean, Virginia; DSDJ, Inc., in Hampton, Virginia; Steven Myers and Associates, Inc. (SM&A), in Newport Beach, California; Proposal Support Center in Towson, Maryland; and Hy Silver and Associates, Inc. (HSA), in Los Angeles, California.

Define and present proposal themes and strategies.

- Ensure adherence to proposal themes within the proposal documents.

  The marketing intelligence that your company's acquisition team has collected for a particular procurement must be translated into the thematic messages that are embedded throughout the proposal volumes. As proposal manager, you will need to set the "marketing tone" for the entire proposal response. Ensure that the documents are designed to be, and ultimately become, sales documents—authenticated by technical, programmatic, contractual, staffing, institutional, and technological strengths, "lessons learned," and innovations.

Organize and conduct the pre-proposal activities and proposal kickoff meetings.

The proposal manager should communicate with the Acquisition Team that has done the direct marketing with the client to ensure that all of the intelligence gathered has been communicated and is understood thoroughly.

Provide needed information to your company's contracts and pricing staff.

Include information regarding direct labor, travel, equipment purchases, property rentals/purchases, *government-furnished equipment* (GFE) and *contractor-furnished equipment* (CFE), subcontractors/consultants to be used, documentation requirements for contract deliverables, proposed cost center, recommended fee/profit margin, and pricing strategies. Develop the technical and management volumes in close coordination with the cost volume.

Review and critique proposal input from the authors.

Whether received in hardcopy form, on diskette, or via electronic mail, a proposal manager must review the sections and subsections of the proposal for a myriad of factors including RFP compliance, thematic continuity, technical accuracy, programmatic appropriateness, risk minimization, and client awareness. Provide writers with interactive, ongoing feedback so that they stay on target from both a technical and schedule standpoint. Help them to understand the importance of their contributions.

Project planning (project organization, schedule, and budgets).

One of the most important aspects of proposal planning is selecting the staff whom you will propose to the government as your company's project or program team. Early staffing selection and management approval of this selection will be extremely beneficial to the entire proposal team. Definitive staffing decisions will make many processes easier, from résumé development and organizational chart generation to cost structuring and management plan preparation. Too many times, staffing decisions are delayed so that technical solutions can be developed and technical writing can be accomplished. Gathering detailed information for each résumé can be extremely time consuming. Every time a staffing change occurs, a new résumé must be developed. Name changes also present editorial challenges in both text and graphics in the sense of finding every instance in which a person's name had appeared and putting in the new person's name.

The proposal manager must ensure that an appropriate project schedule is developed that accounts for all client-required deliverables. This project schedule must be tightly linked with the details of the Management Plan.

- Maintain ongoing communication and liaisons with senior management to select subcontractors, identify key project personnel, review the proposal progress, and obtain assistance as needed. Use every opportunity to brief managers with the proposal status and issues that they can help to resolve.

- Maintain client contact in conjunction with the business development or advanced planning groups.

  Although your company's marketing staff cannot discuss the particular procurement for which a proposal is being prepared with the client, ongoing relationship building and discussions must continue.

- Coordinate subcontractor liaison with the contracts department.

  Ensure that technical information, résumés, project summaries, and cost data are submitted in a timely manner. Many times *small disadvantaged businesses* (SDBs), *disadvantaged business enterprises* (DBEs), and small, *woman-owned businesses* (WOBs) do not have adequate administrative staffs to prepare proposal input for your company (the prime) in electronic format or compatible electronic format. That means additional word processing tasks for your company's publication group.

- Arbitrate Red Team and other review cycle comments.

  Provide written instructions to the Red Team reviewers explaining what the expectations of the review process are and asking that comments be detailed and explicit so that the proposal team can respond effectively. During the recovery period following a review cycle, the proposal manager may have to assign "tiger teams" to ramrod the effort to fix any major deficits identified during the review process. Comprised of a small number of very senior staff, these teams will need to work quickly and effectively to address these major weaknesses. Senior management support for adequate resources is vital at this stage.

- Keep the proposal team members motivated.

    Attend to such needs as meals and special transportation requirements. Do not forget your support staff. They are the ones called upon to produce proposal after proposal on a regular basis. Demonstrate clear, concrete appreciation for a job well done.

- Ensure adherence to the proposal response milestone schedule.

    As proposal manager, you must be *relentless* in adhering to the schedule. Time spent in pre-kickoff strategizing, planning, communicating, and decision making will serve you well downstream in the response cycle. Ensure that even the early proposal response milestones are met. There is no need, indeed it is counterproductive, to take every proposal schedule down to the final hour. Manage the process effectively from beginning to end. It will be exhausting, but everyone will benefit from a well-executed process. And your proposal will have a much better probability of being a winner.

- Review and respond to all amendments to the RFP.

    The government will often release one or more amendments to the original RFP in response to questions from offerors or to make a correction. Each of these amendments must be reviewed carefully because they may, but certainly do not have to, extend the due date of the proposal. In some cases, amendments can have significant impact on staffing, technical approach, costing, page count,[2] and font size, for example. Amendments must be acknowledged in the proposal response, sometimes in your company's transmittal letter, as well as in the completed *Representations and Certifications* (Reps & Certs), Section K.

- Ensure timely delivery of the requisite number of proposal volumes.

---

2   For example, an amendment to a 1996 USACE-Sacramento RFP changed the original technical volume page count from 80 to 65. This change had a significant effect upon the structure of the final technical volume.

Avoid packaging all of the proposal documents in one box when delivering. Divide the copies among several boxes to minimize the loss if a box is dropped or misplaced.

- Ensure that the transmittal letter and cost volume forms have all the necessary signatures.

- Evaluate the performance of personnel assigned to work on the proposal.

  Once the proposal has been submitted to the government, take the time to evaluate the performance of both technical and support staff during the proposal response period. More than likely, you will be serving with these people again. Identifying who requires proposal development mentoring will be beneficial to them and to you. Express your appreciation. Hold a party, recommend people for bonuses, and arrange for proposal team T-shirts or other team-building items to enhance morale.

- Prepare for orals and BAFO.

  Ensure that your entire proposal—text, graphics, and cost data and justifications—are archived in a secure manner and easily retrievable. You will have to refer to your submittal frequently as you prepare for an oral presentation in front of the client or a written BAFO.

Ensure that your presentation team is reinforcing your company's answer to the question, "Why us and not our competition?"

Depending upon what deficiencies (DRs) and items that require clarification (CRs) have been identified in your original proposal, BAFO responses can be involved and time consuming. Deficiencies are significant shortcomings, omissions, or gaps in compliance that must be addressed fully.

In many cases, you as proposal manager will have to gain management support in reassembling the initial proposal team to prepare a solid response. Oral presentations will also require dedicated time to strategize your response and then conduct "dry run" practice sessions to hone the public speaking skills and cooperative interaction of the project team. Individuals who critique these practice sessions should ensure that the presentation team is reinforcing your company's answer to the question, "Why us and not our competition?" Staff should practice a sufficient number of times to present their material effectively while standing calmly looking at their audience. Make sure that the presentation is timed during the dry runs so that all of your material can be presented well in the time allotted by the government. Verify with the government exactly what audiovisual and computer equipment will be provided for you and also inquire as to the size of the room. Perhaps you could visit the presentation room beforehand. If your company is taking its own computer equipment, slide projector, or overhead projector, make sure to take extension cords and spare bulbs. Test the actual equipment you will be taking *before* arriving for the presentation.

Certain government agencies are now *requiring* presentations as part of the proposal response. NOAA and the U.S. Army are two such agencies in which specific organizations may require presentations. Today, with multimedia presentation software such as Microsoft PowerPoint, Gold Disk's Astound, and Macromedia's Action!, presentations can be more professional and alive than ever [1]. Multimedia software, speakers, sound cards, and cabling will be a capital investment for your company; or multimedia-equipped computer systems may be rented for occasional use as well.

- Request and attend the client debriefing.

A debriefings file is an excellent training tool for novice proposal managers within your company.

It is important to understand exactly why your company was not selected for award of a given procurement. Ask questions of the government contracting and technical staffs in attendance at the debriefing. The most important activity is to document and communicate the lessons learned from the loss to other proposal managers and senior management within your company. Maintain a debriefings file and refer to that file before beginning the next proposal. A debriefings file is an excellent training tool for novice proposal managers within your company.

- Overall responsibility for proposal security.

  To help ensure confidentiality, proposals should not be discussed outside the office. Proposal-related *paper*, including graphics, cost tables, and narrative, must be shredded and disposed of when it is no longer needed through a reputable waste management firm. Companies have been known to purchase competitors' trash! And network or PC-level password protection/encryption is essential to guard the *electronic* proposal files.

If at all possible, the proposal manager should have no writing assignments. In small firms, however, this is very difficult to avoid. Yet if the proposal manager is to be effective in that capacity, he needs to attend to a myriad of technical details and human and organizational issues and not be burdened with writing proposal sections.

10.3 The growing importance of oral presentations

During the past several years, the federal government has used oral presentations to *augment* written proposals submitted by the contractor community for competitively negotiated procurements. Oral presentations assisted government source selection teams to understand the written proposal documents. Generally, the oral presentations were not point scored in any way.

Since 1994, various agencies of the federal government including the Department of Energy, Nuclear Regulatory Commision, NASA, Centers for Disease Control, Federal Aviation Administration, Internal Revenue Service, and Bureau of Printing and Engraving, have experimented with selecting contractors on the basis of oral presentations instead of hardcopy or electronic technical and management volumes. And in 1997, the Federal Acquisitions Regulations (FAR) included—for the first time—coverage of oral presentations.

The Procurement Executives Association (PEA), a quasi-official group of senior government executives chaired by Lloyd Pratsch of the Department of State, in association with the Office of Federal Procurement Policy (OFPP), has published a pamphlet that is available on the World Wide Web at http://www.pr.doe.gov. The pamphlet, entitled. "Guidelines for the Use of Oral Presentations," is intended to heighten interest in the oral selection technique. Already, the Internal Revenue Service, Nuclear Regulatory Commission, and the Department of Energy have identified major procurements as targets on which to implement oral presentation components [2].

A major goal for the government is to reduce the level of written materials associated with the procurements process. In addition, oral presentations permit government evaluators to receive relevant information directly from the key members of the contractor's proposed team that will perform the work on the contract. And expect evaluators to address past performance during oral presentations. In this manner, communication and the exchange of information between the government and the contractor can be enhanced. The government is, in effect, conducting job interviews of key staff. Proposal-related costs could be reduced for the government as well as the contractor with the implementation of oral presentations. In addition, procurement lead times can be reduced significantly.

From both a legal/contractual and practical perspective, debate continues over exactly what portions of a contractor's "proposal" would be integrated into the resultant contract. Various federal agencies have videotaped the oral presentations for later review, although there is no regulatory requirement that prescribes that a record of the oral presentation be maintained. Both video and audio tape may become available through the Freedom of Information Act (FOIA). Time limits on oral presentations, number of participants from each contractor, number of presentation materials, and type of presentation materials all vary from agency to agency.

Although no standards for oral presentations have emerged, there is a distinct movement in the direction of such presentations throughout the federal government.

10.4 Attending to the details

Your company may want to consider establishing a petty cash fund to handle unforeseen proposal requirements such as photocopying, color laser copying, photography, graphic enlargements for foldout pages (11 × 17 inches), and small office supplies. The proposal manager should have access to this fund. He or she should also be given keys and access codes to the building and appropriate offices, file cabinets, and personal computers. Evening and weekend proposal work often requires ready access to hardcopy materials, diskettes, hard disk drives, and networks.

At the formal kickoff meeting, it is advisable to generate and distribute a list of key proposal participants, including subcontractors and consultants. The list should provide names and telephone numbers (office, home, cellular, and "beeper").

The proposal manager must ensure timely delivery of the proposal. Often, the mailing address is different from the hand-carried delivery address. For air carriers, such as Federal Express, military bases will need barracks or building numbers and not simply mail stops. Allow sufficient time for contingencies such as inclement weather, breakdown, construction delays, accidents on the road, and delays in public transportation. Know exactly who will deliver the proposal to the government site. Deliver the required number of copies of the proposal document to the correct address. Avoid the use of couriers for proposal delivery because they do not have a vested interest in making certain your proposal documents are delivered undamaged and on time. Finally, obtain and retain a receipt from the government that is stamped with the time and date of your delivery.

10.5 Control of the schedule

Proposals can be challenging to complete, because for most of us they are an effort above and beyond our normal contract-related responsibilities. In your small company, you probably do not have a comfortable depth of staff in either the publications department or in the technical ranks. But adhering to the agreed-upon schedule will help to make the process smoother and help to ensure that the finished product is one that your company can be proud to submit. Winning competitive proposals requires a substantial amount of planning and guidance, both during the actual proposal preparation process and well before the release of the RFP.

Seasoned proposal managers come to realize that, in general, everything about the proposal process will take longer than planned. A novice proposal manager will quickly ascertain that many participants in the process—from the technical writers to the subcontractors to the publications group—feel that they are being pressured unnecessarily. And supporting departments, such as Human Resources or Engineering, may be indifferent to even a well-planned proposal milestone schedule. Subcontractor technical, staffing and cost information is often late or delivered in an electronic format that requires conversion and cleanup.

When generating the schedule for your proposal, be sure to take into account the effects of weekends and holidays, important company social functions, and competing priorities within the publications group. Generating a calendar-type schedule as shown in Figure 10.1 that notes the highlights of the proposal schedule is one effective tool in keeping the overall proposal effort in perspective. A bar chart is another such tool that indicates the many simultaneous proposal activities. Effective implementation of a proposal milestone schedule is challenging and often thankless work. People tend to procrastinate, or want to keep polishing their writing, or have competing priorities. Frequent but brief proposal status meetings coupled with the full support of senior management will definitely assist a proposal manager keep his or her proposal effort on schedule. A word about management support. Senior management involvement is critical to ensure adequate resources are committed at the right times throughout the proposal response lifecycle. Management "buy-in" to your technical solution, staffing plan, management approach, and costing strategy is also pivotal *early* in the response life cycle. However, senior management must also be willing to entrust and empower the appointed proposal manager to orchestrate the proposal response activities. They must allow the proposal manager to take ownership of and responsibility for the entire task. Too many times there is significant micromanagement of the process or senior

Sunday	Monday	Tuesday	Wednesday	Thursday	Friday	Saturday
	1 Release RFP. Copy and distribute RFP to proposal team.	2 Analyze RFP. Generate kickoff package.	3 Analyze RFP. Generate kickoff package.	4 Kickoff meeting. Writing assignments made.	5 Proposal writing. Ongoing communication between proposal manager and writers.	6 Proposal writing.
7 Proposal writing.	8 Preliminary assessment of writing by proposal manager.	9 Proposal manager's comments back to writers.	10 Proposal writing.	11 All proposal sections delivered to proposal mahager by writers in electronic form. Proposal manager reviews all sections and electronic files to Publications Department. Production of Red Team draft.	12 Production of Red Team draft. Distribution of Red Team draft by close of business.	13 Red Team review.
14 Red Team review.	15 Interactive meeting between Red Team reviewers and entire proposal writing team.	16 Red Team recovery period (proposal manager prioritizes sections that require reworking.	17 Red Team recovery period.	18 Red Team recovery period.	19 Proposal manager reviews all sections that have been reworked and provides comments back to writers.	20 Additional writing/ modification.

Figure 10.1 Calendar provides the highlights of the proposal schedule for easy review.

Sunday	Monday	Tuesday	Wednesday	Thursday	Friday	Saturday
21 Additional writing/modification.	22 Proposal manager reviews all modified sections again.	23 All materials provided to Publications Department for generation of final Gold Team draft.	24 Production and editing of Gold Team draft.	25 Gold Team review.	26 Final production.	27
28	29 Photore-production and final Q.C. packaging.	30 Deliver proposal volumes to client.				

Figure 10.1 (continued)

management gets involved late in the response process and significantly revises the staffing plan, costing strategy, or technical solution.

10.6 Training additional staff in proposal management skills

Particularly within small companies, the number of experienced proposal managers is very low, sometimes less than three. One way out of the quandary of always using the same people to manage proposals is to have novice proposal managers directly assist their more experienced colleagues. After two or three assists, perhaps the novice could be appointed as the primary proposal manager with full responsibilities and obligations, but with access to the support of the more senior proposal staff. Outside consultants can also be used for proposal training and mentoring.

In-house proposal management and preparation seminars are another option to be considered. Curricula can be developed in-house, and seminars can be held in the evening or on a weekend. The point is to expand the base of proposal management experience within your company. This helps to prevent personnel burnout and promotes better proposal efforts.

10.7 Finish the job at hand

A proposal manager's duties are not over once the proposal documents have been delivered to the government. The proposal manager should either personally review or assign someone to review the entire proposal to identify, for example, any omissions, critical errors, and areas that require improvement. If the procurement cycle includes discussions and/or a BAFO, the proposal manager and his team need to prepare a meaningful response to written questions as well as rehearse oral responses for the face-to-face meeting with the client. The proposal manager will most likely be a key participant in the oral defense of the proposal.

Finally, the proposal manager should prepare a written "lessons learned" of the entire proposal effort for senior management. Examples of outstanding commitment by members of the proposal team should definitely be noted along with observations, suggestions, and recommendations to improve your company's proposal process. Senior management should then review and act appropriately on the recommendations. "[L]ess successful performers treat every opportunity as a discrete and unique event. These companies spend little time examining and improving their approach.... The proposal process is left unchanged. Each proposal team must relearn how to put together a proposal and experiences the same problems over and over again" [3]. Take the time as a company to learn from each proposal experience and apply those lessons to improving the process on a continuous basis. If ever there was an application for continuous process improvement, the proposal response lifecycle is it!

The successful proposal manager has provided leadership for his or her proposal team throughout the entire proposal response lifecycle.

10.8 Successful proposal managers

The most successful proposal managers with whom I have worked and been associated possess a marked ability to motivate and communicate with people of all levels—from company presidents to technical proposal writers to publications staff to financial and administrative personnel. They command and instill confidence in others by their technical knowledge, business development acumen, self-confidence, and encouragement. Often, they hold advanced degrees in a relevant technical discipline and have an intimate working knowledge of the particular client organization and its business processes. Successful proposal managers know when to push their proposal team, when to empathize, when to encourage, when to be flexible, and when to relax. They have a well-developed sense of humor and can laugh even during the most intense proposal development and production periods. These individuals talk with senior management and are forthright in that communication. Senior management, in turn, better understands the technical, costing, and logistical issues at hand that must be addressed and solved. Finally, successful proposal managers seek work-arounds. They demonstrate a certain "proposal agility." If one way does not produce the desired result, they are very willing to try another and another until they achieve a resolution. Fundamentally, they are leaders, facilitators, communicators, and listeners—and owners of the proposal development process.

END NOTES

1. Wodaski, Ron, "Planning and Buying for Multimedia: Effective Multimedia for the Small Office," *Technique: How-To Guide to Business Communication*, Oct. 1995, pp. 16–25. See also Joss, Molly W., "Authoring Alchemy: Ingredients for Brewing Up a Multimedia Masterpiece," *Desktop Publishers*, Jan. 1996, pp. 56–65.

2. See also Edwards, Vernon J., "Oral Presentations: New Development and Challenges for Proposal Managers," *The Executive Summary: The Journal of APMP's National Capitol Area Chapter*, November 1996, pp. 1, 3–5.

3. O'Guin, Michael, "Competitive Intelligence and Superior Business Performance: A Strategic Benchmarking Study," *Competitive Intelligence Review*, Vol. 5, 1994, pp. 8–9.

Chapter 11

Structuring international proposals

The time and space buffers that used to limit a company's exposure to changes are gone. Business now is vulnerable to anything that happens anywhere instantly.
—Marc J. Wallace, Jr., *Center for Workforce Effectiveness,*
  cited in *Industry Week*

11.1 Overview

The benefits of doing business in the international market can be substantial, however, "[c]ompanies that want to operate globally must have a global mindset" [1]. In the next millennium, your firm's business opportunities will probably be linked very closely to unfamiliar cultures and customs. Advances in information and telecommunication technology have intensified competition in product and financial markets during the past 20 years, as noted by Marina Whitman and Rosabeth Moss Kanter [2]. And new and different trade, economic, and business models are likely to become necessary to understand and implement within your own company. Professor Govindarajan of Dartmouth College's Amos Tuck School of Business Administration suggests that "companies must open themselves to cultural diversity and be ready to adopt best practices and good ideas regardless of their country of origin" [3].

Many large U.S. high-tech firms have doubled their size in the global marketplace. And the Peoples' Republic of China, the Republic of Korea (South Korea), and Mexico are assuming leading economic roles, according to Franklin Root, author of *Entry Strategies for International Markets* [4]. Root sees the transformation of the international economy as "a geographical extension of the industrial revolution that started in Great Britain more than two centuries ago" [4]. He asserts that "[t]o survive and prosper in the 1990s, companies will need to develop new strategies. For the truth is that today *all* business firms—small or large, domestic or international—must strive for profits and growth in an international economy that features enormous flows of products, technology, intellectual and financial capital, and enterprise among countries. In this economy, no market is forever safe from foreign competition. And so even when companies stay at home, sooner or later they learn from hard experience that there are no domestic markets, only global markets" [5].

The losses associated with international business, however, can be devastating. Contract awards can actually result in severe financial losses and liabilities for the selling firm, depending upon the type of the contract and terms and conditions of that contract. Time delays in RFP or tender release dates are frequent, so booking and sales projections should be conservative [6]. A significant challenge associated with conducting international business in many countries (particularly developing nations) is supplying the technical and programmatic expertise as well as the *funding*—all as a package to the host country. Two examples of funding sources are the World Bank, headquartered in Washington, D.C., and the Asian Development Bank, centered in Manila in the Philippines. The World Bank receives its resources from industrialized nations and from tapping the world's financial capital markets. Projects funded by the World Bank are not limited geographically, whereas the Asian Development Bank is a lending institution serving only the Pacific rim [7].

11.2 The importance of the World Bank group

With US$20 billion in new loans each year, the World Bank is the largest global provider and coordinator of developmental and infrastructure assistance, which takes a *financial* as well as technical form. The World Bank is "owned" by its 180 member countries, with the largest industrialized nations—the Group of Seven or G7[1]—controlling about 45% of the Bank's shares. The board of governors and the Washington-based, 24 full-time-member board of directors of the World Bank, however, represent *all* member countries. Of note is that a central tenant of the evolving role of the World Bank is to build it into a world-class *knowledge institution*. In FY98, prototype knowledge management systems (country-specific best practices, lessons learned, success stories) in education and health were established under the direction of the bank's president, James D. Wolfensohn [8]. The World Bank finances more than 200 new projects each year, which involve nearly 30,000 individual contracts. Projects financed in 1998 spanned a broad range from agriculture and education to information technology, urban development, and water supply and sanitation. Specific projects included improving the urban transport services in Rio de Janeiro, Brazil; health-sector reforms in Ghana; improving access to safe, potable water for rural people in Morocco; and enhancing school quality in Pakistan. Together, the International Bank for Reconstruction and Development (IBRD), the International Development Association (IDA), and the Multilateral Investment Guarantee Agency (MIGA) comprise the World Bank Group, which also works closely with the International Finance Corporation (IFC).

The publication *UN Development Business* (UNDB), which is released twice each month by the United Nations (UN) Department of Public Information and sells for US$495/year, contains World Bank and UN procurement information. This includes invitations to bid on World Bank and UN projects as well as information about projects financed by the African, Asian, Caribbean, and Inter-American Development Banks; the European Bank for Reconstruction and Development; and the North American Development Bank. (Established in 1959, the Inter-American Development Bank (IDB) is an international financial institution that finances economic and social development in Latin America and the Caribbean. Its loans exceeded $6.7 billion in 1996.) Sample templates and forms for requests for expressions of interest and invitations for prequalifications and bids can be found on the Web at http://www.worldbank.org/html/opr/consult/tempeng.html, http://www.worldbank.org/html/opr/procure/notices/inviteng.html, and http://www.worldbank.org/html/opr/procure/notices/prenoeng.html, respectively.

Suppliers, contractors, and consultants can learn more about the World Bank's procurement process by attending monthly business briefings at World Bank headquarters in Washington, D.C., as well as in Paris,

---

1   The G-7 members are Canada, France, Germany, Italy, Japan, the United Kingdom, and the United States.

France, or by consulting the *Guide to International Business Opportunities* at http://www.worldbank.org/html/opr/busop/11.html.

UN Development Business provides a Monthly Operational Summary report listing all of the projects being considered for financing by the World Bank; Procurement Notices, that is, invitations to bid/prequalify and submit proposals to supply the products and services required to carry out World Bank-financed projects; and Contract Awards for major projects. Your company can subscribe to this important publication through:

Subscription Department
Development Business
United Nations
P.O. Box 5850 Grand Central Station
New York, NY 10163-5850
Tel. (212) 963-1516
FAX (212) 963-1381
E-mail: dbsubscribe@un.org
UN Development Business can also be accessed for a fee online at http://www.devbusiness.com.

To assess the qualifications of firms and to assist in-country borrowers who are trying to establish a short list of potential contractors, the World Bank maintains an automated database of consulting firms interested in doing business on World Bank-financed projects. This database is called the Data on Consulting Firms (DACON) system, and is open only to consulting companies with five or more staff as opposed to manufacturers or suppliers. Registration cost is only $20. You can contact the DACON Center at:

DACON Center
The World Bank
1818 H Street, N.W.
Room MC10-446
Washington, D.C. 20433

The DACON database on CD-ROM is a valuable tool for identifying consulting firms registered with the World Bank and Inter-American Development Bank. This searchable database provides information on consulting firms, addresses, contact information, year established, staff by occupational group, descriptions of activities, types of services and specializations. It is particularly useful for individual consultants or consulting firms that do not have the financial and technical means to prepare successful proposals on their own. It is available from the UN Publications Office at a cost of US$100.

It is essential to recognize that the borrower, *not* the World Bank, is always responsible for procurement. Borrowers or their respective ministry, governmental agency, or public utility in the borrowing country—often with the help of consultants—prepare the specification documents, establish the procurement schedule, publish the procurement advertisement, evaluate the bids, and award and manage the contract technically as well as fiscally. The World Bank provides financing from its loans for the contracts, but the contract itself is between the borrower and the contractor or supplier. The role of the World Bank is to ensure that the agreed-upon procurement procedures are observed. For any contract to be eligible for financing from a World Bank loan, the procurement must be completed in strict accordance with the procedures established in the loan agreement, which must incorporate those in the World Bank's *Guidelines: Procurements Under IBRD Loans and IDA Credits* publication (January 1995, revised January and August 1996, and September 1997).

Of note is that a private company cannot approach the World Bank directly to request financial assistance for a project in which it and the host country government are interested. Only the government of an eligible World Bank member can request such financing.

11.3 Your company's participation in United Nations procurements

The Inter-Agency Procurement Services Office (IAPSO; http://www.un.org/Depts/ptd/) was established by the United Nations Development Programme (UNDP) in 1978 to assist its partner organizations within the UN system in the most economical acquisition of essential equipment and supplies. Origi- nally based in New York, and designated the Inter-Agency Procurement Services Unit (IAPSU), it moved to Geneva in 1982. In July 1989 the office was relocated to Copenhagen and renamed the Inter-Agency Procurement Services Office (IAPSO).

Interested firms should write to request inclusion in the UN's Supplier Roster. Upon receipt and acceptance of the appropriate questionnaires, the firm's name will be added to the roster and be given equal consideration in the selection process. Competitive bidding procedures are followed, with due regards to servicing and spare parts facilities available in the country of destination. Single purchase orders in excess of US$25,000 are subject to international bidding unless exceptional factors preclude bidding. There is no limit to the number of firms that may submit bids in a single invitation.

Here are the requirements for a complete application:

1. Completed registration form with the requested data;

2. Printout of the form certified by the signature of an authorized person;

3. The firm's most recent audited financial report(s);

4. Information about the firm, its products/services and quality certification, if any;

5. A statement showing the firm's business experience especially with international organizations and national governments.

Each completed application will be evaluated by the UN on the basis of experience, ability to perform, relevance of the goods or services, and financial soundness of the supplier. Applicants may provide one set of catalogs or short-form specifications concerning its products. Return the completed registration form, certified printout, general information and financial report to:

United Nations Procurement Division
Supplier Registration
Supplier Roster Officer
304 East 45th Street, Room FF-245
New York, NY 10017

Submitting a completed application package does not mean that your company will be automatically included in the UN Supplier Database. Registration will depend upon the result of an evaluation. Inclusion of your company in the database does not mean that it will be included in all supplier solicitations. It only means that your company, if included in the database, will be given equal consideration in the process of supplier selection. Suppliers who do not respond to solicitations on two consecutive occasions will be removed from the roster.

11.4 European Bank for Reconstruction and Development (EBRD)

Established in 1991, the European Bank for Reconstruction and Development (EBRD) assists 26 countries in central and eastern Europe and the former Soviet Union (now Commonwealth of Independent States (CIS)) with privatization and entrepreneurship efforts and promotes structural and sectoral economic reforms. The EBRD encourages co-financing and foreign direct investment from both the private and public sectors. The EBRD's *Procurement Policies and Rules*—derived from Standard Bidding Documents (SBDs) developed and in use by the World Bank, the Asian Development Bank, and the Inter-American Development Bank—articulates the procurement guidelines to be followed in EBRD-financed operations and establishes tendering procedures for use by clients when procuring goods and services following open tendering in operations financed by the EBRD. In addition, the EBRD *Procurement Policies and Rules* outlines how tenders are received, opened, and evaluated under "open tendering" procedures for contracts financed by the European Bank for Reconstruction and Development. Invitations to tender, expressions of interest, contract award information, and other essential information regarding EBRD-funded contracts is published monthly by the Bank in *Procurement Opportunities,* which is also available on the Bank's Web site at http://www.ebrd.com. Procurement information is also provided in the *Official Journal of the European Communities* and *United Nations Development Weekly*. A sample entry from *Procurement Opportunities* is shown in Figure 11.1. In Figure 11.1 note that *ECU* refers to the European Currency Unit or "euro," which was introduced on January 1, 1999, in 11 countries of the European Union. During a 3-year transition period, the 11 countries' existing currencies will coexist with the euro, but their exchange rates will be locked together. However, after June 30, 2002, the euro will be the sole currency in the region.

There will be seven euro notes; in different colors and sizes they will be denominated in 500, 200, 100, 50, 20, 10 and 5 euros. There will also be 8 euro coins denominated in 2 and 1 euros, then 50, 20, 10, 5, 2, and 1 cents. The graphic symbol for the euro looks like an E with two clearly marked, horizontal parallel lines across it. The official abbreviation for the euro is "EUR."

The term *EU* in Figure 11.1 denotes the European Union, which now consists of Austria, Belgium, Denmark, Finland, France, Germany, Greece, Ireland, Italy, Luxembourg, The Netherlands, Portugal, Spain, Sweden, and the United Kingdom. This group of European countries has

28 October 1998

Romania

Project name:	Establishment of the National Power Grid Company
Contact:	John Besant-Jones - Fax: +44 171 338 7280
Sector:	Power/Energy Utilities
Description:	To provide advisory services to the Romanian National Power Company for the Establishment of the National Power Grid Company. Services to be provided include: development of commercial arrangements for transmission tarriffs; preparation of a Business Plan; and analysis, design and implementation of programmes for institutional and organisational development.
Estimated cost:	ECU 1,300,000
Funding source:	EU Phare Bangkok facility. Eligibility restricted to EU Member States and Phare recipient countries.
Status:	Consultant to be selected. An invitation for Expressions of Interest will be published.

Figure 11.1
Sample entry from
*Procurement
Opportunities.*

decided to collaborate on a variety of significant areas that encompass a single economic market as well as a common foreign policy and legislative and judicial policies. The European Union came into existence as a result of the ratification of the Maastricht Treaty in November 1993. However, its roots can be traced back to the Treaty of Paris in 1951 and the Treaty of Rome in 1957. Formally, the EU consists of three "pillars," with the European Community flanked by various intergovernmental "pillars." The European Investment Bank (EIB) serves as the European Union's financing institution.

Phare in Figure 11.1 refers to the Phare Programme, the world's largest grant assistance effort for central and eastern Europe. All activities under the Phare Programme are normally undertaken by contracted suppliers by means of competitive tender. You can learn more about Phare at http://europa.eu.int/comm/dg1a/phare/index.htm. The Phare Programme is related to two other initiatives of the Directorate-General 1A (a department of the European Commission)—the *Tacis Programme* and the *Obnova Programme.* Tacis is a European Community initiative for the New Independent States (NIS) and Mongolia, which fosters the development of harmonious and prosperous economic and political links between the European Community and these partner countries. (For further assistance in the United States, contact the Business Information Service for the Newly Independent States, U.S. Department of Commerce, International

Trade Administration, Room 7413, Washington, DC 20230; Tel. (202) 482-4655; FAX (202) 482-3145.) The Obnova Programme is a European Community initiative for the rehabilitation and reconstruction of Bosnia and Herzegovina, Croatia, the Federal Republic of Yugoslavia, and the former Yugoslav republic of Macedonia.

Official Journal of the European Communities

The *Official Journal of the European Communities*, published 5 days each week from Tuesday through Saturday, includes notices of public works and supply contracts as well as invitations to tender from the European Development Fund (EDF). The S-Series Supplement to the *Official Journal of the European Communities* lists all invitations for public tenders. *Tenders Electronic Daily* (TED), http://www2.echo.lu/, is the online version of the printed S-Series Supplement S and it contains calls for tenders, contract awards, and pre-information notices for contracts above the fiscal thresholds from all EU member states. *Tenders Electronic Daily* is an excellent source of information on current worldwide public purchasing. It covers all areas of government spending, namely, supply, service, and public works contracts. TED contracts are from around the globe including the United States, Japan, and the member states of the European Union. The EDF is the principal means by which the European Union provides aid, concessionary finance, and technical assistance to developing countries. The fund was originally established in 1958.

By registering with the Central Consultancy Register (CCR) in Brussels, Belgium, you can make your qualifications and experience known to the European Commission units dealing with the Phare, Tacis, and Obnova Programmes. The CCR is used by European contracting authorities as a source of information about companies being considered for shortlists. A shortlist is a device used in tendering procedures. It is established by the contracting authority for a given project. It is a list of those companies judged capable by the contracting authority of supplying the services required by a particular project. Shortlisted companies are the only ones invited to bid for a service contract. Because different projects need different services, a new shortlist is drawn up for each tender. For practical reasons, a shortlist is limited in size. Being registered in the Central Consultancy Register or submitting expressions of interest thus does not guarantee inclusion in a shortlist. Only the companies considered to be the best qualified for a particular project will be invited to participate in a tender.

Your company can request the forms by e-mail at ccr@scr.cec.be, or contact the CCR administration office directly at:

Central Consultancy Register, B-28 6/84
Rue de la Loi 200
B-1049 Bruxelles, Belgium
Tel. 32-2-295 60 74
FAX 32-2-295 54 31

11.5 Asian Development Bank (ADB)

The Asian Development Bank, a multilateral development finance institution whose capital stock is owned by 57 member countries including the United States, is engaged in promoting the economic and social progress of its developing member countries in the Asian and Pacific region. Since the ADB began operations in December 1966, it has been a catalyst in promoting the development of one of the most populous and fastest growing regions in the world. ADB makes loans and equity investments, and provides technical assistance grants for the preparation and execution of development projects and programs, and also for advisory purposes. It promotes investment of public and private capital for development purposes. You can visit ADB on the Web at http://www.adb.org. Headquartered in Manila, Philippines, the ADB is led by a board of governors and board of directors. In 1999 Mitsuo Sato was serving as the bank's sixth president, a seat he has held since 1993.

Like the World Bank, ADB's mandate to assist developing member countries in the Asia/Pacific region has resulted in the need to maintain an inventory of suitably qualified firms/organizations from its member countries that could act as consultants to provide services required for various projects assisted by the bank. ADB's information on the capabilities and experience of firms/organizations is maintained in the ADB DACON system, the acronym for Data on Consultants. It is a computer inventory used by ADB to record the eligibility of such firms for possible engagement as well as their experience and qualifications for easy retrieval by ADB when needed. To be eligible for registration in the ADB DACON system, a company must have a minimum of five permanent full-time professional staff, must have been incorporated for at least 2 years, and must have completed a minimum of three major projects.

11.6 International market planning

In developing international marketing entry and marketing plans, keep in mind that your company is entering both a new country and a new *market* [9]. The entry mode (export, contractual, or investment) determines the amount of a company's control over the marketing program in the target country. For example, indirect exporting and pure licensing allow little or no control over the marketing program. Your company's international marketing plan should include objectives for sales volumes, market share, profits, and return on investment as well as resource allocations and a time schedule [10]. Market potential and growth as well as risk associated with market entry can be evaluated through such published sources as *Business International,* the *Price Waterhouse Country Information* guide, and Dun and Bradstreet's *Exporters Encyclopaedia.* The U.S. Department of Commerce also published detailed global market surveys for the 20 to 30 best foreign markets for a given industry. There are, however, a limited number of industries addressed. And the Economist Intelligence Unit (EIU) out of London, England, publishes special reports on specific international marketing topics [11]. The EIU, http://www.eiu.com/, is an information provider for companies establishing and managing operations across national borders anywhere in the world. Established 50 years ago in London, the organization now has a worldwide network of offices in London, New York, Hong Kong, Vienna, Singapore, and Tokyo. The EIU produces analyses and forecasts of the business and political environments in more than 180 countries. The EIU is part of the Economist Group, which also publishes *The Economist* newspaper.

Many small companies simply do not have the staff to dedicate to the time-consuming task of international marketing. There are local and regional trade programs designed to assist small businesses with identifying international markets. One such program is the *State University of New York* (SUNY) at Albany's *International Marketing Assistance Service* (IMAS). This service pairs SUNY-Albany graduate students in Marketing with local New York companies to identify and seek out international markets. The IMAS has also conducted several trade missions, which have included trips to Mexico, Chile, Argentina, Venezuela, and Brazil. In New York state, the Department of Economic Development has participated in several trade shows coordinated by the U.S. Department of Commerce that link small companies with representatives and distributors in international countries [12].

Because of the critical shortages of resources in small companies, international business must be a carefully planned and orchestrated process. Small companies successful in this field often have foreign shareholders who know the local environment and its pitfalls and processes. International market entry must become part of your strategic plan and marketing plan if it is to be pursued with the diligence that successful international business requires.

11.7 In-country partnerships

Creating ongoing business partnerships with the host country's key deci-sion makers is essential to long-term marketing, proposal, and contractual success. Partnerships can be strengthened through scheduled in-country visits (be sure to work around major holidays and holy days such as Rama-dan in Islamic countries, Setsubun in Japan, Carnival in Brazil, and Spring Festival (Chinese New Year)) that include social and community activities, reciprocal invitations to the United States, and establishment of offices in country. Well-managed representatives, agents, or distributors who under-stand the language, culture, and procurement laws and processes and are sensitive to the in-country political directions can be invaluable to your company's success. Crosscultural considerations must be taken into account, so it is suggested that these representatives be employees or con-sultants of the buying country's nationality. Also, do not try to cover the whole world—develop a profitable beachhead in a country your company knows already.

In-country partnerships can result in your company defining the tech-nical requirements that will appear ultimately in the RFP or tender.

Your company might also consider establishing formal business part-nerships with several select consulting firms in strategically important countries throughout the world. These firms can then function as in-country partners, providing local support as well as facilitating worldwide coverage for your multinational clients.

11.8 Host country procurement environments

As in domestic U.S. business development efforts, long-term marketing and relationship building on a global scale are critical to your company's contract award success and long-term profitability. Understanding your international client includes his technical expectations, procurement and source selection cycle, business decision-making processes and framework, import-export regulations, cultural environment, language requirements, and political agenda. Detailed understanding extends to potential tax expo-sure, countertrade requirements, work visas required, host country law, U.S. treaties with the host country, quality assurance parameters, delivery and acceptance restrictions, payment parameters, and currency restrictions [13]. In certain countries (e.g., Brazil) and on specific types of procure-ments, your competitors will have the opportunity to openly and thor-oughly review your proposal documents for the purpose of attempting to disqualify your submittal on technical grounds or procedural/administrative technicalities.

11.9 Import-export considerations and technology transfer

Time must be allocated in delivery schedules for any necessary U.S. Department of State or Department of Commerce licensing, such as in the case of advanced electronic equipment. "The international customer community wants to obtain as much technology from U.S. corporations as possible" [14].

11.10 Risk assessment

Your company may want to consider establishing a separate legal entity for the purpose of minimizing corporate exposure in the international arena. Carefully consider the following: taxes and duties, import-export quotas, host country legislation and regulations, language, host country culture, quality and audit standards, countertrade requirements, packaging/labeling standards, political issues and stability, currency exchange rates, foreign direct investment, customs paperwork and protocol, weather, religion, inspection and acceptance guidelines, in-country banking restrictions, and host country law.

11.11 Terms and conditions

Developing restrictive, risk-aware *terms and conditions* (Ts & Cs) is absolutely vital for international proposals. There are many challenges associated with resolving legal problems at an international level. Questions arise as to where to try a particular case, which law should apply, and how to apply foreign laws. To help mitigate the problems inherent in such issues, include clauses in your Ts & Cs that address in detail which specific laws will obtain in any contractual disputes [15].

There are certainly risks associated with international payments, two of which are nonpayment and variations in the foreign exchange rate [16]. Such risk can be mitigated by stipulating cash in advance or an irrevocable confirmed letter of credit. This letter of credit is issued by the importer's bank and then confirmed by a bank in the exporter's (your company) country. Thus, your company does not have to depend on the importer for payment [17]. Be aware, however, that your competitors may be offering easier, more flexible, and therefore more attractive payment terms. U.S.exporters can obtain export credit insurance that protects against the risks of nonpayment for both commercial and political reasons from the *Foreign Credit Insurance Association* (FCIA). This is an association of private insurance companies that operates as an agent of Ex-Im Bank.[2] Having this insurance can allow your company to extend credit on more favorable terms to overseas buyers and thus be more competitive [18].

Ts & Cs will vary depending upon the product or service being provided to the host country or industry. A listing of frequently used line items found in Ts & Cs includes:

- Warranty terms, such as commencement of warranty and warranty for replacement parts;
- Warranty exclusions (for example, improper unpacking, installation, or treatment of equipment; unauthorized attempts to repair, reconfigure, or modify equipment);
- Extended warranty and maintenance;
- Limitations on liability;
- *Force majeure* (protects supplier from casualty or cause beyond reasonable control, such as strikes, floods, riots, acts of governments, and war);
- Pricing for tasks beyond scope of work;
- U.S. government approvals;
- Taxes and duties;
- Dispute resolution (for example, arbitration under the Rules of Procedure of the International Chamber of Commerce, London, England);
- Excusable delays;
- Shipping and insurance;
- Contract termination;
- Acceptance criteria;
- Payment terms, schedule, and currency;
- Training;

2   Export-Import Bank of the United States, 811 Vermont Avenue, N.W., Washington, D.C., 20571; phone: 1-800-565-3946; FAX: (202) 565-3380. The Export-Import Bank of the United States (Ex-Im Bank; http:www.exim.gov) is an independent U.S. government agency that helps finance the overseas sales of U.S. goods and services.

- Documentation (it is critical to specify the language and dialect in which training manuals will be provided).

11.12 Export-Import Bank of the United States assists small businesses

In the past 60 years, Ex-Im Bank has supported more than $300 billion in U.S. exports. The bank's mission is to create jobs through exports. It provides guarantees of working capital loans for U.S. exporters and also guarantees the repayment of loans or makes loans to foreign purchasers of U.S. goods and services. Ex-Im Bank also provides credit insurance that protects U.S. exporters against the risks of nonpayment by foreign buyers for political or commercial reasons. Ex-Im Bank does not compete with commercial lenders, but assumes the risks they cannot accept.

Ex-Im Bank has undertaken major initiatives to reach more small business exporters with better financing facilities and services, to increase the value of these facilities and services to the exporting community, and to increase the dollar amount of Ex-Im Bank's authorizations supporting small business exports. The Working Capital Guarantee Program assists small businesses in obtaining crucial working capital to fund their export activities. The program guarantees 90% of the principal and interest on working capital loans extended by commercial lenders to eligible U.S. exporters. The loan may be used for pre-export activities such as the purchase of inventory and raw materials, or the manufacture of a product. Ex-Im Bank requires the working capital loan to be fully collateralized utilizing inventory, accounts receivable, or other acceptable collateral. Ex-Im Bank's Export Credit Insurance helps U.S. exporters reduce foreign risks. A wide range of policies is available to accommodate many different export credit insurance needs. Insurance coverage protects the exporter against the failure of foreign buyers to pay their credit obligations for commercial or political reasons; encourages exporters to offer foreign buyers competitive terms of payment; supports an exporter's prudent penetration of higher risk foreign markets; and because the proceeds of the policies are assignable from the insured exporter to a financial institution, it gives exporters and their banks greater financial flexibility in handling overseas accounts receivable.

The Bank offers a short-term (up to 180 days) insurance policy geared to meet the particular credit requirements of smaller, less experienced exporters. Products typically supported under short-term policies are spare parts, raw materials, and consumer goods. Under the policy, Ex-Im Bank assumes 95% of the commercial and 100% of the political risk involved in extending credit to the exporter's overseas customers. The special coverage is available to companies that have an average annual export credit sales volume of less than $3 million for the two years prior to application and that meet the U.S. Small Business Administration's definition of a small business.

Ex-Im Bank provides direct loans and guarantees of commercial financing to foreign buyers of U.S. capital goods and related services. Both programs cover up to 85% of the U.S. export value, with repayment terms of 1 year or more. Ex-Im Bank loans are offered at the lowest interest rate permitted.

Small businesses will also find Ex-Im Bank to be supportive when they are confronted by foreign tied aid on a capital project. Ex-Im Bank wants to enable competitive U.S. exporters to pursue market penetration opportunities in dynamic developing countries. The bank has established a Project Finance Division to analyze transactions in which the repayment of the financing is based upon a project's cash flow instead of a bank or government guarantee.

In February 1998, William W. Redway was chosen as the group vice president for Small and New Business within the Ex-Im Bank. The Small and New Business Group provides financial structuring and counseling services to both Ex-Im Bank's foreign and domestic customers. The group promotes Ex-Im Bank's programs and policies by conducting in-house seminars and outreach conferences. The bank also offers briefing programs that are available to the small business community. The program includes regular seminars, group briefings, and individual discussions held both at Ex-Im Bank and around the United States. For a seminar brochure and scheduling information, call (202) 565-3912 or FAX (202) 565-3723.

11.13 Helpful Web-based resources and in-country support infrastructures for small businesses

The World Wide Web provides small businesses with a host of business development sites to support international marketing initiatives.

MERX: Canada's national electronic tendering service

http://www.merx.cebra.com; 1-800-964-6379; 880 Wellington Street Suite #1 (Tower) Ottawa, Ontario K1R 6K7, Canada. MERX is an Internet-based electronic tendering system established and launched in

1997 by Cebra Inc., the Bank of Montreal's electronic commerce company. MERX is designed to improve access, increase competition, and provide a level playing field for small businesses competing for bidding opportunities within the public sector in Canada. At the present time, MERX provides the opportunity to bid on most Canadian federal government contracts. The system also posts opportunities for eight provincial governments: Ontario, Quebec, New Brunswick, Nova Scotia, Manitoba, Prince Edward Island, Alberta, and Saskatchewan. MERX replaced Canada's existing Open Bidding System in October 1997.

In late 1998, MERX announced a new partnership with the Ontario University Purchasing Management Association (OUPMA). As a result of the OUPMA's endorsement of MERX, Canada's National Electronic Tendering Service, 10 new university partners will be joining. They join fellow members, University of Ottawa and Carleton University, in utilizing the MERX service.

The Canadian federal government estimates that they procure between CAN$5 and $6 billion per year. Any Canadian federal government contract over $25,000 must be advertised through electronic tendering. MERX is the central source for these opportunities.

Government Supplies Department (GSD) Hong Kong, China

http://www.info.gov.hk/gsd/index.htm; GSD Headquarters 9/F, North Point Government Offices, 333 Java Road, North Point, Hong Kong. The GSD is the central purchasing, storage, and supplies organization for the Government of the Hong Kong Special Administrative Region, serving more than 80 governmental departments, subvented organizations, and certain nongovernment public bodies. The objective of GSD's procurement service is to obtain, at best value for money and in a timely manner, the goods and services required by the user departments. This is achieved by the use of open competitive tendering procedures, by the use of specifications that are as general as possible, and by the widest possible sourcing to maximize competition.

GSD issues procurement forecasts that are available on the Internet, for example, "Forecast of Major Purchases, The Government of the Hong Kong Special Administrative Region in the next 12 months starting from November 1998." The purchasing activities of GSD are shared among the six buying groups of the Procurement Division. On May 20, 1997, Hong Kong acceded to the World Trade Organisation Agreement on Government Procurement (WTO GPA).

As established in the Stores and Procurement Regulations, government procurement exceeding HK$500,000 (for goods and general services) and HK$1 million (for construction and engineering services) in value is normally done by the use of open and competitive tendering procedures so as to obtain the best value for the money. However, to save administrative costs, for purchases of goods below HK$1 million, this may be arranged through single or restricted tender if it has been established that the existing contractor is a suitable and cost-effective source of supply. Limited or restrictive tendering procedures are only permissible under specified exceptional circumstances. Where the nature of the contract (such as a contract that is time critical or one that requires particularly high levels of skills and proven reliability) dictates that tenders have to be invited from qualified suppliers/contractors, selective tendering or pre-qualified tendering may be used.

Open tendering: Tender invitations are published in the *Government of the Hong Kong Special Administrative Region Gazette*, on the Internet, and, if necessary, in the local press and selected overseas journals. Consulates and overseas trade commissions are also notified, where appropriate. All interested suppliers/contractors are free to submit tenders.

Selective tendering: Tender invitations are published in the *Government of the Hong Kong Special Administrative Region Gazette*, or are sent by letter to all suppliers/contractors on the relevant approved lists of qualified suppliers/contractors established and approved for the purpose of selective tendering. GSD has 4,000 registered suppliers. Suppliers and contractors who wish to apply for admission may contact the GSD. Normally, firms are required to meet certain qualification criteria or technical assessment to ensure suitability. Applications for inclusion may be submitted at any time. Up-to-date lists and the method of application for inclusion in the lists are published in the *Government of the Hong Kong Special Administrative Region Gazette* annually and are reviewed regularly.

Prequalified tendering: Tender invitations are sent by letter to those prequalified suppliers/contractors approved by the secretary for the Treasury. Invitations to apply for prequalification may take the form of open tendering or selective tendering and the respective procedures will apply.

Single or restricted tendering: Tender invitations are sent to only one or a number of suppliers/contractors approved by the secretary for the Treasury or the director of government supplies. This procedure is only used when circumstances do not permit open tendering, for example, on grounds of extreme urgency or security, for proprietary products or for reasons of compatibility.

Established in 1966, the Hong Kong Trade Development Council (HKTDC; http://www.tdc.org.hk/) is an important statutory organization established to promote Hong Kong's trade in goods and services. The TDC's mission is to develop and diversify markets for Hong Kong companies, with special reference to the needs of small- and medium-sized enterprises (SMEs), to enhance the image and competitiveness of Hong Kong's products and services in world markets, to strengthen Hong Kong's role as Asia's premier business and services hub, and to enhance Hong Kong's image as an open market and good business partner. TDC's SME Training Centre has become the permanent home for a year-round program of business training courses and workshops designed to help smaller Hong Kong companies. There are also four SME Service Stations located strategically around Hong Kong to make it even easier to access TDC's wide range of services and publications.

The TDC has a worldwide network of 51 offices in 34 countries and regions that serve effectively as the international marketing arm for Hong Kong's manufacturers and service providers. TDC publishes 20 product/service magazines and trade directories. These are available at the TDC Business InforCentre and SME Service Stations. With a combined worldwide circulation of more than two million, the publications market Hong Kong products and services effectively and make business sourcing easier for international buyers.

Euro Info Centres (EIC)

The Euro Info Centres (EIC; http://www.cdopesaro.com/en/e-eicnet.htm#eic) network is a European Union initiative set up to encourage economic growth and business competitiveness. Its creation stems from the recognition that small and medium-sized enterprises (SMEs) play a vital role in the economy of the European Union. In fact, in 1998 more than 15.8 million businesses in the European Union are SMEs. Their contribution to job creation, growth, and competition had a significant impact on prosperity. Small and medium-sized businesses created 3 million jobs between 1988 and 1993. As information centers with a European outlook, the Euro Info Centres are the preferred partners of SMEs who wish to make the most of opportunities offered by the European market. Organized in a network based in all European Union Member States, their approach is proactive and decentralized, which allows them to give pertinent answers to SMEs' queries.

In many instances, SMEs encounter difficulties when exploiting opportunities offered by the Single Market such as lack of information on European issues, differences between the laws of various Member States, and administrative constraints. Missions of the EIC network inform businesses by offering them targeted information and access to expertise on European matters. EICs assist companies through regular contacts that EICs have established with other network members and the European Commission. The EICs also advise businesses trading abroad by providing assistance with project development and follow-up, and facilitate exchanges within the EIC network by pooling information on laws and business practices in other Member States. The EIC network provides feedback to the European Commission on SMEs' concerns and interests, and on the impact of European programs and legislation. Most EICs have been established in host organizations including, for example, business development agencies, university libraries, or regional chambers of commerce.

British Overseas Trade Bureau (BOTB)

The BOTB (http://www.thebiz.co.uk/botb.htm; 1–3 Victoria Street, London SW1E 6RB, England), located in the Department of Trade and Industry (DTI), advises the British government on international trade and guides the government's export promotion program, including policy, financing, and overseas projects.

The Committee for Middle East Trade (COMET), originally established in 1963, is the British Overseas Trade Board's Area Advisory Group for the Middle East, whose main function is to advise the British government on matters affecting trade between Britain and the Middle East. Through its main committee of senior British industrialists and government officials and its four Area Action Committees, it provides a forum for private-sector/government debate on the promotion of British commercial interests in the area. COMET's second main objective is to make British firms aware of the opportunities for doing business in Middle East markets. COMET publications are available to British-based companies only, and include *Opportunity Middle East*, a bimonthly magazine available free of charge to U.K. companies. This publication contains items of interest on Middle East markets, including news of European Commission funding, business opportunities, and listings of forthcoming events. You can visit COMET at http://www.comet.org.uk/public.htm.

Department of Trade and Industry (DTI) (UK)

The broad public sector in England includes central Government departments, local authorities, the National Health Service, and the academic institutions (http://www.dti.gov.uk/about.html). Generally, all procurement contracts are subject to the Treaty of Rome, regardless of value. The Treaty establishes principles to prevent discrimination against firms from other Member States of the European Union (EU) and promotes the freedom to provide goods, services, and workers throughout the European Community (EC). The Treaty is reinforced by a series of EC procurement Directives. The Directives are implemented into UK law by the Public

Works Regulations 1991, the Utilities Supply and Works Contracts Regulations 1992, the Public Services Contracts Regulations 1993 and the Public Supply Contracts Regulations 1995. These Regulations establish procedural rules which set down nondiscriminatory and transparent criteria for the selection of tenderers and the award of contracts with a value above the relevant threshold.

There is increasing opportunity for small firms to do business with government departments. Departments are not required to buy through central suppliers and are able to save money by purchasing directly from suppliers. They are increasingly aware of the merits of placing contracts with small firms. Government buyers are seeking the best value for their money. To become a successful supplier to this market your firm must: (1) be able to compete with other firms; (2) be able to complete contracts on time and to the required standards; (3) have a sound financial and commercial reputation; and (4) be able to familiarize yourself with Government purchasing procedures. Of note is the fact that in the UK, government organizations have been migrating toward becoming "performance-based" entities, which entails greater autonomy from government-wide rules in exchange for greater accountability for results. This is analogous to the initiatives associated with the National Partnership for Reinventing Government (NPR) in the United States.

Government Departments do not normally deal in component parts or materials, nor do they select subcontractors—they expect their contractors to do this themselves. If your company only produces components, your best opportunities are likely to be as a subcontractor. Names of the main contractors can be found in the information sheets at the back of the booklet "Tendering for Government Contracts," which can be obtained from DTI Publications Orderline at ADMAIL 528, London SW1W 8YT, or at http://www.dti.gov.uk/publications/sme/.

Almost all public procurement contracts for business worth more than a specific threshold amount must be published in the daily supplement to the *Official Journal of the European Communities* (OJ). The daily supplement series of the OJ gives details of current contracts by inviting suppliers to tender, or express an interest in tendering and providing information about contracts awarded. Tender invitations are listed by Directive and provide summary information about each commercial opportunity. Tenders Electronic Daily (TED) is the on-line version of the OJ and provides information on invitations to tender for public contracts. Reports are available to subscribers of this monitoring service on the morning of their publication. TED offers the advantage of being highly selective using subject and country codes to provide users with direct access to notices in their field of interest. It also gives information on some below-threshold opportunities. Many businesses, however, find it easier to use the TED service offered by the Euro Info Centres and other commercial organizations.

English Government departments usually place contracts after a period of broad-based competitive tendering. However, suppliers may not be asked to tender every time. The types of contracts your firm might expect to see span: research and development, production of equipment, supplies for stores, general services, and local purchase orders. "Framework agreements" or "call-off" contracts (similar to Basic Order Agreements or ID/IQ contracts in the United States) are common and are based upon an estimate of a department's total requirement over a stipulated period of time. Orders are placed when the need arises during the period of the contract. Most Government contracts are built upon model documents that may vary from department to department. There may also be variants where special needs are required. It is important to understand the conditions on which these contracts are based. The documents are usually divided into two parts: the first consisting of general conditions as with all contracts; the second containing a series of additional conditions if the contract warrants them.

The completion of the Single Market has created opportunities for firms of all sizes to compete for public-sector contracts across Europe. Government departments, both in the UK and the rest of Europe, now have to advertise certain types of public work. This not only means a greater scope for your company to do business abroad, but also that you may have to compete with other European firms for contracts in the UK.

Specific information on Western European Markets is available from the Department of Trade and Industry's Business in Europe Directorate in London. Information on public-sector contracts across Europe is also available through OJ and the electronic database TED. The easiest way to gain access may be to contact your nearest EIC.

DTI Small and Medium Enterprise Policy Directorate publishes a range of free booklets designed to help small firms grow and survive. These include:

"Tendering for Government Contracts" (URN 98/603)

"Guide to Help for Small Firms"(URN 97/525). This Guide gives details of the help and services available to small businesses with contact points for further information. It covers various topics including sources of help, employing staff, financial help, innovation, and exporting.

"Financing Your Business" (URN 97/762)

"Setting up in Business: A Guide to Regulatory Requirements" (URN 98/763)

"Employing Staff: A Guide to Regulatory Requirements" (URN 97/740)

For most areas of Scotland, business advice is provided by the network of Scottish Business Shops. If your business is based in the Scottish Highlands or west coast islands, you should contact Highlands and Islands Enterprise Business Information Source. For details of the advice services available in Wales, contact the Wales Euro Infor Center (WEIC), which has been very active since its establishment in 1989. The WEIC serves as a local access point for small- and medium-sized enterprises (SMEs) in Wales for European business information. (Learn more at http://www.weic.demon.co.uk/index.html.) In Northern Ireland, an information service is provided by the Local Enterprise Development Unit (LEDU) for firms who employ up to 50 employees and by the Industrial Development Board (IDB) for firms with more than 50 employees.

The Changing Definition of Small Firms

In Europe, there is no single definition of a small firm because of the wide diversity of businesses. Within the UK, the best current description of the key characteristics of a small firm remains that used by the Bolton Committee in its 1971 "Report on Small Firms." This stated that a small firm is an independent business, managed by its owner or part-owners and having a small market share. The Bolton Report also adopted a number of different statistical definitions. It recognized that size is relevant to sector, i.e., a firm of a given size could be small in relation to one sector where the market is large and there are many competitors; whereas a firm of similar proportions could be considered large in another sector with fewer players and/or generally smaller firms within it. Section 249 of the British Companies Act of 1985 states that a company is "small" if it satisfies at least to of the following criteria: (1) a turnover of not more than £2.8 million; (2) a balance sheet total of not more than £1.4 million; and (3) not more than 50 employees. A medium-sized company must satisfy at least two of the following criteria: (1) a turnover of not more than £11.2 million; (2) a balance sheet total of not more than £5.6 million; and (3) not more than 250 employees.

In February 1996, the European Commission adopted a communication setting out a single definition of SMEs—10 employees (microbusiness), 50 (small business), and 250 (medium-sized business), along with annual turnover, balance sheet, and ownership criteria. The Commission will apply this across Community programs and proposals. The communication also includes a non-binding recommendation to Member States, the European Investment Bank, and the European Investment Fund, encouraging them to adopt the same definitions for their programmes.

European Union Small and Medium-Sized Enterprise (SME) Initiative

The European Union Small and Medium-Sized Enterprise (SME) Initiative was introduced in England in December 1997. Its goal is to provide English SMEs, with up to 250 employees, with grants as well as advice to encourage them to adapt to the Single Market and to become more competitive internationally. This enhanced capacity, in turn, will enable the SMEs to increase their export potential. Assistance is offered under a range of measures designed to develop international cooperation and trade. Support will mainly be for the design and implementation of business strategies; cooperation and networking between SMEs in EU Member States; participation in tendering for international public procurement contracts; and training aimed at helping SMEs to export more effectively. Support is limited to projects involving, or having the potential to involve, cooperation with an international SME partner. Project support will be considered until the end of 1999. Scotland, Wales, and Northern Ireland have their own EU SME Initiatives. Your company can contact the Scottish Office, Welsh Office, or the Department of Economic Development in Northern Ireland.

The British Government provides a broad range of export support from DTI and the Foreign and Commonwealth Office, through the Overseas Trade Services (OTS). This support covers basic advice and information services, off the shelf and tailored reports on individual markets and sectors, assistance in visiting markets and overseas exhibitions and other publicity support. DTI manages several campaigns focused on particular countries or regions of the world deemed to have the most promising marketing opportunities. The campaigns raise the profile of the particular region to UK exporters, and strengthen links between those regions and the UK. Activities generally include information about market opportunities, seminars and workshops, and trade missions.

Export Market Information Centre (EMIC)

Located in central London, Export Market Information Centre (EMIC; http://www.dti.gov.uk/ots/emic/What.html.) EMIC, Kingsgate House, 66-74 Victoria Street, London SW1E 6SW, England) is a free, self-service library and research facility provided by the British Department of Trade and Industry (DTI) to enable firms exporting for the UK to research export markets. EMIC is one of a range of British Overseas Trade Services that offers assistance to exporters.

EMIC contains information about business opportunities arising from the US $90 billion per annum funding of Multilateral Development Agencies, such as the World Bank and Inter-American Development Bank.

EMIC maintains a database index of more than 8,000 current MDA-funded projects. These can be accessed by country/region, by individual or business sector, by agency or by any combination of the above. Early notification of project opportunities is given in Aid and Loan notices issued via the DTI's export sales lead service.

EMIC offers British firms the largest collection of overseas trade and telephone directories, including overseas yellow pages, in the UK. The Centre also houses market research reports that cover consumer and capital goods and services and are an excellent way of getting an overview of a market. Country profiles, reports, and forecasts are available to assist UK companies to develop a robust picture of the market and the general economic prospects and conditions. There is an extensive range of country and market sector reports, written by UK Embassy and High Commission staff overseas, are available in EMIC.

European Procurement Information Network (EPIN) (Ireland)

EPIN (http://epin1.epin.ie/default.htm) consists of a WWW server residing in Dublin, Ireland, which has links to a number of procurement information services. For an annual/quarterly subscription, suppliers/customers receive a username and password which enables them to access all public procurement notices from the electricity, gas, water, transportation, telecommunications, local and central government authorities from across Europe and the U.S. The "EPIN Leads" application allows users to query the contract notice database and search for leads of interest to Irish companies stored in the EPIN Central database. The EPIN Central database contains all Contract Notices published in the S Supplement of the *Official Journal of the European Communities*. The "EPIN Industrial Park" application allows suppliers to input their company profile and upload it to EPIN central. The company profile includes information on the products and services a company supplies as well as a brief company description and contact information. The information in the Industrial Park serves as an on-line brochure for your company.

Office of Public Works (OPW) (Ireland)

For more than 160 years, the Office of Public Works (http://www.opw.ie.) has provided the Government in Dublin and the public sector with services in the areas of procurement, property, and construction. The Office of Public Works is a Government Office responsible for the procurement of supplies and services in common use in Government Departments, such as office supplies, print, publications, uniforms, vehicles, furniture, etc.

SIMAP (European Commission)

The SIMAP (Système d'Information pour les Marchés Publics; http://simap.eu.int/. infeurope S.A. 62, rue Charles Martel L - 2134 LUXEMBOURG) project was designed to provide the information system infrastructure needed to support the delivery of an effective public procurement policy in Europe by providing contracting entities and suppliers with the information they need to manage the procurement process effectively. The European Commission launched the project in order to encourage best practice in the use of modern information technology for public procurement. Initially the project is intended to improve the quality of information about European Union procurement opportunities and to ensure that information is made known to all potentially interested suppliers. In the longer term, the SIMAP system will address the whole procurement process, including bids, award of contracts, delivery, invoicing and payment.

European Union companies can consult the SIMAP Web site regularly to learn of joint U.S.-European Union business initiatives. For example, upon the invitation of the State of Texas, and in the context of the Transatlantic Business Dialogue (TABD) and of the Transatlantic Small Business Initiative (TASBI), the Texas Department of Economic Development, the U.S. Department of Commerce and the European Commission are organizing a partnering event scheduled for 1999 between European Union and U.S.-based companies, primarily small/medium enterprises (SMEs).

THEMiS: The System for Regulated Procurement

Public and utility procurement in the European Union is subject to regulation that articulates detailed procedures for awarding contracts. Achilles Information Ltd. is a specialist in this area of procurement, and provides both consultancy and software to help purchasers meet their commercial needs while complying with the legislation. Because of the complexity of the requirements of the EU procurement legislation, and its UK implementation, Achilles has developed THEMiS (http://www.achilles.co.uk/services/themis/themis.html#access), a specialist decision support package that provides the tools to comply with the legislation easily and with minimum cost. THEMiS is updated both with new documents and court cases at least every 4 months. The system also gives access to the latest news and information in the UK, Europe, and beyond, where legislation on public procurement is becoming increasingly important.

Procurement Information Online (PIO) (Germany)

This Web site (http://www.procurement.de/, maintained by the law firm, Arnold Boesen, in Bonn) provides a list of links that potential tenderers in the German procurement market may find to be useful. The following are online databases established for enterprises that want to tender for public contracts. In general, access is for paying subscribers.

Ausschreibungs-ABC: This is a specialized data base for public procurement in Germany. It can only be used by registered paying subscribers. Registration is possible online.

Bundesausschreibungsblatt Online: This is the official and specialized organ for the procurement of public institutions in Germany. Procurement opportunities are published here according to the German *Verdingungsordnungen*. Moreover, there are other institutions such as NATO and institutions of the European Union which publish procurement opportunities here. The database can be used only by subscribers of the *Bundesausschreibungsblatt*. A subscription can be ordered online directly.

bi online: This is a specialized database for public procurement concerning building and construction in Germany. It can only be used by registered paying subscribers. Registration is possible.

In Germany special advice centres *(Auftragsberatungsstellen*; http://www.ihk.de/3ber/diht0613.htm) impart practical experience and information to enterprises which plan to tender for public contracts. Usually, these centres are organized as registered societies or assigned to the chambers of commerce of the several *Länder* (states). Each *Land* (state) keeps a centre of advice to guide enterprises which have a presence in a given *Land*. Some centres already have a site on the Internet, while others can be contacted by mail or telephone/telefax only.

The *Vergabe-News* is a monthly German-language information service updating tenderers and contracting authorities on the latest developments in German and European Procurement Law and Practice. It is edited by the specialized law firm, Arnold Boesen.

Confederation of German Trade Fair and Exhibition industries (AUMA)

Association of the German Trade Fair Industry (AUMA), Lindenstr. 8, D-50674 Köln, Germany. AUMA, in close cooperation with the German Federal Ministry of Economics and also the Ministry of Food, Agriculture and Forestry, prepares the official German trade fair program abroad. Within the framework of this program, the German Federal Government provides financial assistance for German companies' joint participation at foreign events, but also for independent presentations by German industry abroad. These include, in particular, the German investment goods exhibitions, such as TECHNOGERMA Jakarta 1999 and consumer goods exhibitions entitled KONSUGERMA, which was held in April 1998 in Shanghai, China. AUMA's members include the central associations of German business (industry, trade, skilled trades etc.), trade associations of the exhibiting and visiting industries, and the German trade fair and exhibition organizers as well as the companies organizing foreign fair participation.

Das Gepa-Projekt (Marketing Assistance Programs for eastern German enterprises)

The export volume of medium-sized enterprises in the new Federal States of the former East Germany is deemed to be still too small. For this reason, the German Federal Ministry of Economics has entrusted foreign Chambers of Industry and Commerce (http://www.gepa.de/98/englisch/indexeng.html.) and other organizations to carry out marketing assistance programs in markets in western countries. East German medium-sized companies producing capital or consumer goods are meant to benefit from these programs. The marketing assistance program is at the focus of sales promotion abroad. It is used strongly by East German enterprises and has helped many of them to find sales partners and take in export orders. The programs are directed towards markets in Europe and North America. To accompany and support the marketing assistance programs the companies participating in the projects are presented on the Internet–in the framework of the *gepa* project. This gives the foreign Chambers of Commerce and other organizations in the target markets the chance to direct the attention of potential customers or partners to the East German enterprises, their range of products, and individual objectives. The *gepa* Internet platform is presented worldwide through the press, national and international trade fairs, individual events and advertisements to the benefit of the individual companies.

French Committee for External Economic Events (CFME)

The Agency for the International Promotion of French Technology and Trade (http://www.cfme-actim.com/. CFME ACTIM, 14, avenue d'Eylau, 75116 Paris), a nonprofit association based in Paris, develops technical and commercial collaboration between France and her foreign partners. The Agency brings together French business people and their foreign counterparts through such forums as trade shows; distributes information on French products and technology to the international trade press; and develops partnerships between French companies and companies abroad through joint ventures, subcontracting, and technology transfer. With 5,000 members, CFME is part of the French program for export support.

Contracts and Tenders Worldwide (CTW)

10 Main Street, White Plains, NY 10606 Tel. 914.682.2634, FAX 914.682.2635 or 33 St. Thomas Street, Winchester, Hampshire, England SO23 9HJ Tel. +44 1962.866586. www.contracts-base.com.

Suppliers and contracts can subscribe to this service for US$ 250/year. For this fee, your company can advertise its services, browse the Commercial Procurement Database for contract opportunities with large and small firms in the U.S. as well as internationally, retrieve formal invitation for tender (ITT) documents, submit bids, and negotiate and close contracts.

JETRO (Japan External Trade Organization)

JETRO (Japan External Trade Organization; http://www.jetro.go.jp/) is a nonprofit, Japanese government-related organization dedicated to promoting mutually beneficial trade and economic relations between Japan and other nations. JETRO is particularly active in encouraging technology transfers and manufacturing investment to and from Japan. JETRO operates 16 Centers for Industrial and Technological Cooperation (CITEC) in 57 countries throughout North America, Europe, and Oceania to foster exchange between industries of these regions and Japan. The centers provide information on investment and technology and arrange special exchange programs. JETRO also coordinates the Manufacturing Technology Fellowship Program, which gives foreign engineers from these regions firsthand experience with manufacturing technology in Japan. To assist overseas business people in launching their marketing efforts in Japan, JETRO operates Business Support Centers in Tokyo, Yokohama, Nagoya, Osaka, and Kobe. In addition to providing free office space, the centers are staffed with advisors who provide complimentary consulting services on exporting to Japan and doing business there.

Asia-Pacific Economic Cooperation (APEC)

The Asia-Pacific Economic Cooperation (APEC; http://www.apecsec.org.sg/) was established in 1989 in response to the growing interdependence among Asia-Pacific economies. Begun as an informal dialogue group, APEC has since become the primary regional vehicle for promoting open trade and practical economic cooperation in the Pacific rim. Member nations include the Philippines, Australia, Peru, Japan, Canada, the U.S., Mexico, China, and Vietnam. The goal of the cooperation is to advance Asia-Pacific economic dynamism and sense of community. APEC's 18 member economies bring a combined Gross Domestic Product of more than US $16 trillion in 1997 and account for 44 percent of global trade. Malaysia is the APEC Chair for 1998.

High priority is given to the following themes in economic and technical cooperation in six areas: developing human capital; fostering safe and efficient capital markets; strengthening economic infrastructure; harnessing technologies of the future; promoting environmentally sustainable growth; and encouraging the growth of small and medium enterprises.

APEC exhibits a strong commitment to business facilitation and the regular involvement of the private sector in a wide range of APEC activities. Business already participates in many of APEC's working groups and helps shape the policy dialogue in partnership with member economy officials.

Government Electronic Marketplace Service (GEMS) (Australia)

GEMS is an electronic information service (http://www.gems.gov.au/) that offers both buyers and suppliers purchasing advice, policy information, and tender opportunities.

Tender I.N.F.O. (Electronic Tendering and Procurement Network)

An interactive opportunity network (http://www.tenders.com.au/tenders.html.) that has supplied Australian and New Zealand business with information on tenders, expressions of interest, and requests for information since 1990. Information is updated daily.

Australian Trade Commission Online (Austrade)

Austrade (http://www.austrade.gov.au/) administers a rich variety of export assistance programs designed to help Australian companies enter foreign markets. Operating an international network of offices located in 108 cities in 63 countries, Austrade is able to identify potential buyers, accurately match Australian suppliers with interested overseas contacts, and arrange introductions. Austrade maintains an Export Hotline, provides export consulting services, coordinates Australian stands at more than 100 international trade fairs annually, and also offers financial assistance to Australian exporters. In additon, Austrade promotes Australian firms on line in "Australia on Display."

Arabnet: Reaching the Arab World in the Middle East and North Africa

Found at www.arab.net, this Web site offers country-specific business, political, and cultural information about 22 nations in the Arab world in the Middle East and North Africa. For example, Kuwait's Central Tenders Committee, which is attached to the Council of Ministers, administers all public tenders. For open tenders, bidders must have applied for registration as an approved supplier or contractor with the Ministry of Planning. They must also have been classified according to their financial and technical capabilities. U.S.-based companies may participate in a public tender provided that they have a Kuwaiti agent or partner and that the tender is open to nonKuwaiti companies. International firms desiring to prequalify either as consultants or contractors should approach the *client* rather than the Central Tenders Committee. Tender documents may be obtained from the offices of the Central Tenders Committee for a fee. Once a tender is submitted, it is evaluated on the basis of price, conformity to the specifications required, and the level of Kuwaiti involvement in the company as regards the scope of needed work.

Normally when a tender is submitted, a percentage of the price must be put up as a bond. The amount will be specified in the RFP or tender document. Nongovernmental agencies in Kuwait may also put contracts out to tender but these will be handled directly and not through the Central Tenders Committee.

In the case of Saudi Arabia, the Saudi government must receive at least three bids for all contracts larger than one million *riyals* (US$266,667). For construction projects, at least five contractors must be asked to submit bids. A committee of three or more people from the Ministry of Finance & National Economy, or from the government agency responsible for the project, must review the bids. The contract will be awarded according to a vote decided by a majority. These bids are open to the public. Companies with the lowest bids and also meeting all specifications will be awarded the contract. In most cases, the price is estimated by the Saudis and if all bids are significantly larger, then the project will be negotiated. This also applies if the lowest bidder does not meet the conditions of the project.

Tender regulations allow price increases for variations in transportation charges, insurance rates, or the price of raw materials. If all bids significantly exceed the estimate, the government agency may cancel all such bids. The Saudi government insists that bids come reasonably close to practical estimates. Since January 1979, all contracts over 100 million *riyals* have required the personal approval of the King.

Foreign companies wishing to bid on projects supervised or undertaken by government ministries *must be known* to the ministry or agency. A list is compiled of these foreign companies and bidders are selected from this list when projects are available. In order for a company to be properly registered in the Kingdom, a questionnaire must be completed in both Arabic and English. There must also be included the company's latest annual report along with two references and two copies of a list of completed projects. These documents must be submitted to:

Director
Contractor's Classification Committee
Non-Saudi Contractors' Division
Riyadh, Saudi Arabia

In awarding contracts, the Saudi government is required by law to give preference to Saudi companies or to joint-ventures which are more than 50% Saudi-owned. Jubail Industrial City contracts for architecture and engineering, for example, are awarded only to 100% Saudi-owned companies. When a qualified Saudi company is not available, foreigners may then be awarded the contract. The foreign company, however, must have a local address.

Saudi Arabia's offset programme is one approach for generating industrial investment. This policy requires contractors to reinvest a portion of the value of their major contracts within the Kingdom. The government may require foreign-owned consortiums involved in defense contracts to reinvest up to 35% of the value of the contract into high-technology service industries. Foreign contractors must also subcontract 30% of the value of a government contract to local subcontractors.

Stat-USA GLOBUS Information System

STAT-USA (http://www.stat-usa.gov/), an agency in the Economics and Statistics Administration, U.S. Department of Commerce, provides vital economic, business, and *international trade* information produced by the U.S. Government to you so you can make an important decision that may affect your business. STAT-USA, and agency in the Economics and Statistics Administration, U.S. Department of Commerce, collates information you may need that is produced by hundreds of separate offices and divisions of the government. This information is delievered through low-cost subscription services like STAT USA/Internet and the National Trade Data Bank CD-ROM.

11.14 The Unisphere Institute, U.S. SBA, NASA, and International Marketing

The UNISPHERE Institute (http://www.unisphere.com/. 1530 Wilson Blvd. Suite 120, Arlington, VA 22209, Tel. 703.465.4500; FAX 703.465.4530) is a nonprofit organization that assists small and medium high-tech firms in finding international venture partners to pursue commercial opportunities abroad. UNISPHERE facilitates a wide range of commercial services that enable growth-oriented companies to consummate domestic and international transactions generally involving advanced technology. These services include product testing and evaluation, market assessment, strategic partnering, deal structuring and financing. Types of deals closed include licensing, service contract, direct sale, joint venture, strategic alliance, and financing through leveraged leasing, angel investing, and other funding sources. Angel capital is money invested in firms at the earliest stage of development—providing "seed money" to get a company off the ground and early-stage investments to keep a firm operating, often while it's losing money.

UNISPHERE works closely with the Federal Laboratory Consortium (FLC) to identify leading-edge technologies at more than 600 federal laboratories. UNISPHERE and FLC jointly facilitate independent testing, market assessment, and other services needed to commercialize such technologies. The two organizations also provide technology scouting services to assist firms in finding innovations within the federal laboratory system that meet specialized needs. UNISPHERE also assists NASA, the U.S. Department of Transportation, and other federal agencies in commercializing technological innovations developed through agency investments in the Small Business Innovative Research (SBIR) Program. Firms receive Phase I SBIR funds to develop a concept or design. Phase II funding provides for the development of a prototype. UNISPHERE evaluates the commercial viability of these prototypes and, where appropriate, facilitates the deal-making services. (See Section 1.5 for a detailed discussion of the SBIR Program.)

As part of an effort to improve the competitive viability of small disadvantaged businesses (SDBs) in obtaining NASA contracts after those businesses no longer qualify as disadvantaged, NASA has teamed with the U.S. Small Business Administration to support the UNISPHERE Institute.

NASA has contracted with UNISPHERE to match recently graduated 8(a) firms with international partners. It is hoped that through such relationships the SDB firms will expand their business base, financially, and be able to better compete for full-and-open competition NASA procurements.

Market expansion opportunities are sourced through UNISPHERE's global partner network of financial institutions, laboratories, law firms, and consulting firms in the United States and 25 other countries. Partners recommend opportunities, often at firms or labs with new technologies, or at firms needing technologies or strategic partners, to UNISPHERE which in turn coordinates a central matching and deal tracking system. After an opportunity is identified, UNISPHERE and the sourcing partner collaborate closely to find an appropriate match to a technology, strategic partner, and/or financing. A confirmed match is reached when principals on both sides of a potential transaction agree to pursue it. Confirmed matches are referred to an affiliated company, Unifinancial International Inc., which provides the negotiating, structuring, and financing services that may be needed to close the deal.

11.15 British-American Business Council (BABC)

An association of 32 British-American chambers of commerce and business associations, the British-American Business Council (BABC; http://www.babc.org. 52 Vanderbilt Avenue, 20th Floor, New York, NY 10017 Tel. 212.661.5660; FAX 212.661.1886.) is based in major cities throughout the United States and the United Kingdom. It is the largest transatlantic business organization, representing more than 4,000 companies and 10,000 executives. The BABC's objectives are to support the unique business partnership between the U.S. and the UK and to ensure its continuing vitality; to help its member associations provide business development and business intelligence services to their member companies; and to provide a broader, transatlantic business network for these member companies. Founded in 1993, the BABC now includes two associate organizations in Canada and Mexico. The commercial interactions between the U.S. and UK represent the most substantial business relationship between any two countries in the world, including more than $320 billion in two-way investment and some $60 billion annually in two-way trade. The UK is the largest overseas investor in the U.S. and receives more than 40% of U.S. investment in Europe. U.S. companies provide employment for about a million people in the UK, and UK companies employ about a million people in the U.S.

The BABC has been playing a more active role in supporting a positive environment for the further growth in transatlantic business. It has established an ongoing dialogue and partnership with the U.S. and UK governments and participated actively in a number of business initiatives. These include USA Engage, the business coalition formed to support the renewal of "Fast Track" authority for the U.S. President, and the Transatlantic Business Dialogue, the high-level U.S.-European business partnership formed to improve the two-way flow of goods and services. Since April 1997, the BABC has also launched a number of new initiatives to support its member associations, including establishing a searchable, password-protected Online Membership Directory (www.babc/members) of the member companies of all the BABC's business associations, thereby providing these companies with access to contact information about all the other member companies of the BABC. The Council has also facilitated BABC-wide business networking and now maintains an Online Calendar (www.babc/calendar) of the events being organized by all BABC member associations.

The BABC is governed by a Board of Directors, on which all its member associations are represented, and an Executive Committee. The BABC's Secretariat is based in New York and led by its Executive Director.

11.16 U.S. Trade and Development Agency (USTDA)

The U.S. Trade and Development Agency (USTDA; http://www.tda.gov/; 1621 North Kent Street, Suite 300, Arlington, VA 22209-2131 Tel. 703.875.4357, FAX 703.875.4009) assists in the creation of jobs for Americans by helping U.S. companies pursue overseas business opportunities. Through the funding of feasibility studies, orientation visits, specialized training grants, business workshops, and various forms of technical assistance, TDA enables American businesses to compete for infrastructure and industrial projects in middle-income and developing countries. TDA, through the Trade Promotion Coordinating Committee (TPCC), works closely with the U.S. Department of Commerce, The Export-Import Bank, the Overseas Private Investment Corporation, and other export promotion agencies to advance American business interests abroad. Of note is that small businesses win nearly one-third of TDA's program budget for U.S. companies.

11.17 U.S. Agency for International Development (USAID)

Established in 1961 by President John F. Kennedy, the United States Agency for International Development (USAID; http://www.info.usaid.gov/) is the independent government agency which provides economic development and humanitarian assistance to advance U.S. economic and political interests overseas. USAID publishes a Forecast of contracting opportunities for businesses to use in their planning efforts. The Forecast is divided into two sections—one for USAID/Washington and the other for USAID/Missions. The Washington Forecast is updated every quarter and the Mission Forecast report is updated twice a year. USAID also issues Procurement Information Bulletins to assist companies.

Under a cooperative agreement with USAID's Office of Business Development (BD), the International Executive Service Corps (IESC; http://www.iesc.org) delivers an array of business development services to assist small and medium-sized businesses in developing countries and emerging democracies through joint and co-ventures with counterpart U.S. companies, including transfer technology and other business development activities. Technology Assistance Centers (TAC) are USAID-funded business development operations designed to help private sector companies in developing countries access the technology and expertise needed to compete effectively in local and global markets. BD works with missions to develop an in-country TAC that is linked to network of partner organizations to expand technical assistance services in country. TAC's serve as a mechanism for organizing and assisting developing country firms—especially small and medium-scale enterprises (SMEs) seeking access to U.S. technology and expertise—as well as a source of business opportunities (trade leads) for U.S. SME firms seeking to market their products and services abroad.

The Global Technology Network (GTN; http://www.usgtn.org) is operated and administered by the Office of Business Development (BD). The primary objective of BD is to leverage Agency resources by establishing partnerships and networks with the private sector in support of USAID's global economic development mission. GTN is a trade lead/business matching program that facilitates the transfer of U.S. technology[3] and services to address global development problems. This is accomplished by

---

3   Note, however, that the International Traffic in Arms Regulations (ITAR) controls the export of defense articles and defense services. ITAR addresses certain commodities such as computers, electronic equipment and systems, encryption devices, and information security systems adapted for military use.

matching a country's development needs with U.S. firms equipped to provide the appropriate technological solutions. GTN focuses on the following sectors: agricultural technology, communications and information technology, environment and energy technology, and health technology. GTN trade leads (available at http://usgtn.org/pages/new leads.html) are electronically transmitted from the field to GTN/Washington, where they are matched and disseminated to U.S. firms registered in GTN's database. GTN is operational in 33 countries.

USAID's Global Bureau's Office of Business Development serves as the central point of contact for U.S. firms interested in doing business in developing countries and emerging economies worldwide. Through USAID's regional business outreach offices and public and private partnerships, the Office of Business Development offers a wide range of outreach and marketing activities and services. These activities and services are specifically intended to increase U.S. private enterprise participation in USAID international development programs, while opening up new market opportunities for U.S. product and services firms—especially small/medium-scale enterprises. Technical consultants interested in submitting their résumé to USAID should register in the Agency's Technical Consulting Services (TCS) database.

USAID Regional Business Outreach Offices

The Office of Small and Disadvantaged Business Utilization/Minority Resource Center (OSDBU/MRC) is USAID's advocate for U.S. small businesses, small disadvantaged businesses, and women-owned small businesses. OSDBU/MRC ensures that their U.S. entities receive consideration and access to USAID-financed procurement of goods and services. OSDBU/MRC is the initial point of contact for these U.S. entities and offers in-depth information about Agency programs.

In general, USAID publishes both intended procurements and awards of procurements in the *Commerce Business Daily*. Publishing this information is required by the Federal Acquisition Regulations (FAR), and it allows U.S. small businesses to identify business opportunities and potential joint venture and subcontracting partners. The FAR requires that prime contracts exceeding $500,000 ($1M for construction of public facility) have a Small Business Subcontracting plan, if the Contracting Officer has determined the procurement has components that can be subcontracted, and if the awardee is other than a small business.

FAR 19.704 and FAR 52.219-9 require the Small Business Subcontracting Plan to express subcontracting goals in terms of both percentages and dollars for small businesses, for small disadvantaged businesses, and for women-owned small businesses. The Plan is also required to identify the contractor's employee who will administer the Subcontracting Plan, with detail of his/her duties; the procedures the contractor will employ to afford a fair and equitable opportunity for subcontracting for small businesses, for small disadvantaged businesses, and for women-owned small businesses; the assurance that records will be maintained for review by the contracting agency and the U.S. Small Business Administration (SBA); and, the assurance that FAR-required subcontracting reports will be submitted to the contracting agency.

The subcontracting reports are used by both the contracting agency and the Federal Procurement Data Center to measure the contractor's achievement of the subcontracting goals expressed in the Small Business Subcontracting Plan.

A prime contractor's performance in implementing a Small Business Subcontracting Plan and in achieving goals expressed therein are part of past performance evaluations that are considered for future awards.

For small companies in the United States, resources for international business planning include:

Breaking Into The Trade Game: A Small Business Guide (http://www.sba.gov/ OIT/info/Guide-To-Exporting/all.html). This publication is the product of a private/public sector initiative between the U.S. Small Business Administration and AT&T. The U.S. Small Business Administration's (SBA) Office of International Trade (OIT) developed this *Trade Guide* as an information tool to assist American business to develop international markets. This Guide will help answer questions and take the mystery out of exporting. SBA employs 76 District International Trade Officers and 10 Regional International Trade Officers throughout the United States as well as a 10-person international trade staff in Washington, D.C.

(http://www.libraries.rutgers.edu/rulib/socsci/busi/intbus.html provides a comprehensive listing of international business resources.)

The National Trade Data Bank (NTDB; http://www.cd-rom-guide.com/ cdprod1/cdhrec/010/344.html) is a CD-ROM database produced by the U.S. Department of Commerce. It contains more than 160,000 documents, including books, reports, serials, and statistical tables on overseas markets, export and import statistics for thousands of products, how-to guides for exporters, general information on foreign countries, and more than 60,000 foreign business contacts for exporters. The information on the NTDB comes from 26 U.S. Government agencies, such as the Central Intelligence Agency, Department of Commerce, Department of Energy,

Department of Labor, Federal Reserve System, International Trade Commission, and Small Business Administration. Previous trade leads and the Export Yellow Pages also are included.

Price Waterhouse, a major accounting firm, publishes a series that they call *Price Waterhouse Information Guides*. Most of the volumes are entitled "Doing Business in (country)"; they focus on business conditions, accounting procedures, and tax rules.

(http://www.nafta.net/SmallBiz.htm is a NAFTA-sponsored small business information resource.)

The World is your Market: An Export Guide for Small Business by the U.S. Small Business Administration and AT&T.

A Basic Guide to Exporting (1989; http://www.smartbiz.com/sbs/ books/book153.htm) by the U.S. Department of Commerce in cooperation with Federal Express.

United States International Trade Commission (USITC)

The USITC (http://www.usitc.gov/) is an independent, quasi-judicial federal agency that provides objective trade expertise to both the legislative and executive branches of government, determines the impact of imports on U.S. industries, and directs actions against certain unfair trade practices, such as patent, trademark, and copyright infringement. USITC analysts and economists investigate and publish reports on U.S. industries and the global trends that affect them.

International Trade Development Organizations

(http://agis-usa.org/global-biz/trade.htm.) Excellent listing of trade organizations in Asia, Central and Eastern Europe, Russia and the Newly Independent States (NIS), Western Europe, the Middle East, Africa, Latin America, and North America.

We now turn our attention to innovative and successful proposal production and publication techniques.

END NOTES

1. From "Test Your Global Mindset," *Industry Week*, 2 November 1998, p. 12.

2. Whitman, Marina, and Rosabeth Moss Kanter, "A Third Way?: Globalization and the Social Role of the American Firm," *Washington Quarterly*, Spring 1999.

3. "Test Your Global Mindset," p. 12.

4. Root, Franklin R., *Entry Strategies for International Markets*, revised and expanded, New York: Lexington Books, 1994, p. 1.

5. Ibid., p. 19.

6. Flagler, Carolynn, "The ANY Aspects of International Proposals," *Contract Management*, March 1995, p. 13.
     Franklin Root suggests that "[f]or most companies the entry strategy time horizon is from three to five years, because it will take that long to achieve enduring market performance." Root, Franklin R., *Entry Strategies for International Markets*, revised and expanded, New York: Lexington Books, 1994, p. 22.

7. Conceived in 1944 at the United Nations Monetary and Financial Conference held in the United States, the World Bank is a multilateral development institution whose purpose is to assist its developing member countries further their economic and social progress. The term *World Bank* refers to two legally and financially distinct entities: the International Bank for Reconstruction and Development (IBRD) and the International Development Association (IDA). The IBRD and IDA have three related functions: to lend funds, to provide economic advice and technical assistance, and to serve as a catalyst to investment by others. The IBRD finances its lending operations primarily from borrowing in the world capital markets. IDA extends assistance to the poorest countries on easier terms, largely from resources provided by its wealthier members. (See *The World Bank Annual Report 1994*, Washington, D.C.: The World Bank.)

8. *The World Bank Annual Report 1998*, Washington, D.C.: The World Bank; http://www.worldbank.org/html/extpb/annrep98/overview.htm.

9. Root, Franklin R., *Entry Strategies for International Markets*, revised and expanded, New York: Lexington Books, 1994, pp. 25–26.

10. Ibid., pp. 41–42, 44.
     An entry mode is "an institutional arrangement that makes possible the entry of a company's products, technology, human skills, management or other resources into a foreign country." Root, Franklin R., *Foreign Market Entry Strategies*, New York: AMACOM, 1987,

p. 5. See also Erramilli, M. M. Krishna, and C. P. Rao, "Choice of, Foreign Market Entry Modes by Service Firms: Role of Market Knowledge," *Management International Review*, Vol. 30, 1990, pp. 135–50.

11. Douglas, Susan P., C. Samuel Craig, and Warren J. Keegan, "Approaches to Assessing International Marketing Opportunities for Small- and Medium-sized Companies," *Columbia Journal of World Business*, 1982, p. 27.

12. Farrell, Michael, "Many Small Firms Want to Do Business Internationally But Lack the Wherewithall," *Capital District Business Review* (Albany, N.Y.), Vol. 22, Sept. 1995, p. 15.

13. Flagler, Carloynn, "The ANY Aspects of International Proposals," *Contract Management*, March 1995, pp. 14, 16, 19.
    A common area of tax liability exposure relates to sending technical staff into the buying country to provide technical support and training. This type of technical support is often subject to taxes.
    Countertrade refers to a mandated purchase requirement imposed on the foreign seller by the buying community and/or its government.

14. Flagler, Carolynn, "The ANY Aspects of International Proposals," *Contract Management*, March 1995, p. 17.

15. See McCubbins, Tipton F., "Three Legal Traps for Small Businesses Engaged in International Commerce," *Journal of Small Business*, Vol. 32, July 1994, pp. 95–103.

16. Root, Franklin R., *Entry Strategies for International Markets*, revised and expanded, New York: Lexington Books, 1994, p. 97.

17. Ibid., p. 98.

18. Ibid., p. 99. Regional marketing offices are located in Chicago (312-641-1915); Houston (713-227-0987); El Segundo, California (213-323-1150); Miami (305-372-8540); and New York (212-227-7020).

Chapter 12

Proposal production/publication

PUBLICATION OF A SET of superior-quality proposal volumes is a *professional-level*, time-intensive, dedicated effort. It is not a lower level clerical function that can be accomplished adequately and consistently by marshaling the secretarial and administrative support available within your company. Even small businesses benefit from a dedicated core publications staff of at least one desktop publisher/word processing operator and one graphic artist.

Senior management is well served to recognize and support the publication professionals they have on staff, even if they number only two or three people. Support is most meaningful in the form of senior management's judicious, targeted involvement in assisting the proposal manager maintain the proposal response schedule at every milestone. Otherwise, publication staff are faced with compensating for schedule slippages along the way. And when schedules are missed on a regular basis and significant levels of night and weekend publication time become routine, publication staff morale can be at risk. Turnover among publication staff can be highly detrimental to your company's ability to prepare outstanding proposal documents every time.

Proposal efforts are not complete when technical and management volumes are written and reviewed, and the costing is completed after careful integration with the technical and management approaches and staffing plan. There remain the following critical production-related activities:

- Editing of both text and graphics;
- Creation/generation of tables of contents, lists of figures and tables, and compliance matrices;
- Proofreading of text and graphics (pay extra attention to proposal covers; they make a first impression and you want it to be positive);
- Electronic and hardcopy configuration management and version control;
- Word processing/desktop publishing;
- Generation of graphics/preparation of photographic prints for photocopying,[1] including insertion of graphics, figures, and tables into the documents, either electronically or manually;
- Coordination of outside printing activities (for example, covers, spines, tabs, and color pages);
- Photoreproduction and in-house color printing;
- Collation;
- Assembly/binding (such as three-ring vubinders, Velobinding, GBC, and plasticoil), including placing covers, spines, and tabs into binders;
- FINAL quality checking;
- Packaging;
- Completion of specific bid forms and packaging labels; labeling;

---

1 For the best reproduction quality without scanning technology, black-and-white as well as color prints should be "statted" and "screened" prior to being pasted down and photocopied. Most printing companies can prepare your photos for photocopying for a nominal charge. And certain photocopiers come equipped with a specialized capacity to reproduce prints with reasonable quality. With a flatbed scanner, both black-and-white and color prints can be scanned directly into such applications as WordPerfect and Adobe PhotoShop. In a two-step process, these same prints can be scanned and imported into CorelDRAW!, Adobe Persuasion, and Adobe PageMaker. Note, however, that flatbed scanning technology has become so inexpensive (less than $400) that your company should strongly consider acquiring a scanner for proposals as well as for contract deliverables. Scanner manufacturers include Hewlett-Packard, Canon, Epson, and Agfa.

- Shipping/delivery: Delivery is more personal in the international and commercial arenas. A senior person from your company should perform this function;
- Obtaining time- and date-stamped receipt.

It is recommended that the proposal publication be a centralized corporate function. Continuity of publication staff promotes uniformity of the image of the proposal documents from proposal to proposal. A core group of staff should become intimately familiar with, and crosstrained in, the document preparation process. Policies and procedures are much more easily implemented in a centralized working environment.

12.1 Internal documentation standards

To ensure that your company's proposal and contract deliverable documents have a similar "look and feel," it would be beneficial to develop a corporate *handbook of format and style, publications guide*, or analogous guidebook. The publications group, in conjunction with the business development staff and senior project management staff within your company, should be involved in developing and maintaining the format and style standards that will help to ensure the presentation of a uniform company image for both proposals and deliverables. This guidebook should address some or all of the following topic areas:

- Editorial style;
- Technical writing guidance;
- Technical report formats;
- Correspondence (memos, transmittal letters, etc.);
- Proposals, including résumés and project summaries;
- Marketing presentations and brochures.

Always follow the formatting guidance provided in Section L (instructions to offerors) of the RFP when preparing your federal proposals. However, in-house written documentation standards will assist technical writers, proposal reviewers, and publication staff to work toward a consistent style and structure of presentation.

A useful technique for each proposal effort is to have someone in your company with editorial capability develop a listing of RFP-specific terms that will appear in your proposal so that writers and publication staff will capitalize, hyphenate, and abbreviate those terms in accordance with the particular government agency's parlance.

12.2 Document configuration management and version control

Once sections of a proposal are written, it is critical to control the internal *release* of and *changes* to those sections throughout the proposal lifecycle. The centralized publications group should consider maintaining a master hardcopy proposal book for each volume in three-ring binders. Such binders allow for quick page replacements. One set of master proposal books should be made available to the proposal manager, and another should be under the direct control of the publications group. The pages from a third and final master copy may be hung on the cork-lined walls of your secure "war room" facility to allow for comment and review.[2] Then, as new proposal sections are created and existing ones modified with the authorization of the proposal manager, the publications group should be responsible for generating the change pages and inserting them in the master proposal books and hanging them on the walls of the war room. Establishing and maintaining a "living" master proposal document that iterates during the course of the proposal response lifecycle will be of valuable assistance to all proposal contributors.

Your company's publications group should maintain electronic and hardcopies of the proposal documents at each review stage. For example, when the publications group produces the Red Team draft, all electronic files should be copied to a storage medium appropriate for the data volume of the proposal. Mass storage devices and removable cartridge drives and media include IOmega ZIP® drives (100 MB) and JAZ drives (1 and 2 GB) for PCs and Macs; Imation LS-120 SuperDisk drives; and SyQuest Technology, Inc., SyQuest, Sparq, and Syjet cartridges. Your firm might also consider tape backup systems, compact disk (CD), or rewritable magneto-optical (floptical) cartridges. One hardcopy should be retained in the exact form and format of the Red Team draft. Then, if a computer hard drive fails, the network server crashes, or electronic files are corrupted in some way due to power surges or electrical storms, the last major draft version will be available, *unmodified*, in both electronic and hardcopy forms. It is

---

2   The technique of affixing documents to the walls of the war room is slowly being phased out, in part due to file sharing made possible through computer networks and the Internet.

advisable to back up all electronic proposal files to secure storage media at least once each work day.

A key aspect of proposal document configuration management is the use of time, date, file name on the hard drive, and diskette/CD-ROM/ magneto-optical rewritable disk number headers on *all* draft proposal pages. Headers can be in a small font and, of course, must be removed prior to submittal to the client. An example of a useful header for a page in the Technical Volume of a proposal to the Federal Aviation Administration is

14:25 Friday, March 10, 2000 c:\proposal\A-1-3.FAA *CD: Tech-A.1*

Such headers allow easy identification of when a particular page was last modified and on which storage medium the backup file for A-1-3.FAA resides.

Another beneficial configuration technique is the use of colored paper for various proposal drafts—for example, pale blue paper for the Blue Team and pale red for the Red Team.

And finally, do not discard draft proposal sections until after the proposal has been submitted. You can never be sure that you will not want to refer to an earlier version of the proposal. For security, keep all such materials in locked storage and shred them when the proposal is submitted. Remember to retain a complete hardcopy and electronic copy of the final version of the proposal submitted to the government for use during BAFO and orals preparation.

12.3 Freelance and temporary publication staff

Many companies, including small businesses, respond to fluctuating publication workloads with outside support from freelance, part-time on-call, or temporary staff. Your company's publications group manager will be well served to develop and maintain a listing of prequalified local freelance graphics artists, word processing operators, editors, and proofreaders to call upon during peak periods. Freelance staff, if called upon consistently, will be able to develop an understanding of your company's internal documentation standards, proposal development processes, and technical and support staff. In the case of freelance staff, make certain that they sign a nondisclosure agreement so that they cannot legally share proposal-sensitive information with any other company or individual.

12.4 Incorporating technical brilliance up to the last minute

One of the most challenging aspects of responding to an RFP is incorporating the best materials into the final document within a very limited timeframe. Technical and programmatic input, tailored boilerplate plans, marketing intelligence, nuances of corporate image, résumés, project descriptions (summaries) of past performance, cost and pricing data, and legal opinion—all of these elements must be brought together quickly and effectively. However, precisely because the proposal preparation cycle must be a *controlled, choreographed process* of planning, analysis, writing, multiple review, and publication, potentially brilliant technical or programmatic ideas and knowledge might not be incorporated into the final document. There simply may not be enough hours to integrate the change(s) into the text from a publication standpoint. Technical, management, and cost volumes of a proposal are "ecological" in the sense that a change in a system design concept affects cost, and an alteration in the WBS and bid task list affects the program plan and perhaps the cost as well. These wide-ranging changes take time to identify and make consistent across the proposal volumes both in the text and graphics. Unfortunately, the publication staff invariably are caught in the crossfire of last-minute changes. Their task is to generate a well-presented, smooth-reading, consistent, and integrated set of documents, which in final form always takes more time to produce than in draft form. Deciding between *structure* (generating a polished, final document) and *chaos* (ongoing free-thinking that can result in significant technical and programmatic enhancements) is a challenging pivot point for the proposal manager, but one that can actually work to the benefit of your company's proposal.

The balancing act for the proposal manager is to allow the entire proposal process to remain fluid enough to accommodate evolutionary change while simultaneously maintaining a firm commitment to milestone schedules and completion of action items. Regular brainstorming sessions can prove valuable, and building in extra space in the proposal documents will facilitate the easy incorporation of new ideas late into the production process. Human and organizational dynamics come into play in graphic relief during the proposal process. Management commitment of sufficient human and material resources, bonus and incentive programs, effective crosstraining, and the infusion of a winning attitude all can be brought to bear on the often arduous schedule of responding to federal procurements.

It seems imperative to me for the proposal manager to direct the publications group to generate as complete a document for each volume as possible as early in the proposal response lifecycle as they can. Not that final formatting needs to be accomplished, because this can be counterproductive given today's high-end desktop publishing software applications. What is important is seeing and working with a complete rough draft so that technical gaps, RFP compliance issues, and program management inconsistencies are identified *early*.

12.5 Graphics are an integral part of your proposal

Appropriate use of graphics, photographs, line art, and clipart (noncopyrighted art available on CD-ROM, for example) of all kinds will increase the evaluators' interest in and positive response to your company's proposal. It is essential not to submit a boring, lackluster proposal to any potential client organization. Well-designed graphics can convey complex information in an easily understood format. See Figure 12.1 for an example of a graphic appropriate for an executive summary. And in page- or word-limited proposals, graphics can present significant quantities of

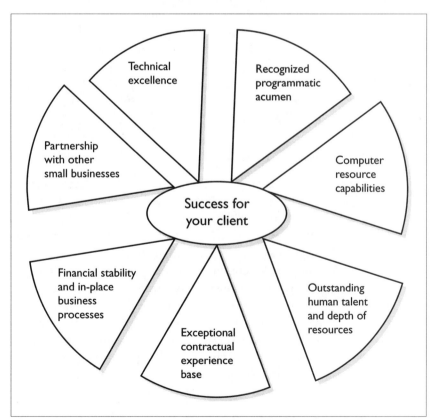

Figure 12.1
Sample graphic that introduces your company's primary thematic storyline to your client.

information in very limited space. Most proposals benefit from and many, in fact, require graphical presentations of staff skills, contractual experience, client-contractor interfaces, company organization, project organization, and project milestone schedules. Photographs might be added to personalize your company's résumés, particularly if your client knows your proposed key staff. Appropriate photographs of your company's facilities, specialized computer equipment, engineering and manufacturing centers, OTS products, and design prototypes can also greatly enhance your proposal's sales value. Photodocumentation brings projects and products to life, and also serves to "break up the text" for easier reading and fuller comprehension.

A word of caution about graphics found on the Web. The majority of photos and graphics on the Web are designed to look good on your monitor, which can display a maximum resolution of 72 pixels per inch (ppi). Printed material requires at least 150 dpi, with 300 dpi preferred for high-quality hardcopies. Do not assume that you can capture *noncopyrighted* photos and graphic images from the Internet and use them in your proposal. Plan ahead and purchase industry-specific photo CDs (e.g., Photo-Disc in Seattle, Washington; http://www.photodisc.com) and electronic clipart (e.g., Corel Gallery, which includes 140,000 clip art images).

One powerful software tool that can convert, capture, compress, organize, enhance, view, and print graphics files is IMSI's HijackPro. This application supports more than 85 graphics formats, including raster (e.g., .gif, .jpg, .pcx, .psd, and .tif), vector (e.g., .DXF, .TXT, and .PIC), and metafile (e.g., .EPS, .CDR, .PCT, .WMF, and .WPG) types. See http://www.imsisoft.com.

12.5.1   Action captions

Fundamentally, proposals are *sales* documents. In any sales document, whether it is a brochure or mailer or proposal, there are significant *messages* conveyed through photographs, artwork, and words. In proposals, these messages are called *themes*. Think of a theme as a statement that you make in a proposal that explains exactly why the client should select your company for contract award and not select your competition. In short, "Why you?" themes should always be substantiated and authenticated with facts. One great place to use theme statements is in the captions for figures, tables, photographs, and other graphic presentations. Let's look at the difference between simply identifying what a graphic or photograph is and using the caption as a forum to convey important sales messages that have been tailored to the specific procurement.

- Instead of a "project organizational chart," you might use "*Our company's project team is configured to manage multiple USACE work assignments efficiently.*"

- Instead of "map of local company locations," you might use "*The proximity of our company's offices to Fort Campbell will afford rapid response to changing requirements.*"

- Instead of "capabilities matrix," you might say "*Each one of our eight proposed senior technical staff holds an accredited degree in a relevant engineering and scientific discipline.*"

- Instead of "laboratory staff," you might consider "*Our well-trained laboratory staff are dedicated to and trained in quality-oriented USEPA procedures that ensure appropriate chain-of-custody handling.*"

- Instead of "appropriate H/S procedures," you might use "*We actively practice a comprehensive Health and Safety program designed to meet the requirements of 29CFR 1926.65 for hazardous waste site work.*"

- Instead of "scope of our support services," consider using "*We provide cost-effective, innovative support services across a broad spectrum of environmental regulatory acts.*"

In each of these examples, we have taken a flat statement and added positive, dynamic sales messages. We have introduced benefits of our approach and capabilities to the client. These sales messages, or themes, should be developed early in the proposal response cycle so that the entire proposal writing team can incorporate them into their narrative and into the captions they develop for their graphics. If a particular RFP stipulates that figure and tables shall be named in a particular manner, you must adhere to that guidance.

12.5.2 Configuration control of graphics

Configuration control of the graphics in your proposal is made difficult because many graphics are submitted to your publications group without captions (for figures) or legends (for tables) and their number within the proposal volumes may change several times depending upon placement within a specific section or subsection. The proposal manager can assist with this configuration management issue by insisting that writers submit graphics concepts with appropriate numbering and clear linkage to specific outline sections. Adding a file path name in a very small font directly into the graphic file (so that the path specifications print with the graphic image) will also help when these graphics iterate or shift within the document.

12.6 Role and structure of your publications group

A small company's publications group is often tasked with multiple responsibilities, such as proposals, contract deliverables, presentations, and marketing brochures. The core staff of this group should be cross-trained in a variety of operations, software applications, communications protocols, and hardware platforms. And it is critical that the publications group be able to accept, incorporate, and manage "outside" assistance. The group's policies and procedures, software, and hardware should not be tailored so that temporary/freelance employees or other in-house staff cannot support the group effectively and efficiently during "crunch" (crisis) documentation periods. In addition to an active list of local freelance staff, the manager of your company's publications group should consider maintaining lists of local photography/visual imaging vendors and photoreprographics houses that offer pickup and delivery services as well as rapid turnaround.

The continuity of publications group core personnel is very important for smoothly operating proposal efforts. Proposal managers benefit from seasoned, competent documentation staffers. Having established successful proposal production departments from the ground up for four federal contracting firms, I can testify to the benefits of human continuity within the publications group. Continuity, crosstraining, and a positive attitude are salient elements in proposal publication success. Loss of skilled and crosstrained staff can cut deeply into productivity and inflate overall proposal publication costs significantly.

In order to facilitate a smooth proposal operation, your company will need to determine the document *throughput capacity* of the publications group. How many new proposals of average size (for your firm) per month can be handled adequately with the core staff? If your bid/no bid process causes additional proposals to enter the publications "pipeline" each month, will additional staff, computer equipment, and floor space be needed? Keep in mind that an informed, aggressive bid/no bid or down-select process is the checkpoint that controls proposal document flowthrough. Your company should avoid overloading your publications group and your business development infrastructure with proposal efforts that have low-win probabilities. That practice wastes B&P money and can be very detrimental to morale. Preparing and submitting proposals on a "law of averages" basis—the more we submit the better the chance of winning—is a devastating and debilitating practice in both human and financial terms.

Keep in mind that editing and proofreading must be integral parts of the proposal publications process, not merely "nice-to-have" services. The publications group can be a good source of editorial support.

12.7 Software and hardware compatibility, standards, and recommendations

In today's era of desktop publishing—*integrating* sophisticated text configurations and complex graphics and creating multicolumn page layout designs on a personal workstation—computer platforms, peripherals, and software applications are important considerations for any company that prepares proposals. The ongoing explosive proliferation of third-party hardware "compatibles" and multiple variations of software applications argues strongly for your company adopting a uniform standard for hardware platforms, software applications, storage media, and peripherals.[3] This standard is best adopted as early in your company's corporate life as possible. The same version of operating software, such as Microsoft Windows 98 or NT, should be installed on all your company's computers. As much as possible, the default settings should be consistent from machine to machine, and the printer drivers and fonts should be the same.

In addition to software and hardware compatibility, computer and peripheral *redundancy* and *maintenance* should also be considered. A limiting factor in the publications process is document printing. Complex graphics, for example, can take many minutes to print just one copy. Your publications group should be adequately equipped with printers that have expanded *random access memory* (RAM) to print complex graphic images, Pentium-class computers with high-storage hard drives (minimum of 2.0 GB) and expanded RAM (minimum of 32 MB, with 128 MB preferred), and color *video graphics adapter* (VGA) monitors (15 in or larger, with 21 in preferable). A four-hour response time maintenance agreement for all your proposal publication computer equipment is highly recommended.

When establishing hardware and software standards for your company, keep in mind the various external organizations, vendors, and

---

3 Printers manufactured by the same company, for example, can vary in the number of words printed per page. The Hewlett-Packard LaserJet IIISi and IVSi printers are cases in point.

freelance support with which your system at the hardware, software, and storage media levels must interact on a regular basis. For example, perhaps several of your major federal clients require contract documents be delivered in Microsoft Word 97 rather than WordPerfect 8.0 for Windows, or your top two graphics service bureaus[4] use SyQuest 200 MB cartridges (or Bernouilli cartridges[5]) for transferring large files rather than CD-ROM or magneto-optical disk, or three of the firms with which you team most frequently operate in a Macintosh Power PC environment rather than with IBM-compatible PCs, or the majority of the known freelance staff in your immediate area are most competent and proficient in WordPerfect.

Significant computer-based experience during the past 11 years suggests strongly that whatever system platforms your company selects should be open architecture—designed for and fully capable of responding to the exponential pace of software and peripheral advancements. Your *central processing units* (CPUs) should be equipped with as much RAM as is necessary to operate and execute all of your software applications and printing tasks in an efficient manner. Printers should have expanded RAM to accommodate graphics files as well as integrated text and graphics files. Storage media must be capable of handling ever-increasing data volumes and of rapid random retrieval. Today, an average-sized proposal—replete with full-color cover and integrated text and graphics—can extend to 200+ MB. One proposal alone would fill a SyQuest drive, and only 6 or 7 would fit on a 1.3-GB floptical disk! A single graphics file that contains several scanned photographs can be 30+ MB, the equivalent of more than 20 3.5-in high-density diskettes [1]. One final note: *Back up files regularly. Your proposal process probably cannot sustain major data losses.*

From a software applications standpoint, proposals do not generally lend themselves to being prepared with high-end, full-scale desktop publishing packages such as Ventura, PageMaker, and Interleaf. This is because of the very tight production timeframes. Ventura, for example, is best-suited for the publication of books and other lengthy documents that are "frozen" (approved with no changes) and then have "styles" (page layout parameters) applied to them. In the IBM-compatible world, WordPerfect for text and CorelDRAW! for graphics (imported into WordPerfect as Windows metafiles .WMF) have proven to be appropriate over the long term. By keeping text in WordPerfect, which many technical

---

4   Service bureaus are for-profit organizations that provide such support as typesetting, electronic file conversions and *raster image processing* (RIP), 35-mm slide generation, and prepress film generation.

5   IOmega has also released JAZ™ 1-GB cartridges and ZIP 100-MB cartridges for archival purposes and large file transfers.

and administrative staff can manipulate, the documentation process can receive more in the way of ad hoc support from throughout your company and the local freelance community. In addition, technical changes can be accommodated until a much later stage in the production lifecycle. The high-end packages also have an associated steep learning/proficiency curve.

A very recent development in computer platforms is the prototype workstation with a motherboard that offers users the option of starting up under Microsoft Windows or the Macintosh MacOS operating system [2]. Assuming that this converged hardware reference platform prototype is produced and mass marketed successfully, users could purchase one machine that would offer the flexibility of running applications for either operating system while providing total hardware compatibility.

12.8 Electronic proposal submittal and evaluation

In an effort to streamline the evaluation and source selection processes, certain federal agencies are now requiring both hardcopy and electronic copy (such as diskette and CD-ROM) proposal submittals. It is already commonplace that cost data are submitted on diskette in Lotus or Excel, for example. The technical and management narrative are migrating in that direction as well. Be certain to provide your proposal in the specific version of the particular software application requested on the correct size and density of diskette.

The *Windows Proposal Evaluation Tool* (WinPET), which is maintained by the U.S. Army's *Information Systems Selection and Acquisition Agency* (ISSAA), reduces the time spent on evaluating large procurements [3]. Developed by User Technology Associates, Inc., WinPET facilitates information flow among the government's evaluation team and documents each member's narrative comments. WinPET can import text from a variety of commercial word processing applications.

The U.S. *Air Force Material Command* (AFMC) is moving in the direction of electronic evaluation for all major system procurements [4]. Electronic evaluations are already being used for many smaller procurements. Each AFMC center is in the process of defining and developing its own methods for electronic evaluation. As of yet, there is no initiative to develop a common electronic proposal evaluation process or standard proposal evaluation software tool that is shared across all AFMC centers. For example, the U.S. Air Force *Space and Missile Systems Center* (SMC) is refining a process based upon Microsoft Access data base software, while the AFMC center at Wright Patterson Air Force Base in Ohio is building its process around Lotus Notes. SMC is now building an electronic proposal evaluation facility at Fort MacArthur in San Pedro, California.

In the case of the recent *Evolved Expendable Launch Vehicle* (EELV) procurement, SMC reduced the time to evaluate proposals from 180 days to 52 days. This was accomplished in part through the use of Microsoft Access electronic templates. The electronic process was only a mechanism to store, manage, and retrieve data; this process did not electronically manipulate data that influenced the final selection. Contractors delivered their proposals on CD-ROM in *Portable Document Format* (.PDF),[6] which is application-independent and suitable for both text and graphics. Proposals in .pdf that are prepared on Macintosh or IBM-compatible PC platforms are readable on either platform. Adobe Acrobat software then permits government evaluators to zoom in on compressed graphics and read large pages (11 × 17 inches). Electronic proposal submittal and evaluation will help the government reduce the overwhelming amount of paper that resulted from manual source selection processes.

Increasingly, federal solicitation documents are being issued and disseminated in electronic form. For example, the Department of Transportation's Operations Systems Center IRM Technical Services Project was distributed to potential offerors on floppy disk and simultaneously made available to offerors to download from the Coast Guard file server via the Internet. That solicitation was 500 pages in length and was disseminated to 400 contractors. The Coast Guard saved approximately $10,000 in hardcopy reproduction, postage, and labor costs. The U.S. Navy and Department of Commerce are also releasing both RFP documents and subsequent amendments via its Web site.

12.9 Important documentation tips

Proposal documents are presented very attractively in double-column format, which provides the added bonus of 10% to 15% more text per page, depending upon the font selected. Conserving the "white space" at the end of lines of text is part of the reason why double-column formatting provides for more words per unit of space. This space savings is very

---

6 The file format for documents created using Adobe Acrobat is PDF. PDF documents allow people to share formatted documents across different platforms. To create a PDF document, you use Adobe Acrobat Exchange and print the existing document to a file. The PDF file can be viewed using the Acrobat Reader, which is distributed free of charge.

1.0   TECHNICAL APPROACH

Proposal documents are presented very attractively in double-column format, which provides the added bonus of 10-15% more text per page, depending upon the font selected. Conserving the "white space" at the end of lines of text is part of the reason why double-column formatting provides for more words per unit of space. This space savings is very important for page-limited proposals.

Proposal documents are presented very attractively in double-column format, which provides the added bonus of 10-15% more text per page, depending upon the font selected. Conserving the "white space" at the

> **This space savings is very important for page-limited proposals.**

end of lines of text is part of the reason why double-column formatting provides for more words per unit of space. This space savings is very important for page-limited proposals.

1.1   Understanding the Technical Requirements

Proposal documents are presented very attractively in double-column format, which provides the added bonus of 10-15% more text per page, depending upon the font selected. Conserving the "white space" at the end of lines of text is part of the reason why double-column formatting provides for more words per unit of space. This space savings is very important for page-limited proposals.

Proposal documents are presented very attractively in double-column format, which

provides the added bonus of 10-15% more text per page, depending upon the font selected. Conserving the "white space" at the end of lines of text is part of the reason why double-column formatting provides for more words per unit of space. This space savings is very important for page-limited proposals.

Proposal documents are presented very attractively in double-column format, which provides the added bonus of 10-15% more text per page, depending upon the font selected, as shown in Figure 1-1. Conserving the "white space" at the end of lines of text is part of the reason why double-column formatting provides for more words per unit of space. This space savings is very important for page-limited proposals.

Figure 1-1.

Proposal documents are presented very attractively in double-column format, which provides the added bonus of 10-15% more text per page, depending upon the font selected. Conserving the "white space" at the end of lines of text is part of the reason why double-column formatting provides for more words per unit of space. This space savings is very important for page-limited proposals.

1.1.1   Software Applications

Proposal documents are presented very attractively in double-column format, which provides the added bonus of 10-15% more text per page, depending upon the font selected. Conserving the "white space" at the end of lines of text is part of the reason why double-column formatting provides for more

Figure 12.2   Sample double-column proposal page format.

important for page-limited proposals. See Figure 12.2 for an example of a double-column proposal page format.

When compiling a list of acronyms, perform a global search for a beginning parenthesis "(". If acronyms have been defined in the text, and they should be, this technique should identify most of them. The acronym list for your proposal can, like the compliance matrix, be put on a foldout page so that the government evaluators can have constant and easy access to the acronym definitions.

For lengthy proposals, customized tab pages should be prepared to serve as page dividers for ease in locating important sections. Tabs are ordered by the "set" and by the "cut." Ten sets of seven-cut tabs means 10 complete sets of tabs each of which has seven tabbed pages. Many higher-end photocopier machines can now produce tabs, even laminated ones, right in your office.

The human and organizational dynamics of the proposal process are probably the most important elements of all. Chapter 13 will highlight these very real issues.

12.10 Virtual proposal centers, intranets, and extranets

An *intranet* is an internal information distribution system that uses Web technology to allow an organization to access specific information in a form and format designed just like an Internet home page. However, the general Internet community cannot access your company's intranet site. In effect, an intranet is a separate system with limited access[7] using a Web browser identical to Internet Web browsers such as Netscape Navigator and Microsoft Internet Explorer. The organization develops its own internal home page, with links to various sites or locations containing information within the site. On the other hand, an *extranet* incorporates encryption technology to allow secure access to an intranet over the Internet (see Figure 12.3). The Federal Express do-it-yourself tracking system represents an example of a sophisticated extranet. Now how are these systems applicable to, and useful for, proposal development?

An intranet can be used to create a "virtual proposal center" within your company, which in turn can be used to manage, support, and simplify the proposal development process for the firm. A virtual proposal center enables your entire proposal team to have a common electronic work

---

7   As opposed to a file server, electronic mail, or groupware such as Lotus Notes or Microsoft Exchange messaging software.

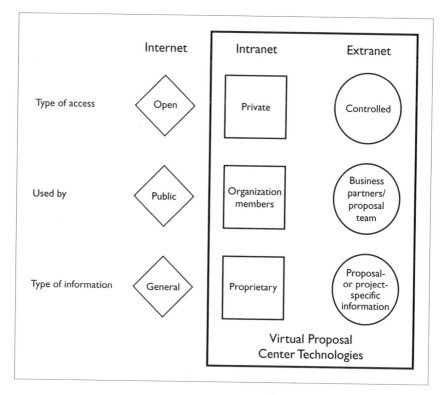

Figure 12.3 Comparison of electronic networked data warehouses.

center with access to an expandable repository of important automated proposal tools, project- and marketing-related information and knowledge base, and proposal templates required to help develop a successful proposal. A virtual proposal center facilitates real-time status checks for every proposal participant, despite geographical location. Intranet and extranet technologies become invaluable time and B&P cost-savers with geographically dispersed, complex teams. Both travel and administrative time and costs can be reduced. Schedules and action item lists can be shared. And appropriate levels of security and access control can be built into your system. Companies such as intr@vation, inc., in Los Altos, California (http://www.intravation.com) offer off-the-shelf software applications designed to build virtual proposal centers within your firm.

12.11 Using freelance proposal writers to maintain technical productivity

Carefully managed freelance "writer/interviewers" can be of invaluable assistance to your proposal manager at select phases of the proposal response life cycle (see Figure 12.4). If deployed on a "just-in-time,"

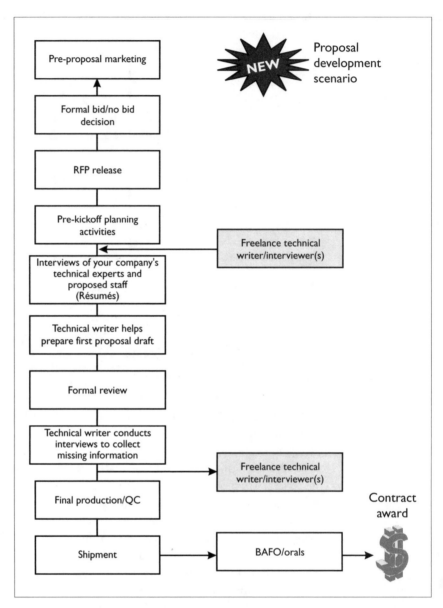

Figure 12.4
Carefully managed freelance writers can help you to control B&P costs.

as-needed basis, individual assignments should last from 2 or 3 days to 2 or 3 weeks. The point here is not to add to B&P costs, but rather to maintain your company's technical staff at maximum productivity and billability. The writer/interviewers would be trained by your proposal manager to extract appropriate technical and programmatic information from your firm's professional and managerial staff. Following the interview process, the freelance writers would then translate "technical-ese" into marketing-oriented, high-impact narrative that conveys the benefits and value of your approach and credentials to the client. Interviews can be conducted in person, by telephone, or via carefully constructed e-mail queries, and can be worked in before and after core business hours or during lunch breaks. The goal is to *minimize* the time that any billable staff person is diverted from direct client support, while simultaneously controlling the proposal schedule and generating integrated, well-crafted proposal text.

Effective freelance writer/interviewers bring an understanding of the documentation process and possess technical/professional writing or English/communications skills. Real-world experience with well-managed and trained freelance specialists serving in this capacity demonstrates an astounding *40% reduction in overall B&P costs!*

Sources of freelance writer/interviewers include local colleges and universities as well as the following organizations.

Association of Proposal Management Professionals (APMP)
http://www.apmp.org/seeking.htm#lm. Proposal professionals list their expertise and availability at this important Web site.

Washington Independent Writers (WIW)
220 Woodward Building
733 15th Street NW
Washington, DC 20005
Tel. (202) 347-4973
http://www.washwriter.org

Independent Writers of Chicago (IWOC)
7855 Gross Point Road
Unit M
Skokie, IL 60077
Tel. (847) 676-3784
http://www.iwoc.org/

Independent Writers of Southern California (IWOSC)
P.O. Box 34279
Los Angeles, CA 90034
Toll-free phone: 877-79-WRITE
http://www.iwosc.org/

Cassell Network of Writers (CNW)/FFWA
P.O. Box A
North Stratford, NH 03590
Tel. 1-800-351-9278 or (603) 922-8338

World Wide Freelance Directory
http://www.cvp.com/freelance/. Comprehensive listing of independent consultants and firms in all areas of expertise.

Society of Freelance Editors and Proofreaders
Mermaid House
1 Mermaid Court
London SE1 1HR, United Kingdom
http://www.sfep.demon.co.uk/

European Federation of Freelance Writers
P.O. Box 1116
1227 Carouge
Geneva, Switzerland
http://www.eurofed.org/

END NOTES

1. In 1994, the average file size was approximately 18 MB. In late 1995, that had grown to an average of 73 MB. See Hevenor, Keith, "Storage Product Buyers Guide," *Electronic Publishing*, 1 August 1996, p. 10.

2. Hevenor, Keith, "Signs of Life," *Electronic Publishing*, 1 Aug. 1996, p. 4.

    The PC holds the large majority of the mass market, while the Mac holds the majority share of the prepress and publishing market.

3. Hall, Dane, "Electronic Proposal Evaluation: How One U.S. Government Agency Does It," *The Executive Summary: The Journal of the APMP's National Capital Area Chapter*, June 1996, pp. 6, 8–9.

4. Information about the AFMC's electronic proposal evaluation initiative was obtained from Davidson, Paul, "U.S. Air Force SMC Innovations in Electronic Procurement," *APMP Perspective* I, No. 1, July/Aug. 1996, pp. 7, 12, 13, 15.

Chapter 13

Human and organizational dynamics of the proposal process

The classic mechanistic model ... fails to engage to the fullest
extent an organization's most important asset—its people...
—*The PeopleWise Organization*, Kepner-Tregoe, Inc.

C *ritical thinking in the workplace.* In 1996, Kepner-Tregoe, Inc., a
Princeton, New Jersey-based organizational management consulting firm, conducted its fourth in-depth, audited survey of more
than 1,400 managers and hourly employees. What was being examined in
this important study was how much an organization's collective thinking
is being tapped to resolve the big strategic and operational issues as well
as the day-to-day impediments to competitiveness and profitability. *The
results?* Sixty-two percent of workers surveyed believed that their organizations are operating at half—or less—of the brainpower available to them.

These perceptions were reflected in responses by the managers as well. More than 40% of workers said they do *not* feel valued by their organization. And yet delighting clients and pleasing stakeholders "are the hallmarks of organizations in which people can—and want to—contribute their very best" [1].

13.1 Modifying our thinking to win

Your company's entry into the arena of *competitive* procurements brought with it the requirement for a fundamental shift in thinking and business-related behavior throughout the ranks of management as well as the professional and support staff. Senior management, for example, must appreciate the significance of proposals and proactively support the development and enhancement of the processes associated with the proposal response infrastructure. In addition, each division of your company must make full use of the collective human and contractual expertise resident throughout the entire firm. Divisions cannot operate effectively in isolation from each other or from corporate direction. The dynamics of genuine *teamwork* take on accentuated meaning in the competitive marketplace. People at all levels within the company must work together with clearly defined objectives, for example, to collect and assess marketing intelligence, write proposals, review proposals, and publish proposals. In too many companies, a small cadre of technical and support staff produce the significant bulk of the proposals. That practice does not encourage team building, knowledge sharing, and a wider sense of ownership for the company's success.

Proposals do not follow a fully *democratic* process. Companies must recognize that effective proposal management follows from informed, albeit authoritative decision making at the appropriate time. A proposal effort can become mired if people are not instructed to heed the guidance of the proposal manager. On the other hand, the proposal manager must listen to team members and make informed judgments. And senior management must clearly and repeatedly reinforce the role and authority of the proposal manager, the criticality of sharing knowledge internally, and the importance of meeting the schedule milestones.

There is no more important corporate activity than proposal development for a contracting firm.

An entrepreneur—the very person(s) who founded the company—can often be an unwitting impediment to the development, implementation, and enforcement of rational, formal, repeatable business development processes. Not because they do not want their company to succeed. Quite the contrary. However, they know that they built their company to its current revenue base and position within the small business contracting community on savvy, intuition, risk taking, and "shoot-from-the-hip" decision making. Successful marketing and proposal development in the competitive arena are, on the other hand, built upon a methodically applied and continuously enhanced set of business practices and processes. And that is an entirely different mind set than that which founded and nurtured the fledgling company.

13.2 Building a competitive work ethic

It is likely that your company's very existence and future growth depends upon winning 25% to 40% of the formal proposals you submit. Proposal development is an absolutely essential corporate activity within your organization. Every resource in your company must be made available to the successful completion of each proposal. There are many details that must be attended to in the course of preparing a proposal that do not require the direct, hands-on attention of the proposal manager. Such activities include electronic searches of the résumé and project description subdirectories or data bases to determine appropriate staff to meet position descriptions and to ascertain relevant company contractual experience. Technical editing and tailoring of résumés, project descriptions, and the technical and management narrative would be of invaluable assistance. In addition, administrative support in the form of photocopying, meal arrangements for evening and weekend work, and "text-entry" word processing are each vital to the overall success of the proposal preparation process.

No one person can address all of the details, for example, of technical content quality, writing consistency, and compliance with the RFP. Therefore, your company would benefit from more people being trained and available on an as-needed basis to assist in the preparation of the proposal in addition to the technical authors, proposal manager, and publications group. The burden of preparing winning proposals must be shared among all professional and support staff. Instill this work ethic among all your staff through regular briefings on new business prospects, competitors, and new technologies. A rotating schedule of administrative support being on call to support proposal activities after core business hours is one mechanism for ensuring broader involvement and appropriate levels of support for the proposal team and publications group. Everyone's future depends upon winning! The acquisition team is an excellent vehicle through which to build a winning attitude for a given proposal effort. And when your company expands, you might hire technical and programmatic staff who possess strong proposal-related backgrounds, either from competitors or good business colleges and universities. Successful proposal experience and the mindset that fosters proposal success can be highly synergistic. I have personally witnessed a small company, which had very few people trained in the art and science of proposal development and management, become far more sophisticated and successful with the addition of key technical and programmatic staff who had proposal backgrounds. This knowledge influx helped to underwrite and authenticate the ongoing proposal development training programs that had been initiated in the company previously.

13.3 Strong link between project performance and proposal success

The integral connection between superior project performance and proposal success with that particular client cannot be overemphasized. Past performance and key staff are two of the most significant evaluation factors in most support services proposals. Your client or potential client wants to ensure that your company has performed to expectation on contracts of similar type, size, technical scope, staffing level, complexity, and geographic area. Solid references from COTRs and COs are critical to your company being awarded the next contract. Project schedule slippages, low-quality deliverables, and cost overruns on your current or past contract work translate into hurdles that have to be overcome during the bid/no bid, marketing, and proposal processes. Verify all of your project-related references *before* your proposal is submitted to ensure that you will receive a positive reference from the government points of contact. Also ensure that telephone numbers, mailing addresses, and e-mail addresses are current.

Client staff have demonstrated on numerous occasions that they have enduring memories—for superior as well as poor performance. Perceptions are paramount in contracting. Just as your company networks and shares information, so do your client counterparts. If your company has allowed significant schedule slippages or incomplete deliverables to be submitted, you will be remembered for the wrong reasons. Those negative perceptions can take several contract cycles to overcome. They can also seep throughout the client's organization; so that if your performance was below average at a particular Air Force installation in Texas, a potential client at an Air Force location in California may have heard negative comments. The key is to leverage outstanding contractual performance into exceptional project experience in your proposals. Project and proposal processes and successes are closely linked, as shown in Figure 13.1.

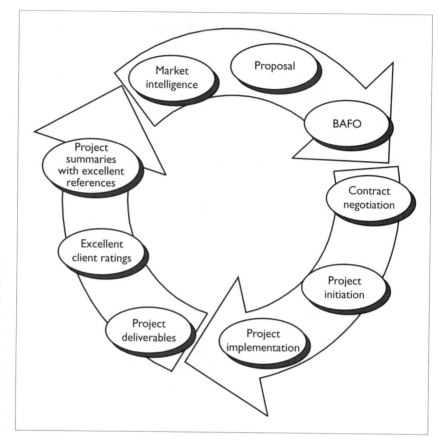

Figure 13.1
Proposals and project deliverables each require planning, technical effort, senior review, contractual involvement, and production expertise.

13.4 Proposals can be fun!

Successful proposal efforts can be *fun*, in addition to being a lot of hard work. The level of fun is directly dependent upon how well a variety of people can interact during the arduous proposal response lifecycle. Publications staff, for example, must interact effectively and courteously with technical contributors. A graphics artist might need to spend some time talking with one of the technical writers in order to understand fully what that author wanted to convey through a particular graphic concept or to help that author conceptualize a graphic. That interaction need not be confrontational in either direction. The writer should want to ensure that his graphic is well presented and be pleased that another professional is attending to that detail. And the artist should want every graphic to be accurate and convincing and not become overly frustrated with the "back-of-the-envelope" "stick figures" that the writer may have submitted origi-nally. Communication and crossfertilization of ideas within the Proposal Team should be actively encouraged by the proposal manager.

Fun can take the form of some office-appropriate humor injected into the proposal process in liberal proportion. And a special meal involving the proposal team members is a nice perk, too. The proposal manager should help make the proposal team members feel appreciated and good about what they are doing for the company. As the days and weeks of the proposal response lifecycle unfold and the proposal team becomes more tired and worn, pleasant humor, nice meals, meaningful compliments, and effective communication assume even larger importance.

[U]nlike traditional raw material—which is inspected, warehoused, bar-coded, and audited—corporate knowledge is scattered, hard to find, and prone to disappear without a trace.
—Thomas A. Stewart, *Fortune* magazine

13.5 Maximizing human intellect

Converting human intellect into useful products and services, including proposals, is becoming increasingly critical to any business's success [2]. " 'We now know where productivity—real and limitless productiv-ity—comes from.' " " 'It comes from challenged, empowered, excited, rewarded teams of people. It comes from engaging every mind in the organization' " [3]. There is an accelerating recognition among corpora-tions that their future profitability depends upon intangible assets such as "creativity, flexibility and the speed with which they can share new ideas and information. In a literal sense, modern corporations *are what they know*" [4]. Major international corporations such as Skandia AFS (Swe-den), Xerox Corporation, Hewlett-Packard Co., British Petroleum, and Dow Chemical Co. are spending significant resources on knowledge iden-tification and management. And the United States *Securities and Exchange Commission* (SEC) hosted a conference in April 1996 that focused on accounting for and reporting intangible assets. Alan Webber, a founding editor of the Boston-based business magazine, *Fast Company*, has noted that " '[t]he world of business is realizing that ultimately, what matters is the quality of the people in the organization, and how they treat their cus-tomers and how creative they are in coming up with new products and new services' " [5].

Another significant transition among such major corporations such as Microsoft, Compaq, Sun Microsystems, and Hewlett-Packard is the paradigm shift away from "business as a battlefield" to that of business as a "complex ecosystem in which the company that wins is the one that is quickest to adapt. Employee individuality and diversity are honored and encouraged" [6]. In order to adapt to new market circumstances and translate them quickly into meaningful opportunities, your employees simply cannot be viewed and treated as mere "mindless troops," "crank-turners," or "line items on an accounting spreadsheet." The originators of business process reengineering (BPR)—Michael Hammer, James Champy, and Thomas Davenport—have come to recognize the real-world failure of their concept of "fundamental rethinking and radical redesign of business processes to achieve dramatic improvements in critical, contemporary measures of performance, such as cost, quality, service, and speed." In essence, they forgot the people [7]. As Michael Verespej cautions, "We must remember that knowledge workers keep their tool kits in their heads. We must learn to lead people, not contain them. We must learn to listen"[8].

With small companies in particular, leveraging the collective intellect is integral to business acquisition achievement. Providing work environments that are conducive to the generation, exchange, respect, and appropriate application of ideas will pay dividends in morale, employee retention, and revenue. Pleasant physical facilities, progressive human resources policies, an "open door" management culture, and concrete incentive programs, for example, all participate in inspiring and harnessing the best in and from your employees.

Dr. Douglas Soat, a psychologist who consults to businesses regarding human resources issues, has identified two critical motivators for both technical and nontechnical staff that result in excellence on the job. The first is "setting challenging but realistic expectations," and the second is "demonstrating a true concern for employees" [9]. I can certainly attest to the validity of the latter. I have seen my role in proposal development during the past 11 years to be that of "facilitator"—a person to "grease the wheels" of the proposal response process as well as the management decision-making mechanism. How can I help my staff do their jobs more effectively and efficiently? What training and additional equipment do they require? Do they need a ride to public transportation after working very late at night? I have arranged for and purchased meals for my staff and have had special breakfasts at which I have brought in the food from home. I have talked with my employees, at their request, about personal issues, not offering any solutions, but simply providing an attentive ear. In addition, I have also pursued and received out-of-cycle raises, bonuses, and gift certificates for my staff in an effort to demonstrate in a tangible way their value to the proposal process.

Your company will be well served to recognize the value of competent, dedicated, team-, and solution-oriented support and technical staff in maximizing performance and engineering business development success.

On a regular basis, I have worked hand-in-hand with my staff when we had to stay 26 hours straight (with breaks and meals) to finish a proposal or provide support over a weekend or on a holiday. My staff knows that I will be there with them to the end to get the job done. The concern for quality is lived each day. They also know that I will share their accomplishments on and contributions to proposals and contract deliverables with senior management. Their successes are documented and reappear in detail in the narrative of their performance appraisals. I have extremely high standards of quality and productivity, and these are communicated to my staff regularly and in a variety of forms. I ask a lot of them and attempt to give a lot back to them as well. In every employee relationship, I have attempted to foster trust in the person by trusting him or her and treating the person with dignity and respect. I recognize the value of competent, dedicated, team- and solution-oriented support and technical staff in maximizing performance and engineering business development success.

13.6 Proposal professionals as "change agents"

Professional proposal staff can become integral participants in the positive transformation of an organization and its business processes [10]. The detailed analysis and articulation of a company's identity as well as its business processes, operational functions, relationships with clients and vendors, success stories and best practices, and internal resource- and knowledge-sharing mechanisms—the very activities that proposal staff do on a day-to-day basis—can be of direct benefit to the company in ways that extend far beyond proposal development and production. Your professional proposal staff can be among the most effective facilitators of knowledge exchange across your organization's divisional, geographical, and hierarchical boundaries. The "excellent ideas, streamlined operations, cost efficiencies, and technical superiority" [11] that are shared internally with and by your proposal staff can be of immense benefit, not only for the next contractual competition, but also to each element of the company.

13.7 Wellness in your proposal process

Fundamental health and wellness involve nutrition, exercise, sleep, and mind/body interrelationships. Employees who manage stress well tend to be more healthy, productive, cooperative, and creative. By supporting employee stress resilience, proposal managers may improve the quality of the work environment and reduce their own stress.

Senior management should investigate ways to realistically apply flex time in order to allow employees to participate in wellness programs that help them "recharge" during the work day. Proposal managers should actively encourage staff to take breaks away from their workstations, during which they might read, walk, or exercise in some other way. Supply fresh fruits, vegetables, other low-fat, low-sodium nutritious foods, and spring water to the proposal team on a regular basis. Ensure that staff are given some dedicated time during the *height* of the proposal life cycle to get out of the office physically and mentally and share time with family and friends. The proverbial "all-nighter" is not only counterproductive to the overall proposal process and product, but physically and mentally detrimental to the *people* who are the proposal team.

Proposal managers must make every effort to support the proposal team in a proactive manner with words and actions of encouragement and reliable guidance. Provide a private forum for staff to express their feelings in appropriate ways rather than letting those feelings fester and build to the point of an inappropriate emotional outburst sometime later. Help everyone know and feel that they are appreciated and respected, and are an integral part of an exciting proposal process.

During the marketing life cycle, your company will want to control costs very carefully. Monitoring and managing bid and proposal costs is discussed in Chapter 14.

END NOTES

1. "Minds at Work: How Much Brainpower Are We Really Using?: A Research Report," Princeton, NJ: Kepner-Tregoe, 1997.

2. Quinn, James Brian, Philip Anderson, and Sydney Finkelstein, "Managing Professional Intellect: Making the Most of the Best," *Harvard Business Review*, March/April 1996.

3. Quinn, Judy, "The Welch Way: General Electric CEO Jack Welch Brings Employee Empowerment to Light," *Incentive*, Sept. 1994, p. 50.

4. Hamilton, Martha A., "Managing the Company Mind: Firms Try New Ways to Tap Intangible Assets Such as Creativity, Knowledge," *Washington Post*, 18 Aug. 1996, p. H1.

5. Ibid., p. H5.

6. James, Geoffrey, "It's Time to Free Dilbert," *New York Times*, 1 Sept. 1996, p. F-11.

7. Mariotti, John, "Nursery-Rhyme Management," *Industry Week*, 5 May 1997, p. 19.

8. Verespej, Michael A., "Only the CEO Can Make Employees King," *Industry Week*, 16 November 1998, p. 22.

9. Soat, Douglas M., *Managing Engineers and Technical Employees: How to Attract, Motivate, and Retain Excellent People*, Norwood, MA: Artech House, 1996, p. 119.

10. McVey, Thomas W., "The Proposal Specialist as Change Agent," *APMP Perspective*, May/June 1997, p. 13.

11. Ibid., p. 3.

Chapter 14

Controlling bid and proposal costs

S
MALL BUSINESSES SPEND on the order of $1K to $2K for every $1M of total contract value. This expands to a range of $30K to $50K for a $5M procurement, or approximately 1%. Proposal consultant Hy Silver and many others have suggested that 5% of total contract value be spent on marketing, B&P, and IR&D. Decision drivers regarding marketing, B&P, and IR&D expenditures will, of course, vary according to the strategic importance, contract type (cost-reimbursement or fixed-price), fee/profit margins, and your company's incumbency status regarding the particular business opportunity. Small businesses often cannot spend 5% of total contract value due to cash flow, line of credit, and contract backlog parameters. It certainly does require money to win proposals, but the careful application and control of that money can be leveraged into contract award.

14.1 What does it cost to get new business, and how are those costs recovered?

If a company wants to capture $5M of new business during Fiscal Year (FY) 2000 and it has a win percentage on proposals of 30%, then it has to estimate the costs associated with pursuing $15M of new business.

All companies, including those competing in the federal marketplace, need to develop strategies for becoming knowledgeable about and then capturing new business. Within the small business contracting community that serves the federal government, a relatively narrow subset of professional staff is involved in identifying, tracking, and capturing new business. This staff may include business development/marketing, proposal development, and division-level management professionals. Coordinating short- and long-range opportunity tracking can be challenging in small businesses because of staffing limitations and also because *centralized* control of business development in such firms is often an evolutionary development rather than a fact of life present from the point of the company's founding. Many times, division managers are primarily responsible for "growing" their respective divisions and have been accustomed to overseeing their own business acquisition activities. Corporate marketing staff may then assist division managers with particular marketing activities.

Marketing and B&P costs can only be projected, monitored, and controlled once a company has determined the extent to which it wants to expand each operating division in accordance with an agreed upon mission statement and strategic plan. The importance of extending the mission statement and strategic plan down to the competitive procurement level is the key to success. Procurements within the appropriate dollar range and contractual type must be ascertained to be available before growth projections can be made realistically. In-depth knowledge of specific client organization planning, programming, budgeting, and procuring processes, cycles, and decision-making pathways is imperative to make this determination. If, for example, a company wants to capture $5M of *new* business during its FY00 and it has a win percentage on proposals of 30%, then it has to estimate the costs associated with pursuing $15M of new business. (Start by listing the potential marketing opportunities, identifying the best opportunities, calculating associated B&P costs, and totaling the numbers.) Fifteen million dollars of viable marketing opportunities in this company's specific LOBs must then be available during FY00. This small company should plan to spend $150K during FY00 in marketing and B&P

costs. These cost estimates should be tracked, monitored, evaluated, and modified on an ongoing basis.

Marketing, B&P, and IR&D costs are recoverable through G&A charges to the federal government. Upon invoicing the government for services rendered, a contractor will typically collect G&A and other indirect costs associated with operating a business.

14.2 Tracking B&P expenditures

Analyzing direct labor and ODCs associated with preparing a proposal as well as the associated postproposal activities *in advance of a bid/no bid decision* and then tracking these costs carefully during the entire proposal lifecycle will be beneficial to your company for several reasons. First, accumulating *real B&P data* over time will allow for increasingly accurate B&P projections as new estimates are calibrated against these real data. Early B&P analysis and estimation will help ensure that adequate resources are made available to prepare an outstanding proposal. Figure 14.1 and Tables 14.1 and 14.2 present templates that can be used to capture and track B&P data. These data should be entered electronically into a spreadsheet or data base software application to access easily and prepare comparative graphics for trend evaluation and future projections.

Date prepared: March 11, 1999

Analyzed by: M. Hendrick, Proposal Manager

| | **Projected Spending** | | | | | | | |
| Month | **April 1999** | | **May 1999** | | **June 1999** | | | |
Employee Name	Hours	Dollars	Hours	Dollars	Hours	Dollars	Total Hours	Total Dollars
B. Frank	15	$525	35	$1,225	25	$875	75	$2,625
G. Sinclair	5	$125	45	$1,125	10	$250	60	$1,500
J. Landis	10	$200	25	$500	5	$100	40	$800
Monthly Total	30	$850	105	$2,850	40	$1,225	175	$4,925

Table 14.1 Direct Labor Analysis—Software Engineering Procurement for the FAA

B&P Initiation Form

B&P initiation date: _____   B&P internal tracking number: _____

Client: **FAA**
Procurement title: **Software engineering**
Division responsible for contract (if awarded): _____

Acquisition manager: _____
Proposal manager: **M. Hendrick**

Due date of proposal: _____
Prime: _____   Sub: _____
Teaming partners: _____
                   _____
                   _____

Location of performance contract: _____
Type of contract: _____
Expected period of performance: _____ (base and option years)
Persons/year staffing level: _____
Estimated contract dollar value: _____

Estimated B&P costs (please attach the direct labor analysis and ODC forms)
      Hours required: **175**
      Direct labor: **$4,925**
      Overhead: _____
      ODCs: **$825**
      Total: _____

Start date of B&P effort: **April 1999**   Ending date of B&P effort: **June 1999**

Signatures below indicate that the staff have (1) full awareness of this procurement, (2) have reviewed the items above for accuracy and completeness, and (3) accept the B&P hours estimate as reasonable for a procurement of this dollar and level of effort.

Signature lines: _____

Figure 14.1   B&P initiation form.

14.3 Business development bonus policy

B&P expenditures increase the indirect rates that your company must submit on competitive procurement solicitations. Proposal managers are the individuals charged with developing, defending, monitoring, and controlling B&P budgets. Success in the highly cost-competitive federal

	Month Projected						
Date prepared: March 11, 1999							
Analyzed by: M. Hendrick, Proposal Manager							
Include types of expenditures such as: documentation, travel, training, advertising, equipment purchases, lease agreements, supplies, software.							
	Amount	Amount	Amount	Amount	Amount	Amount	Total
Type of expenditure	April	May	June				
Local travel	$75	$150	$50				$275
Position Advertising	$550	0	0				$550
	$625	$150	$50				$825

Table 14.2   ODC Analysis—Software Engineering Procurement for the FAA

marketplace of the new millennium demands that acquisition and proposal development costs be kept to a minimum. Therefore, to the fullest extent possible and keeping within Department of Labor guidelines, proposal development efforts must be accomplished during nights and weekends. And the reality as I have observed it and lived it on both coasts is that many major U.S. contracting/consulting firms nationwide also minimize B&P costs by unpaid night and weekend work. As a small company, your professional, *exempt* employees should keep B&P charges to an absolute minimum. If they do not, you will have difficulty in expanding your company's revenue base; to do so you must incentivize them and communicate regularly with them about the good job they are doing.

In recognition of the extended effort of your professional-level and support staff, your company should strongly consider developing and implementing a clearly defined business development bonus policy. The policy should be designed to provide incentives and extra monetary compensation for those employees who have provided outstanding support for proposals or business development. You certainly want to retain your technical and support talent base.

In a recent reward and motivational survey conducted by Ms. Jo W. Manson, then of the Andrulis Research Corporation in Arlington, Virginia, it was determined that during the actual proposal response lifecycle, special meals were used as a motivator by 25% of the respondents. After a proposal had been submitted, expressions of appreciation, bonuses, and time off were used as motivational tools.[1] Ms. Manson reminds us that "we are in the people business and, no matter how automated and process driven we may become, we must still rely on well-motivated staff to produce our winning proposals."

According to research by Shari Caudron that appeared in *Industry Week* magazine, "[C]ompanies are starting to extend their incentive programs to nonsales employees with the hope of generating renewed enthusiasm and commitment" [1]. Incentives have begun to take the form of select merchandise, travel, catalog merchandise, cash, jewelry, and honorary titles.

14.4 Stretching limited marketing funds

Most small companies face limited staffpower for marketing and business development activities. One mechanism to augment and supplement your company's marketing staff is to utilize the services of any of a wide variety of marketing consulting firms, such as Federal Sources, Inc. in Virginia; INPUT, Inc., which has North American offices in Vienna, Virginia, New York, and San Francisco; and Environmental Information Ltd., headquartered in Minneapolis, Minnesota. Consulting companies can provide market research and trend analyses, competitor analysis, marketing opportunity identification, procurement informational data7bases, intercompany information sharing, strategic planning support, and proposal planning and guidance. Spending dollars for specific, well-directed marketing and planning support from an outside source can help to stretch your company's internal resources and actually help them to be more efficient. Taking the marketing information from outside firms and assessing, distributing, and archiving it in an easily retrieved manner is critical in order to maximize its value to your company.

Let us consider several *specific* ways in which outside marketing consulting and research firms can assist your company. They can be used to identify and conduct initial screening of potential teaming partners for an important procurement. Such firms can also help substantiate the funding for a particular procurement. They can be employed to collect competitor newsletters and annual reports. Often, these marketing research firms offer a basic suite of services such as a regularly updated automated database and hardcopy notebooks of marketing information. This standard suite is,

---

1   The survey was distributed to 400 members of the *Association of Proposal Management Professionals* (APMP) and resulted in a 25% response rate. Expressions of appreciation noted included letters, verbal praise, win parties, and publicity within the company.

in turn, augmented for *additional costs* with such services as on-demand telephone hotline support, Web-based information access, news release support and access, and senior-level consulting on a one-on-one, personal level.

Direct experience has shown that a company is well served to judiciously employ the services of several outside marketing consulting firms. Although there will be overlapping information provided, each firm contributes potentially important analysis and data that another may not.

Accurate, current marketing intelligence must be built into your company's proposal. We will now examine proposal writing techniques that are effective in incorporating that intelligence into a solutions-oriented, compliant set of knowledge-based sales documents.

END NOTE

1. Caudron, Shari, "Spreading Out the Carrots," *Industry Week*, 19 May 1997, p. 20.

Chapter 15

Tried and true proposal writing and editing techniques

Proposals are knowledge-based sales documents

Accuracy, brevity, and clarity

The essence of effective proposal writing is responding to the RFP requirements while *completing* the process begun during marketing of convincing the client that your team and approach are the most appropriate and cost effective. Remember that proposals are, first and foremost, *sales documents*. They are not technical monographs, "white" or position papers, or user's manuals. Conceptualizing and developing the technical solution is an engineering or scientific problem. Packaging that solution in a combination of crisp, convincing narrative and high-impact graphics is a *sales* problem (see Figure 15.1).

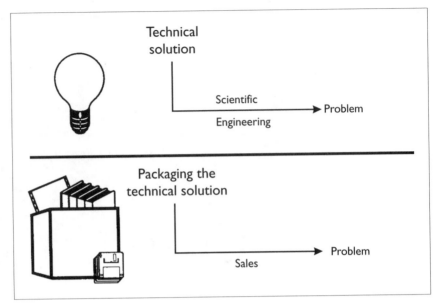

Figure 15.1
Packaging the
technical solution
is a sales problem.

Your company's proposal manager must articulate clearly to his or her technical writing team exactly what the expectations and acceptance criteria are. He or she must foster ongoing communication and feedback not only to himself but also among all of the writers. This will help to ensure a consistent approach and minimize rewriting.

Most proposals for a given procurement look and read essentially the same. The challenge is to incorporate well-substantiated information in your proposal that *only* your company can say. Identify precisely what will separate, or *discriminate*, your company from the competition. If your company is the incumbent contractor for a particular project and has performed favorably, use the names (and photos) of your incumbent staff people throughout your proposal. Reassure the client that they will be working again with the same competent technical staff who will be supervised by the same responsive management team. Write to a level of technical detail that exceeds what nonincumbents could gain from your project monthly reports (obtained through the FOIA),[1] published articles, and marketing conversations with government staff. Demonstrate that your company

---

[1] FOIA requests should be made to the appropriate agencies in written form. Letters should contain only one specific request. The reason for this practice is that an agency could decline to process one item of a multi-item request, and then the entire request would be returned to the contractor and the process would have to begin again. It is suggested that you begin your request with, "Pursuant to the Freedom of Information Act, 5 U.S.C., Section 552 as amended, Company XYZ hereby requests a copy…." Keep an electronic log of each FOIA request, and place followup calls to the agencies after two weeks has elapsed.

understands the technical risks as well as the success criteria. Strive to have your proposals not be boring.

Contributors should keep in mind the *ABCs* of proposal writing.

- *Accuracy:* It is not enough, for example, to tell your client that your company has saved time and money on projects of similar size, contract type, technical scope, and geographic area. Tell the client exactly *how much time* you have saved, how you accomplished this, precisely *how much money*, and the reasons why. Authenticate your claims with concrete, accurate information. Quantify your claims whenever possible with reference to your company's ACASS ratings,[2] contractual evaluation reports prepared by your clients, or FASA-driven[3] client "report cards" for contracts in excess of $1M.

- *Brevity:* Keep your writing *crisp.* Do not assume that volume will make up for quality. Too many times I have heard technical staff say, "Give them 100 pages, and they'll just assume the correct responses are in there." Guess what? No, they (the evaluators and their support staffs) will not. Unless your company demonstrates through subheadings, graphical pointers, chapter-level or sectional tables of contents, and cross-reference matrices that it has responded completely and appropriately to all elements in Sections L, M, and C, you will not score the maximum points available.

- *Clarity:* Assist the client's evaluators to understand, for example, exactly how your company proposes to manage this new project, how you will integrate your subcontractor staff on the job, how your proposed project manager can access corporate resources, and how you will manage tasks in geographically diverse locations. Amplify your crisp, direct narrative descriptions with clear, easily interpretable graphics.

---

[2] The ACASS, or Architect-Engineer Contract Administration Support System, is a database of selected information on architect-engineer (A-E) firms nationwide. ACASS provides required information for federal government selection committees to assist them in the process of awarding A-E contracts. The ACASS system was developed in response to the FAR as well as the Brooks A-E Act. The use of ACASS for assessing past performance history is now mandated for all DoD agencies in the Defense Federal Acquisition Regulation Supplement (DFARS). Captured in the ACASS database are the past performance evaluations for contractors for the past 6 years, as well as current SF254 information supplied by individual contracting firms. The ACASS database resides at the U.S. Army Corps of Engineers Western Processing Center in Portland, Oregon.

[3] As required by the Federal Acquisition Streamlining Act (FASA), a contractor's past performance in government contracts is now considered a relevant factor in the award of future contracts.

Ten pages of thematically integrated, well-constructed, and compliant prose are much preferred to 50 pages that contain most every technical detail the writers happened to know. Technical "data dumps" glued together produce a very uneven and most likely noncompliant proposal. Keep your ideas focused on the client's requirements as stated in the RFP, and keep your sentences short. This author has seen sentences in proposal narrative that have extended more than 80 words!

Focus on translating the features of your technical and management approaches and into benefits to the client and his project, his budget, his schedule (see Figure 15.2). Demonstrate that your company will help him succeed in his mission within his organization. This is an aspect of proposal writing that demands concrete, current, and in-depth marketing intelligence and sharing of this knowledge internally. Without this, your proposal will describe your staff, facilities, contractual experience, and technical capabilities in a vacuum. There will be little or no direct linkage of your company's capabilities with your client's perceived requirements.

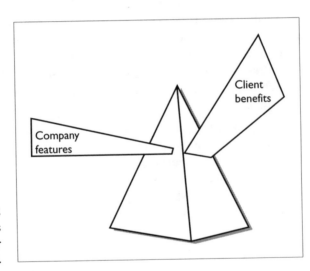

Figure 15.2
Translate features into benefits for your clients.

15.2 Active voice adds strength and saves space

Proposal writers should make every effort to employ the active voice. "We accomplished XYZ," not "XYZ was accomplished by us." The active voice adds strength to the proposal presentation and can save as much as 14% in terms of space as compared with passive voice depending upon the font that is used.

Contributors should attempt to vary sentence and paragraph structure. If you examine a proposal and most of the paragraphs on a page begin with

"The" or your company's name, the narrative requires editing to infuse variety.

Writers should also try to employ strong and descriptive verbs, adjectives, and adverbs (see Figure 15.3), such as in the following examples.

- Our incumbent staff of seasoned professionals all exceed the stipulated requirements.
- Over the course of three years of successful service to the Air Force Space Command, our senior programmers developed and implemented effective and cost-efficient...
- Our company offers field-tested prototyping experience...
- Fifteen analysts embody a knowledge base and legacy of experience that is unmatched...
- Lessons learned from award-winning performance-based IT support will be shared...
- We are a people-oriented firm with a very low turnover rate as measured against the industry standard in the Washington metropolitan area...
- We will deliver highly defensible results through the conduct of all activities in strict compliance with our sound QA/QC program...
- We will maintain technical continuity on this important Department of State program through proactive incumbent hiring...
- Our technical understanding is greatly enhanced by our past performance on two critical contracts for the Office of Personnel Management...
- With an extensive inventory of ADP equipment that includes Sun workstations to support off-site data processing, our company...
- Four years of progressive support of the VA validates our company's commitment to...
- We are committed to joining into a partnership with NOAA during the life of this contract...

15.3 Guide the client's evaluators through your proposal

Formulate ideas precisely, concretely, and simply. To include every technical detail each writer knows about a certain topic will probably not result in a successful proposal. This is where bullet drafts come into play—to guide your writers in preparing a thematically consistent response of

High-impact descriptors
• Rigorous development approach
• Principled management
• Seasoned field technicians
• Creative networking solutions
• Comprehensive community relations plan
• Results-oriented project team
• Innovative sampling methodologies
• Demonstrated success in providing superior-quality engineering solutions
• Responsive organizational structure and seasoned program leadership
• Proven business processes
• Technology-driven core competencies

Figure 15.3 High-impact descriptors for proposal text.

uniform technical depth. Proposals must appear as if written by a single, well-organized individual. Your proposal narrative must contribute to fast and easy comprehension by a variety of client evaluators who are reading your document with technical, contractual, and programmatic frames of reference. Use frequent subheadings and/or graphical icons to break up text and to facilitate the evaluation of your proposal.

Contributors should write from the *general* to the *specific*, from the easy-to-understand to the more difficult to comprehend. Proceed from an overview of each major topic to the finer grained technical and programmatic details. Always ask yourself the question "What's next?" [1]. Writers should identify and discuss tangible *benefits* to the client of your company's technical and management approach; examples:

1. When we build and implement solutions, we offer value engineering recommendations that result in significant cost savings for our clients.

2. We have established clear lines of authority and channels of communication to ensure accountability for this important FAA program at all levels within our organization and to expedite decision making.

3. We will incorporate our formalized and widely deployed operating procedures along with automated applications into the operational environment to ensure that state-of-the-art services are being provided that meet the requirements of Department of Treasury end users.

4. Our company has been successful in transitioning five other projects by focusing on the prospective employees and their career goals, and by offering a superior benefits package.

5. Our single-vendor contract management approach will eliminate program accountability complexities, reduce costs, and minimize overall management risks.

6. Our demonstrated record of careful cost controls on Army and Air Force programs will result in best value support for the Navy throughout the life of this important contract.

7. With an established regional presence in Texas since 1995, we have performed 16 projects in the Dallas–Ft. Worth area in the past 4 years.

8. Complementing our presence in Connecticut and our technical knowledge base with Coast Guard programs is the collective technical, programmatic, and human talent of our teaming partners...

9. To effectively manage and conduct technical work under each task order assigned, we have created the position of task leader. The task leader will serve as the single point of contact with the DHHS...

10. Our outstanding and well-established lines of communication among our team and with the Air Combat Command will help to ensure...

There is a natural tendency to dwell on familiar ground. Often, this is of the least significant as far as the RFP requirements are concerned. Demonstrate how the *features* of your approach translate into benefits. Use FACTS—support all statements with concrete, quantified (if possible) examples. Include:

- What your company will do for the client;
- How your staff will do it;
- Why you will do it that way;
- What you did in the past, that is, previous or current contractual experience. Highlight the relevant lessons learned and successes from similar contractual experience.

Summarize the content of each proposal section in the first paragraph of discussion. Write for a variety of readers, including the "skim" or executive-level reader. To be a winner, a proposal must contain concise, understandable, and closely related thoughts. Identify the critical technical areas. In discussing them, use a level of detail that exceeds what a nonincumbent could use, but do not drown the reader in jargon or equations. Your proposal should be built upon *solution-oriented writing*. Writeups should be risk-aware and solution-oriented and demonstrate an understanding of the evolutionary changes likely to occur over the life of the contract.

Make your company's compliance with the requirements and responsiveness to your client's evolving needs apparent to the evaluator over and over again. Use RFP terminology exactly (or in shortened form) for your proposal headings and subheadings. Employ the RFP terminology as a point of departure for further *original* writing. Do not simply recite the RFP verbiage in the actual narrative of your proposal. A DoD COTR once told me that he felt personally insulted when contractors, both large and small, merely replaced the words "the contractor shall" that appear in the SOW, with "our company will." Such an approach demonstrates no technical understanding whatsoever.

Define acronyms and abbreviations the first time they are used in each major proposal section. Not all evaluators see the entire technical and management proposal volumes, so redefinition of acronyms is certainly acceptable assuming page-count constraints allow. Avoid sectional or page references in the text because references must be changed each time the section or page numbers change. References such as "See Section 5.1.1 on page 5-34" should not be used. Let the compliance matrix and table of contents assist the evaluators in locating specific pages or sections. In addition, avoid fifth-order headings (for example, Section 4.3.1.3.1).

Proposal writers should attempt to think graphically as well as in written form. Every proposal section should have a figure or table associated with it that is referenced and discussed clearly in the text. Figure captions should appear centered beneath the appropriate figure. Table legends should appear centered above the appropriate table.

15.4 Action captions

A great place to use theme statements is in the captions/legends for figures, tables, photographs, and other graphic presentations. Let's look at the difference between simply identifying what a graphic or photograph is and using the caption as a forum to convey important sales messages that have been tailored to the specific procurement.

- Instead of "Project Organizational Chart," you might use "Our project team is proficient in each of the USACE technical requirement areas."

- Instead of "Map of Our Local Locations," you might use "The proximity of our offices to Fort McCloud will afford rapid response to changing requirements."

- Instead of "Capabilities Matrix," you might say "Each one of our four proposed senior technical staff holds an accredited degree in a relevant engineering and scientific discipline."

- Instead of "Our Laboratory Staff," you might use "Our well-trained laboratory staff are dedicated to and trained in quality-oriented procedures that ensure appropriate chain-of-custody handling."

- Instead of "Appropriate H/S Procedures," you might use "We actively practice a comprehensive health and safety program designed to meet the requirements of 29CFR 1926.65 for hazardous waste site work."

- Instead of "Scope of Support Services," you might use "We provide cost-effective, innovative support services across a broad spectrum of financial regulatory acts."

In each one of these examples, we have taken a flat statement and added positive, dynamic sales messages. We have introduced benefits of your approach and capabilities to the client. These sales messages, or themes, should be developed early in the proposal response cycle so that the entire proposal writing team can incorporate them into their narrative and into the captions they develop for their graphics. Please note, however, that it is always critical to be totally compliant with the RFP. If a particular RFP stipulates that figure and tables shall be named in a particular manner, you must adhere to that guidance.

15.5 Methods of enhancing your proposal writing and editing

Toward the goal of keeping your proposal writing highly focused, consider following the stages presented below [2].

1. Collect the technical and marketing intelligence information you need for your section(s). Focus on the basic features of those materials.

2. Identify the information in those materials that is relevant to the RFP requirements for your particular section and which

supports the strategic win themes that have been notated for that section.

3. Extract, organize, and reduce the relevant information. Begin to compose the relevant information into sentence form in accordance with the proposal outline that was part of the Kickoff Package. Slant your writing toward your audience, that is, the client evaluators of the proposal. Be informative but brief. Be concise, exact, and unambiguous. Use short, complete sentences.

4. Refine, review, and edit the relevant information to ensure completeness, technical accuracy, and the inclusion of theme statements.

One of the best ways to polish the writing in your company's proposals is to read it aloud to a colleague or yourself. Awkward sentence constructions quickly become obvious. The flow of the narrative can be smoothed and refined through this process of reading.

A longer term key to effective proposal writing is *reading* many different types of materials across a variety of disciplines and media—such as professional and scholarly journal articles, public relations brochures, newspapers, textbooks, novels, and other works of literary fiction. An enhanced functional vocabulary is one result. Another is learning how to use the same word in a broad spectrum of contexts. Reading in this manner can also produce an appreciation of nuance, that is, using precisely the most appropriate word for the given application.

There are many courses available through local colleges and universities that can be of assistance in enhancing your staff's technical and business writing capabilities.

In terms of editing proposal narrative, the following checklist can prove to be very useful [3]:

1. Check for completeness.

2. Check for accuracy.

3. Check for unity and coherence.

4. Include effective transition statements.

5. Check for consistent point of view.

6. Emphasize main ideas.

7. Subordinate less important ideas.

8. Check for clarity.

9. Eliminate ambiguity.

10. Check for appropriate word choice.

11. Eliminate jargon and "buzzwords."

12. Replace abstract words with concrete words.

13. Achieve conciseness.

14. Use the active voice as much as possible.

15. Check for parallel structure (for example, all "ing" words and all "ed" words).

16. Check sentence construction and achieve sentence variety.

17. Eliminate awkwardness.

18. Eliminate grammar problems.

19. Check for subject/verb agreement.

20. Check for proper case (such as upper case and initial caps).

21. Check for clear reference of pronouns.

22. Check for correct punctuation.

23. Check for correct spelling, abbreviations, acronym definitions, contractions (avoid them), italics, numbers, symbols, for example.

24. Check for correctness of format.

15.6 Government-recognized writing standards

There are several writing, editorial, and proofreading standards recognized by the federal government. Among these are the United States Government Printing Office *Style Manual* (1984); the Supplement to the United States Government Printing Office *Style Manual*, Word Division (1987); the *Gregg Reference Manual* by William A. Sabin, Eighth ed. (1996); and MIL-HDBK-63038-2 (TM) (1 May 1977). The consistent application of one such standard is important to implement within your company's publication or editorial group. Pages 5 and 6 of the GPO *Style Manual* provide a useful reference for proofreader's markings.

15.7 Additional sources of writing guidance

Judith A. Tarutz, *Technical Editing: The Practical Guide for Editors and Writers*; (1992); Philip R. Theibert, *Business Writing for Busy People* (1996); *The Chicago Manual of Style*; Edward T. Cremmins, *The Art of Abstracting* (1982); C. T. Brusaw, G. J. Alred, and W. E. Oliu, *Handbook of Technical Writing* (1976); *Words Into Type*; R. A. Day, *How to Write and Publish a Scientific Paper* (1979); and Herbert B. Michaelson, *How to Write and Publish Engineering Papers and Reports* (1982) each offer insight and direction for clearer and more concise writing. Edward Cremmins' work is exceptional in helping one to understand how to convey information accurately and robustly while simultaneously conserving the number of words used.

We will look at the critical information management processes that link directly to business acquisition success.

END NOTES

1. Miner, Jeremy T., and Lynn E. Miner, *A Guide to Proposal Planning and Writing*, http://www.oryxpress.com/miner.htm.

2. Adopted from Cremmins, Edward T., *The Art of Abstracting*, Philadelphia, PA: ISI Press, 1982, pp. 17, 73.

3. Adopted from Cremmins, *The Art of Abstracting*, p. 85.

Chapter **16**

Packaging and managing proposal information effectively

16.1 **Overview**

Proposals are authenticated information in carefully packaged form. Organizing centralized, fully operational, and *evolutionary* informational data systems are imperative to your company's business development success. Maintaining frequently used proposal materials in compatible electronic formats as well as in easily accessible hardcopy form will help your proposal planning and preparation process significantly. These materials include résumés, project descriptions or summaries, proposal boilerplate or "reuse material" such as configuration management and health and safety plans, and previously submitted proposals. Proposal managers require ready access to the latest[1] company and marketing intelligence information. But be sure to store all company-sensitive marketing and proposal information in a secure physical area and within a secure partition of an electronic network.

In light of the EC/EDI trajectory within federal acquisition discussed in Chapter 6, computer-based file management and information exchange assumes increasingly important proportions.

16.2 The all-important résumés

Your company's human, technical, and programmatic talent is the basis for your success to date. Staff are a particularly important consideration in the case of support services contracts, because the client is, in effect, purchasing human expertise in, for example, pollution prevention, electrical engineering, systems design, office automation, facilities management, and *integrated logistics support* (ILS). Indeed, a company's staff being presented in the best possible light is valid for most every federal government procurement. This means that résumés have to be customized or tailored for each proposal. Tailoring is not altering or misrepresenting a person's experience or education in an unethical or illegal manner but rather involves *highlighting relevant* experience, publications, certifications, and education to the exclusion of other information. Some RFPs require that each person proposed sign a statement that the information contained in his or her résumé is accurate. Adequate time within the proposal response schedule needs to be built in to secure those requisite signatures. Smaller companies need stellar résumés; they function like "name brands."

At times, the RFP will provide a very detailed résumé format, or indicate in narrative form how the issuing government agency wants to receive the résumés. When résumé format, content, and/or page count are stipulated, you must comply fully. However, many times you will have a degree of autonomy to structure the résumé.

A technique for helping evaluators quickly understand your staff's capabilities is to include an "Experience Summary," "Benefits to the Client," "Basis of Team Selection," or "Relevance to the Project" box on the first page of each résumé required by the RFP. Within that summary box,

---

1   Ensuring that proposal managers are provided with the *latest* company information is an ongoing effort in configuration management. For example, a comprehensive written overview about your company that is used in most every proposal needs to contain, for example, the most up-to-date information on contract awards, company commendations and success stories, and corporate organizational structure and leadership. If this business overview is updated for a given proposal, the corporate library boilerplate file that contains this same overview must be revised as well.

highlight the particular individual's expertise and accomplishments relevant to the specific position description and/or technical requirements of the procurement—this might include education, management experience on specific contract types, number of staff managed, publications, certifications, professional society memberships, work on similar projects, and years of successful experience in an operational environment. Three to four "bulleted" items are sufficient. If you do not use an "Experience Summary" box, you might consider including a "Benefits to the Project" section in each résumé. Do not assume that the evaluators will take the time to carefully review each résumé and extract the relevant details that correspond to the position description. If you make their job difficult, you simply cannot expect to receive the maximum points possible.

Résumés should focus on relevant technical and programmatic accomplishments and excellence. They should accentuate *results*—increased production rates, improved quality, on-schedule performance, cost savings, and the implementation of innovative techniques and technologies. As appropriate, they should be client specific, site specific, and geography specific. Résumés should use transferrable, action-oriented lead words to describe an individual's activities and contributions (see Figure 16.1). These types of words help the evaluators to understand the relevance of your staff's current and past experience to the new project at hand. They facilitate the transfer of talent and capability from one project to another.

Figure 16.1
Use action-oriented words to transfer capabilities from past contracts to the current opportunity.

Action-Oriented Words for Staff Résumés

• Manages	• Negotiates	• Leverages	• Brings	• Field-tests
• Performs	• Reviews	• Translates	• Provides	• Establishes
• Implements	• Writes	• Offers	• Facilitates	• Co-creates
• Conducts	• Designs	• Serves	• Enforces	• Assists
• Directs	• Supports	• Oversees	• Selects	• Plans
• Coordinates	• Builds	• Delivers	• Trains	• Employs
• Prepares	• Drives	• Ensures	• Operates	• Studies
• Initiates	• Leads	• Participates	• Troubleshoots	• Examines
• Develops	• Determines	• Maintains	• Audits	• Assesses
• Investigates	• Applies	• Practices	• Analyzes	• Consults
• Inspects	• Characterizes	• Deploys	• Compiles	• Surveys

Review the following example.

- Wrong way: "Mr. Jones managed an investigation at Fort Baltimore, Maryland."
- Appropriate way: "Mr. Jones managed a lead contamination groundwater investigation at Fort Baltimore, Maryland. He negotiated technology-based cleanup criteria with the Maryland Department of the Environment in the decision document."

One of the most important information management activities your company can undertake is to create, update, archive, and make available for retrieval a résumé for each member of your technical, management, and support staff. Utilize the *same software application and version* that you use to produce your proposals, and format each résumé in a style that will be compatible aesthetically with the rest of your proposal text. Store hardcopies of all résumés in alphabetical order in three-ring notebooks for ease of access. Establish an electronic subdirectory of all company résumés so that they might be searched (using Isys Query from Odyssey Development, Smart Text Finder from Spinning Technologies, Search 32 from Anet, or dtSearch from DT Software) for a combination of appropriate keywords based upon the position descriptions in the RFP. And you would be well served to generate and maintain a genuine data base (using database software such as Microsoft Access) of résumé information to facilitate rapid searches to meet position description parameters or prepare tables/graphics of information about your technical and professional staff. Many times RFPs will request the number of staff with specific degrees, certifications, or levels of experience. Appendix C illustrates a form that can be used to capture the employee information used to populate the database.

Creating, wordprocessing, editing, proofreading, and updating résumés is a time-consuming and ongoing task. But without résumés readily accessible, much of your proposal response time will be consumed preparing them. And having more information than is required by the average RFP—such as grade point average, name and address of high school, previous supervisors and telephones, security clearances, years of supervisory experience, and professional references (including telephone numbers and e-mail addresses)[2]—is always preferable to having to search for that information under the pressure of preparing a proposal.

16.3 Project descriptions (project summaries)

A cumulative, regularly updated hardcopy and electronic file copy of your company's contracts, both as prime and sub, is a vital building block in the information foundation.

Create project descriptions at the time of contract award and update them at the completion of specific tasks, or semiannually, and then again at contract completion. As with résumés, storing hardcopies of each project description in a three-ring notebook allows ease of access. In the front of the binder, include a current list or matrix that includes all project description titles, contract/purchase order numbers, and client agencies in some logical order. It would be advisable to put your "boilerplate" project descriptions in the format required by your major client(s). One such format is presented in Figure 16.2. Specific information can always be added or removed in the tailoring process. Project or division managers should oversee and approve the updating process to ensure completeness and accuracy. As with the proposal response process, the updating of résumés and project summaries will require top-down management support, otherwise it will rarely be given the attention it deserves.

Project descriptions or summaries should focus on project successes—cost savings to the client, schedule adherence, awards (including repeat business from the same client), value-added service provided, and application of innovative technologies. *Be specific,* and *be quantitative* in your narrative descriptions. Highlight applicable lessons learned, such as the development of appropriate timesaving techniques or the application of certain automated project management tools.

As client evaluation factors for award are weighted increasingly toward past performance, well-written and photodocumented project summaries become more and more valuable. Clients buy from companies that have performed similar work and can demonstrate that fact effectively and appropriately in their proposals. You might consider investing in a matrix

---

2   To this listing might be added date of hire, current approved job title, total years of professional experience, maximum number and type of staff supervised, technical publications written, knowledge of MIL and DOD standards as well as industry-specific regulatory guidelines such as technology export compliance rules, experience working for or supporting specific federal agencies, general technical areas of expertise, and computer-related experience (subdivided to include such items as hardware; hardware operating systems; *Computer-Aided Software Engineering* (CASE) methodologies and products; application development tools and languages; communications protocols; communication controllers, hubs, and hardware; network operating systems; databases; system tools; and software).

Title:
Name of client:
Address of client:
Contract number:
Contract type:
Contract value (with options):
Period of performance (with options):

COTR:
Telephone:
E-mail:

Contracting officer:
Telephone:
E-mail:

Company point of contact (POC):
Brief description:
Detailed description:

Method of acquisition (competitive/non-competitive):
Nature of award (initial/follow-on):
Technical performance/accomplishments:
Schedule performance (adherence to program schedules):
Cost/price management history:

Termination history:
Client performance evaluation highlights:
Average number of personnel utilized per contract year:
Percentage turnover:
Incumbent capture:
Incumbent retention rate:

Figure 16.2
Effective project
summary format.

camera to take client-approved photographs on site for your company's projects.

Sometimes in small businesses, because of the corporate culture or the pathway of company growth, one division may be reluctant to use staff and contractual experience from other divisions when responding to a procurement. This is one example of the change of thinking that needs to occur as a small business enters the competitive arena. Each division of your company must harness fully the collective human and contractual expertise and knowledge base resident in the entire firm.

One important caveat: Be certain to obtain your client's permission to prepare an in-depth project summary about your company's support for that client organization. In particular, private-sector clients can be extremely sensitive about having the scope of work and specific points of contact made public. It is always the best policy to confirm what you intend to do with your client point of contact.

16.4 Proposal boilerplate ("canned" or "reuse" material)

Translate boilerplate into client-focused text.

As your company prepares more and more proposals, it would be well to extract certain "boilerplate" sections from past proposal documents, copy the text *and graphics* (see Figure 16.3) from these sections, and archive them in a central corporate library or proposal data center. Boilerplate must still be *tailored* for each application, but it represents a rich collection of "fodder" for the writers and planners of future proposals. Examples of boilerplate files include:

- Management plans;
- Technical overviews of your company's core competencies;
- Health and safety plans;
- Technology transfer plans;
- Professional compensation plan;
- Description of business processes and infrastructure (e.g., cost accounting, scheduling, tracking tools, invoicing procedures, purchasing mechanisms);
- *Configuration management* (CM) plans;
- Phase-in/phase-out plans;
- Design control plans;
- Software *quality assurance* (QA) plans;
- Data management procedures;
- Make-or-buy policy;
- Employee recruitment and retention plan;
- Company resources and facilities;
- Previous executive summaries;
- Client quotations and letters of recommendation;
- Computer equipment and resources;
- Overviews of support for specific clients;
- Engineering or manufacturing capability;

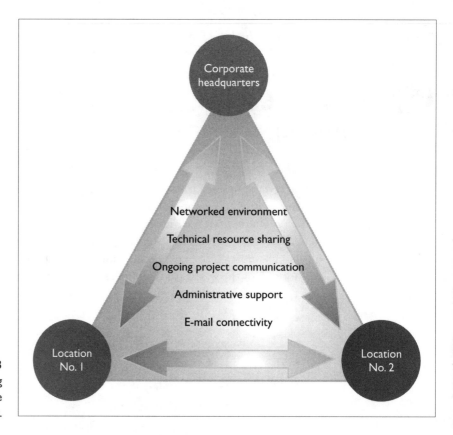

Figure 16.3
Integrated staffing
and resource
management.

- In-house training programs;
- Software development capability;
- Conflict of interest plan;
- Documentation capabilities;
- Small business subcontracting plan;
- Company financial condition (e.g., lines of credit);
- Security clearances (facility and individual);
- Company registration, certifications, and bonding.

Proposal writers should not have to start "from scratch" on every proposal effort. You might create an electronic "proposal toolbox" so that proposal writers can consult the latest reuse material under one electronic umbrella.

16.5 Marketing targets

As your company grows, it will reach a point at which it needs a systematic, formalized mechanism to track its business opportunities. A variety of excellent data base software packages, such as Microsoft Access or Lotus Notes [1], can be employed for this purpose. Database software allows for the generation of customized business reports and provides a quickly retrievable information source, for example, for planning and bid/no bid decisions. General access to the PC on which marketing targets reside should be controlled carefully through key lock and/or network password protection.

16.6 Corporate library

An important step in the transition from the 8(a) environment to the competitive world is to establish a centralized library or proposal data center in which previously submitted proposals and presentations, RFPs, proposal debriefings files, résumés, project summaries, technical materials, FOIA-requested materials (competitor proposals and monthly progress reports for specific projects), professional and trade journals, potential teaming partner information, public relations materials, and photographs reside. The library is best administered by the business development group.

Cataloguing materials can be done using a variety of alphanumeric schemes that best fit your needs and uniquely identify each informational entity in the library. For example, all proposal documents could be catalogued with the prefix "1," followed by a two-digit calendar year designator, a three-digit federal agency code of your choosing, and so forth. Be prepared to dedicate at least one large office with lateral and vertical file cabinets to your company's library. The volume of hardcopy information multiplies very rapidly. Using *magnetic*-backed drawer labels in the library is particularly advantageous. As your company's collection of materials increases in certain areas, the labels can be switched easily from drawer to drawer.

Personal experience has shown that storing a duplicate set of *diskette, CD-ROM, or tape copies* in appropriate containers along with the paper copies of proposals and other library documents is very beneficial. The original storage medium for a given document can reside with the publications group, but the duplicate medium can be used for electronic searches to locate specific verbiage, numbers, for example, that would be difficult to find manually. And having a backup of electronic files is always a good idea. An alternative to a duplicate set of stored diskettes, CD-DOMs, or tapes is network storage, such as an intranet or in Lotus Notes. Intranets allow companies to connect their own people as well as their vendors. "Participants using browsers inside and outside the company share a common interface and communications environment" [2]. All of a company's business processes can be executed online in a secure manner. Important intranet standards include TCP/IP and HTML. "[T]hink of the intranet's design just as you do the physical organization of your company. If there's a locked door in front of an office, lock that door to the virtual office. If you have a secretary acting as a gatekeeper in front of certain offices, require passwords to access those files in the virtual office. Model the virtual world on your real world" [3]. A significant advantage of an intranet is the "increased productivity that results from quick, companywide communications and data sharing.... Some of the many functions that intranets provide include repositories for volatile company information; easy access to company handbooks, guidelines, and forms of all kinds; and real-time calendaring and scheduling services" [4]. Resources for more information regarding intranets include www.lotus.com, www.microsoft.com, www.netscape.com, www.novell.com, and www.oracle.com.

As data volumes of proposal-related files increase due to high-end desktop publishing software and the use of integrated text and graphics, the average proposal can no longer be stored effectively and efficiently on high-density (1.44 MB) floppy diskettes. With files such as style files, graphics files, text files, and mirror files, a midsized set of proposal volumes produced with desktop publishing software can range from 40 MB to more than 200 MB. Storage media are now more appropriately CDs, removable drives, flopticals, magnetic tape, and multi-gigabyte hard drives.

Effective information management is a critical factor in achieving proposal and marketing success. It is essential that this activity within your company receives the attention and support of senior management. Although it appears administrative and mundane, nothing could be farther from reality. Even very large companies have not done a good job in documenting and tracking their human and contractual talent and experience. Your company, being small at present, has the opportunity to start on the right footing. Build an appreciation for the integrated nature of externally directed business development, Internal Sales support, information management, and corporate image.

To be effective, the information management activity within your company must receive the support of senior management.

16.7 Proposal "lessons learned" database

To ensure that your company will derive both short- and long-term benefits from both proposal successes as well as losses, it is advisable to develop and maintain an automated proposal lessons learned database (see Table 16.1) using commercially available software applications such as Lotus Notes or Microsoft Access. Proposal managers should be responsible for providing the specific information for the lessons learned database when proposals are won or lost. Once the data have been captured, they should be subject to senior-level review to help ensure impartiality.

By carefully tracking the reasons for wins and losses, you can begin to discern *patterns* in your business development performance with regard to specific client organizations, particular lines of business, certain types of contract vehicles, and proposal costing strategies. Are you consistently receiving low scores on ID/IQ contracts on which you submit proposals? Is your win percentage on Navy proposals particularly low in the Pacific Northwest? Does the lack of an established office near Wilmington, Delaware, seem to impact your chances of being selected for contract award for particular DuPont projects? Are your management plans missing the mark far too consistently? Those patterns must be discerned and communicated to appropriate business development and operations staff within your company, and then, most importantly, translated into prudent shifts in marketing, proposal development, and operational strategies and processes. In addition, through this database you can begin to generate a solid basis for estimating B & P costs for specific types and sizes of proposal efforts. This is critical for efficient staff and equipment resource planning and allocation, as well as more accurate B & P fiscal projections.

It is worth noting that your company must strive—from the top downward—to develop a business culture that allows for appropriate and constructive *introspection*. The lessons learned database will be most valuable when internal company politics do not skew the interpretation of the information that the database contains. For example, let's say that your company received a low score under the "Key Staff" evaluation factor because your proposed project manager had allowed a previous contract for the same client to run significantly over budget and behind schedule. Although this fact does not place this PM candidate in the best possible light, it is important to record the causes for your loss as completely and accurately as possible, and then act upon the patterns discerned in the causes. For

Client name

Brief scope of work

Place of project performance

Contract type (ID/IQ, CPFF, FFP, T&M, etc.)

Proposal due date

Internal proposal B&P number

Contract value to our company

Status of proposal (won or lost)

Proposal manager and other key staff

Was a debriefing conducted with the client organization?

Date of debriefing

Has our company worked for this client before? (yes or no)

What was our relationship with the client before the RFP was released?

Was our proposal determined by the client to be in the competitive range (shortlisted)?

Did we conduct advanced marketing with this client?

Winning contracting firms

Winning bid amounts ($)

Our bid ($)

B&P budgetΔ: estimate versus actual direct labor and ODCs

Specific reasons why we were not selected (check all that apply):
❏ Technical approach
❏ Lack of understanding of client mission and requirements
❏ Past performance
❏ Contractual experience
❏ Staff qualifications
❏ Management approach
❏ Company size
❏ Financial condition
❏ Direct labor costs
❏ ODCs
❏ Subcontracting plan
❏ Teaming arrangements
❏ Geographic location of staff, projects, and/or offices
❏ Missing information
❏ Late delivery of proposal
❏ Noncompliance with RFP-driven proposal structure, format, and/or page count

Additional relevant information

Client's source of feedback

Table 16.1
Sample Information
for a Proposal
Lessons Learned
Database

example, management plans in your future proposals may have to reflect specific strategies that your firm has implemented to monitor and control costs and schedule effectively and successfully.

16.8 Applying IT solutions: Evolutionary informational data systems

As your company grows, staff in various office locations need easily accessible and usable proposal-related information that can be modified to reflect local and regional requirements. Ongoing advances in information and communications technologies continue to facilitate data and information sharing across distributed locations and networks. Scenarios ranging from virtual proposal centers to CD-ROM data sharing have come into usage.

16.8.1 Lotus Notes scenarios

One powerful automated solution to this ever-growing requirement is a Lotus Notes-based domain of information we will refer to as *proposal building blocks*. Lotus Notes groupware provides a searchable, sortable document storage and management system for group use. As illustrated in Figure 16.4, the proposal building blocks are represented by icons on a Lotus Notes screen on a PC. From their desktops and with no third-party interaction, your field office or corporate staff can click on these icons to access proposal guidelines and key internal points of contact for support along with sample outlines, résumés and project summaries, technical narrative, and standard graphics. Each icon should have a HELP function associated with it to further enhance this system's self-containment. The building blocks should be designed to guide proposal managers and technical staff through the proposal-building process.

Drawing upon Lotus Notes' replication feature, electronic files can be updated, amended, or deleted from the proposal building blocks domain by a corporate-based systems administrator in a controlled manner and made available to your staff nationwide in real time. Everyone in the company will be able to access the latest materials. Such a system facilitates *localized* customization of proposals that are built upon a common, corporate-approved source domain of information. Distributed office locations can have access to the latest narrative on your company's business processes, résumés of recent new hires, key success stories from ongoing projects, and the most current letters of commendation from your clients.

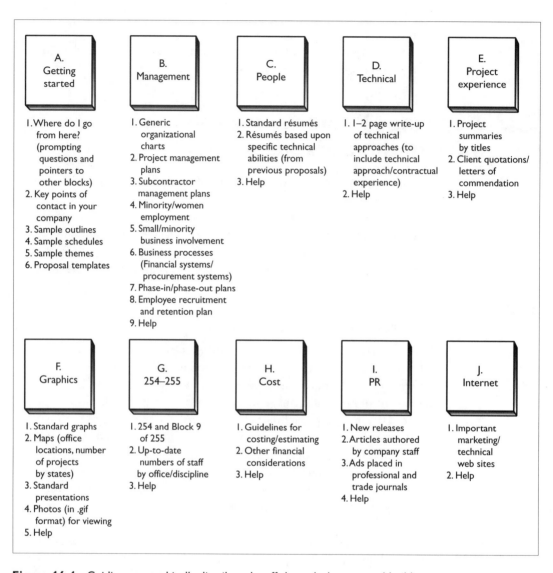

Figure 16.4 Guiding geographically distributed staff through the proposal building process.

16.8.2 CD-ROM scenarios

To address these same information requirements in very small companies, critical proposal files, which are password protected and/or encrypted, can be loaded onto a CD-ROM or removable drive (e.g., ZIP, JAZ). This technique can also be useful for staff who travel frequently and need rapid access to proposal information. Data security is the primary concern with this scenario of information sharing.

16.8.3 **Intranet scenarios**

An *intranet* is an internal information distribution system supported by a server or multiple servers that uses Web technology and graphical user interfaces. An intranet can be used to make the information depicted in Figure 16.4 available to your staff at multiple geographic and office locations. Search engine software such as the ht://Dig system (http://htdig.sdsu.edu/ main.html), developed at San Diego State University, and InfoMagnet from CompassWare Development (http://ipw.internet.com/search/InfoMagnet.html) can be employed to find and filter relevant information on your company's intranet.

16.9 **Leveraging federal performance appraisal systems to your company's benefit**

An important element of your company's proposal information repository is the evaluations and appraisals conducted by your clients about your project support. Make certain to request and archive the most recent appraisal results from the following three federal evaluation systems, as appropriate.

CCASS—the Construction Contractor Appraisal Support System— is an automated database of performance evaluations on construction contractors. The system provides past performance information for federal government contracting officers to aid them in their process of evaluating construction contractors' past performance. CCASS is a tool to track the performance of construction contractors throughout the DoD and federal government. It was developed for use in the process of determination of responsibility for construction contracts on sealed bid solicitations. CCASS was later mandated for use by all DoD agencies in the Defense Federal Acquisition Regulation Supplement (DFARS), Subpart 236.2. The DFARS implements the regulations within the Federal Acquisition Regulation. These requirements include that performance evaluations be prepared for each construction contract of $500,000 or more; or over $10,000 if the contract was terminated for default; that the performance evaluations be retained for 6 years; and that past performance evaluations may be reviewed as part of the determination of responsibility prior to award. Contractors should aggressively demonstrate in their proposals how they have addressed and corrected past faults.

The Service & Supply Contract Appraisal Support System (SSCASS) and the A/E Contract Appraisal Support System (ACASS) comprise the other two performance evaluation systems currently in use by the DoD.

Your company can certainly trumpet your excellent scores accrued under these federal performance appraisal systems in your proposals, as shown in Figure 16.5.

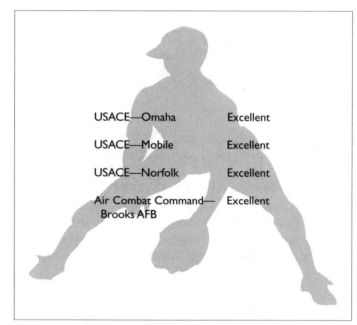

USACE—Omaha   Excellent

USACE—Mobile   Excellent

USACE—Norfolk   Excellent

Air Combat Command—   Excellent
Brooks AFB

Figure 16.5
Project
prowess—our
ACASS ratings are
consistently in the
strike zone.

End Notes

1. Elbert, Bruce, and Bobby Martyna, *Client/Server Computing*, Norwod, MA: Artech House, 1994.

2. Blankenhorn, Dana, "11 Lessons from Intranet Webmasters," *Netguide,* Oct. 1996, p. 82.

3. Ibid., p. 89.

4. Tittel, Ed, and James Michael Stewart, "The Intranet Means Business," *Netguide,* July 1996, p. 122.

Chapter 17

Leveraging business complexity in a knowledge-based economy

Knowledge = Productivity

17.1 Turbulent transition toward knowledge-based business

In merely 50 years, "the transistor, whose modest role is to amplify electrical signals, has redefined the meaning of power, which today is based as much on the control and exchange of information as it is on iron or coal. The throbbing heart of this sweeping global transformation is in the tiny solid-state amplifier invented by [John] Bardeen, [Walter] Brattain, and [William] Schockley. The crystal fire they ignited during those anxious postwar years has radically reshaped the world and the way in which its inhabitants now go about their daily lives" [1]. When processed individually and collectively, information is interpreted, congealed, repackaged, and produced into personal and organizational knowledge. And it is knowledge that has risen to prominence as the currency of the global economy late in the twentieth century. The timeworn aphorism that "cash is king" is undergoing emendation, and employees are being reconceptualized as knowledge or gold-collar workers who provide and possess valuable, strategically advantageous, and renewable intellectual capital. As British research associate Michael Lissack asserts, "In the future information will be the load-bearing structure of organizations." Economies of scale—bigness—are being replaced by economies of speed in this digital age. And Swedish Professor Johan Roos of the International Institute for Management Development (IMD) in Switzerland speaks of the "knowledge economy" and "knowledge landscapes."

The transition from the money-capital and management skill-capital eras to the intellectual-capital era has been concretized by a host of discrete decisions and events, including the creation of careers based upon knowledge management, measurement of (metrics for) knowledge-creating processes and intangible assets, and implementation of knowledge-sharing business cultures. Noteworthy among these events was when Skandia Assurance & Financial Services (AFS)—a major Stockholm-based insurance company—appointed Lief Edvinsson as the corporate world's first director of intellectual capital in 1991. Other companies and organizations at the vanguard of recognizing and acting upon the reality, and the value, of intellectual capital and "intangible assets" (e.g., corporate brain power, organizational knowledge, client relationships, innovation ability, and employee morale) include the Canadian Imperial Bank of Commerce; Booz-Allen & Hamilton; DuPont; and Hughes Space & Telecommunications, which have launched considerable efforts to understand and enhance intellectual capital management. In addition, Buckman Labs, a U.S.-based biotechnology firm, has created a Knowledge Sharing Department. Buckman Labs incentivizes those employees who contribute to its knowledge-sharing culture through financial rewards and management positions. Chevron has engineered a best-practice database to capture and make available the company's collective experience with drilling conditions and innovative solutions to technical problems on site. Sweden's Celemì company published the world's first audit of intangible assets in its 1995 Annual Report. Celemì's Intangible Assets Monitor focuses on its customers (image-enhancing customers, from whom testimonials are valuable; brand names; trademarks); their internal organization (patents, computer systems, management infrastructure); and their staff (competencies, flexibility). Skandia published the first ever annual report supplement on intellectual capital. Pfizer of Switzerland has created competence models for recruiting executives that include knowledge building and sharing as important criteria. WM-data of Sweden, a fast growing information technology (IT) company links non-financial indicators to strategy, and considers financial ratios of little use for management. The Swedish telecommunications company Telia has published an annual Statement of Human Resources for each of the past 8 years. This Statement includes a profit-and-loss account that visualizes human resources costs and a balance sheet that shows investments in human resources [2].

During a speech before the American Accounting Association's annual meeting held in Dallas, Texas, in August 1997, Michael Sutton—Chief Accountant of the United States Securities and Exchange Commission (SEC)—noted that "historically, accounting has been strongly influenced by the reporting needs of a manufacturing-based economy" [3]. In April 1996, the SEC had convened a symposium on intangible assets in Washington, D.C., during which invited participants from prestigious business, academic, and government organizations discussed issues related to the measurement of intangible assets by preparers of financial reports, concerns about disclosures related to intangible assets, and the experience of U.S. and overseas trendsetters with regard to the accounting and disclosure of intangible assets. As of January 1999, however, the SEC has not provided any guidelines or issued any directives vis-à-vis intangible assets for direct application in American corporations.

Converting human intellect into proposals, and ultimately into useful products and services, is becoming increasingly critical to any business' success [4]. "'We now know where productivity—real and limitless productivity—comes from. It comes from challenged, empowered, excited, rewarded teams of people. It comes from engaging every mind in the organization'" [5]. There is an accelerating recognition among corporations that their future profitability depends upon intangible assets such as "creativity, flexibility and the speed with which they can share new ideas and information. In a literal sense, modern corporations *are what they know*" [6]. Major firms such as Xerox Corporation, Hewlett-Packard Co., and Dow Chemical Co. are spending significant resources on knowledge identification and management. Alan Webber, a founding editor of the Boston-based business magazine *Fast Company* has noted that "[t]he world of business is realizing that ultimately, what matters is the quality of the people in the organization, and how they treat their customers and how creative they are in coming up with new products and new services"[7].

Another significant transition among such major corporations such as Microsoft, Compaq, Sun Microsystems, and Hewlett-Packard is the paradigm shift away from "business as a battlefield" to that of business as a "complex ecosystem in which the company that wins is the one that is quickest to adapt. Employee individuality and diversity are honored and encouraged" [8]. To adapt to new market circumstances and translate them quickly into meaningful opportunities, employees simply cannot be viewed and treated as mere "mindless troops," "crank-turners," or "line items on a traditional Excel or Lotus accounting spreadsheet." With small companies in particular, leveraging the collective intellect is integral to business development achievement and superlative, long-term client support. Providing work environments that are conducive to the generation, exchange, and respect of ideas will pay dividends in morale, employee retention, and financial revenue. Pleasant physical facilities; progressive human resources policies; an "open door" management culture; articulated, achievable proposal win and performance incentive programs; and so forth all participate in inspiring and harnessing the best in and from employees.

17.2 How to communicate effectively on your knowledge landscape

The command-and-control model... remains part of the baggage carried by many of our best executives.
—*The PeopleWise Organization*, Kepner-Tregoe, Inc.

Profound redefinition of the images we hold and the language we use to define ourselves as well as our companies, product lines, colleagues, competitors, proposals, business processes, and the entire economic spectrum is required to generate and sustain business health within our increasingly knowledge-based economy. Metaphors and language directly influence the manners in which corporate and project managers perceive the world and, consequently, the approaches by which they manage their companies, projects, and proposals. Transformed iconography, or "word pictures," can expand the domain of possible and realizable interactions and approaches. Business metaphors that include concepts of "war," "a race," "survival of the fittest," or "a jungle" and machine- and clock-like images (e.g., "go full steam on a project," "crank up the pace of activity," "C^3 or command, control, and coordinate," and "reengineering") will do little to enhance the fitness of American companies on a worldwide knowledge landscape that is subject to nondeterministic change, plateaus of relative stability, and coevolution. Open, complex systems such as company organizations demand concomitant open, flexible architectures of understanding and interaction that have matured far beyond the linear, top-down paradigms that served to generate business success from the Industrial

Revolution to the mid-twentieth century. Frederick Taylor's enduring principles of scientific management no longer provide valuable enhancement of human productivity.[1]

"Metaphors allow the transfer of bands of information where other means only transfer smaller bits" [9]. In effect, new words and metaphors constitute new linguistic domains. These new domains promote the conceptualization of avant garde models and discourse that in turn can guide business decision making on an economic landscape contoured functionally and in real time by human thoughts and ideas. Johan Roos and David Oliver assert that "contemporary management issues will increasingly be hindered by the lack of appropriate language to describe emerging organizational phenomena." This is precisely the point at which complexity theory may, in fact, offer senior managers and project managers a robust, multidimensional mode of understanding and decision making.

Within a complexity framework, a company might be conceptualized as a "complex adaptive organization," a shifting "constellation" (à la Henry Mintzberg) with independent and semiautonomous organizations. Machine-like, linear, cause-and-effect must be transformed to the nonlinear ecological interconnectedness of living organisms [10]. Rather than being fiscal liabilities, employees might be perceived to provide valuable intellectual capital and intangible assets within a structured organization and environment. The complex role of management then shifts to visualizing, articulating, measuring, leveraging, expanding, and creating options for the knowledge resources of the firm. Dr. Touraj Nasseri, a consultant and strategist for industry and government, suggests that mapping a company's intellectual capital and employing it productively to achieve strategic objectives are valuable initiatives for senior management to pursue [11]. Nasseri goes on to say that "[t]he potential inherent in well managed intellectual capital extends its impact well into the future as it adapts, renews, and replaces capabilities so that strategies remain responsive to rapid change and much uncertainty" [12]. "Management needs to develop systematic processes that stimulate languaging throughout the company so that, over

1   Frederick Winslow Taylor, M.E., ScD. (1856–1915), conceptualized a system that he termed *scientific management* to address "the great loss which the whole country is suffering through inefficiency in almost all of our daily acts." In his work *The Principles of Scientific Management* (1911), Taylor asserted that "the remedy for this inefficiency lies in systematic management, rather than in searching for some unusual or extraordinary man." Taylor's form of industrial engineering established the organization of work on Henry Ford's assembly lines in Michigan, which were the pride of the industrialized world at that time. However, many of Frederick Taylor's principles were misinterpreted or misapplied, and too often were translated into time-and-motion studies to extract more work from people for less pay. The migration away from labor-intensive mass production operations in American business and manufacturing has neutralized the effectiveness and applicability of Taylor's principles.

time, an internal lexicon is formed that can provide the foundation for an effective corporate identity" [13]. "Languaging," according to Professor Johan Roos, is the art of word choice. And Karl Sveiby introduces the term *knowledge management*, "the art of creating value from an organization's Intangible Assets." Traditional command-and-control management functions must now give way to mapping and acquiring intellectual capital, communicating a clear vision for the firm, developing and implementing a business lexicon with shared meanings and nuances to foster generalized understanding of strategic goals, and facilitating the rapid assessment and multidirectional flow of knowledge throughout the organization. Caveats and pronouncements issued from senior management will not be dispersed and internalized throughout the knowledge organization.

In the knowledge economy, companies should focus on adapting, recognizing patterns, and building "ecological" webs to amplify positive feedback rather than trying to achieve "optimal performance." Managers should attempt to discern interrelationships and recognize patterns rather than conduct forecasting based upon rationalist (read: Tayloresque) causes and effects. Hierarchical management schemas that rely heavily upon the paternalistic authority that marked the Industrial Era are best transformed into partnership models for the Knowledge Era that incorporate values such as cooperation, caring, creativity, empathy, connectedness, mutualism, and compassion. A sustainable competitive advantage is now maintained on a nonlinear, unpredictable economic landscape (i.e., a "rugged" knowledge terrain) through leveraging the collective knowledge base of a company and fostering a "learning organization" rather than simply effecting financial capital investment in infrastructure and institutional legacy systems.

At a practical level, sustainability at the end of the twentieth century follows from nurturing a business culture, infrastructure, and managerial approaches that encourage "self-organization." (Biologists use the term *self-organization* to describe organisms that continually adapt to their environment without losing their basic identity.) Fundamentally, there are no accurate predictions that pertain to a business landscape that is subject to multidimensional, emergent change and on which companies co-evolve, that is, mutually affect the development and direction of each other.

Michael Lissack correctly and profoundly observes that "[t]he descriptive metaphor that everything is changing and thus an organization must be poised to adapt to change says nothing about what to do next or about how to convert conflict into cooperation." What does the "everything is changing" metaphor mean at the level of hour-by-hour, day-to-day managerial decision making? Functionally, very little, according to this author's observation during the course of the past 12 years in the aerospace, defense, IT, and environmental marketplaces. Many American companies continue to operate according to a cluster of timeworn nineteenth-century paradigms, superimposed with a smattering of late twentieth-century management metaphors and buzzwords such as value innovation, total quality management (TQM), business process reengineering (BPR), continuous process improvement (CPI), and quality circles. The parental, autocratic, or mechanistic models of management, inextricably coupled with and often driven by manufacturing-related paradigms of accounting and asset evaluation, remain deeply entrenched and closely guarded. Business performance is still measured largely in financial terms, rather than collectively from a knowledge development perspective, infrastructure perspective, and customer perspective, as well as a financial perspective [14]. However, this situation presents the opportunity to think, and therefore act, in fundamentally different ways. Collectively, American companies are fully capable of inspired vision.

17.3 Envisioning supple business models

Creating open-architecture business models that can function effectively (i.e., converge on reality rather than diverge from real-world data and experience) and evolve within a knowledge-based economy will involve a shift in approach that spans several orders of magnitude. With information and knowledge becoming the "currency" of American business in the late twentieth century, the organizations that can share this new currency "ecologically" in an effort to arrive at effective and appropriate solutions for both their external and internal clients in the shortest possible time and with the least amount of resources expended will be those firms that occupy positions, or *optima*, above the knowledge landscape. From these vantages, companies will be less likely to be displaced by new technologies, reconfigured marketing paradigms, social and demographic shifts, and political and legislative climate changes.

Let us reconsider the notion of *clients*, for example. Within a traditional company, clients are perceived as the buyers or procurers of products or services being offered. In a knowledge-oriented company, clients include both external elements such as buyers, key subcontractors, regulatory agencies, and government institutions, as well as internal clients. The latter might include professional, project management and support staff; operational units; sales and marketing as well as proposal development elements; financial and accounting modules; senior management; and so forth. The currency of knowledge must flow in meaningful, multidirectional pathways among all of these "clients." Each client is critical to the sustainability and performance (i.e., positive energetics) of the company, and each requires, and in turn provides, vital information and knowledge to the other much as enzymes are shared within a living cell to perform a host of vital maintenance, replicative, and growth functions.

The traditional organizational diagram that most American companies use to illustrate internal structure will have little applicability in the knowledge-based economy. Organizational charts are by nature highly linear, pyramidal, and top-down in direction. However as Figure 17.1 depicts, businesses in the late twentieth century should strongly consider an ecological model with knowledge constituting the currency or fuel that is transferred among clusters of entities we will call: Clients, Business Process and Meta-Processes, Employees, and Finances. Table 17.1 suggests how traditional line and staff organizations can be reconceptualized according to this new model.

In this model, every cluster and module within the learning organization is afforded a pivotal complex of roles in ensuring the sustainability of

Traditional Organizational Element/Function	Reconceptualized Cluster in a Complex, Learning Organization	Reconceptualized Functions/Comments
Chairman of the Board Chief Executive Officer (CEO) President	Meta-Processes	Map intellectual capital, communicate a clear vision for the firm, develop and implement a shared business lexicon to foster generalized understanding of strategic goals, and facilitate the rapid assessment and multidirectional diffusion of knowledge throughout the organization. In effect, senior management must leverage and manage the intellectual capital of the company as well as conduct strategic planning activities.
Risk Management	Business Processes	Develop business continuation strategies that permit the transfer of core business-specific knowledge to a new management team.

Table 17.1 Reconceptualizing Organizational Elements Toward Knowledge-Based Framework

Traditional Organizational Element/Function	Reconceptualized Cluster in a Complex, Learning Organization	Reconceptualized Functions/Comments
Operations Manufacturing	Business Processes (now called "Process Owners")	Leverage and empower the intellectual capital of knowledge workers to generate localized, real-time solutions that can then be cataloged and transferred to other operational modules within your company.
Finance, procurement, travel, and facilities	Finances	Assess and report intellectual and structural capital in monetary terms. As Dr. Nasseri notes, "physical assets owe most of their value to intellectual capital" [15].
Human Resources	Business Processes (now called Human Capital Development)	Acquire, retain, train, foster continuing education for, and incentivize human intellectual capital. Develop compensation policies appropriate for knowledge workers. Ensure that employees are always climbing a "learning curve."*
Information Services (IS)	Business Processes	Facilitate the rapid assessment and multidirectional flow of knowledge throughout the organization via data warehousing, data mining, and data filtering supplemented by ongoing training.
Sales & Marketing Secondary Marketing Business Development Advanced Planning	Business Processes	Assess and communicate the economic landscape, develop and share knowledge of clients, conduct benchmarking[†] and competitor assessments, define client and product lines (congealed knowledge), and design and develop proposals.

* Far from being a detriment, a "learning curve" within a learning organization is a prerequisite to continued sustainability.

[†] Benchmarking is a rigorous process for linking competitive analysis to your company's strategy development. Benchmarking is a method that measures the performance of your "best-in-class" competitors relative to your industry's key success factors. It also is a mechanism for determining how the best-in-class achieve those performance levels and competitor assessments, define client and product lines (congealed knowledge), and design and develop proposals.

Table 17.1   (continued)

Traditional Organizational Element/Function	Reconceptualized Cluster in a Complex, Learning Organization	Reconceptualized Functions/Comments
Legal Support (contracts, insurance, litigation, compliance and regulatory support) Auditing	Business Processes	Congeal intellectual capital as patents, software source code, published papers and books, marketable user manuals, research papers, productizable business processes and protocols, etc.
Project Management[‡]	Business Processes (now called "Knowledge Integrators" or "Knowledge Managers" rather than Project Managers)	Serve external and internal clients; implement client-focused vision; empower staff for success; and foster accountability, commitment, and ownership. Serving external clients extends to assembling appropriate knowledge sources from within the organization and matching those sources to external client requirements on budget and within the required timeframe. Internal client servicing includes reporting requirements.
Administrative Support	Business Processes	Produce congealed knowledge products in the form of proposals, contract deliverable reports, marketing presentations, annual and other SEC-required reports, client correspondence, etc.
Internal Research & Development (IR&D) Product Development	Business Processes	"Innovation in a company is nourished and driven by knowledge-based capabilities and by management systems that leverage the capabilities." [16].
Health & Safety Quality Control/Quality Assurance	Business Processes	

[‡] Too often in American business, the functional reality is that project managers are technical staff who also happen to sign time sheets. Their focus is on technical nuts-and-bolts, frequently to the exclusion of client interaction, staff empowerment, knowledge sharing, and so forth.

Table 17.1   (continued)

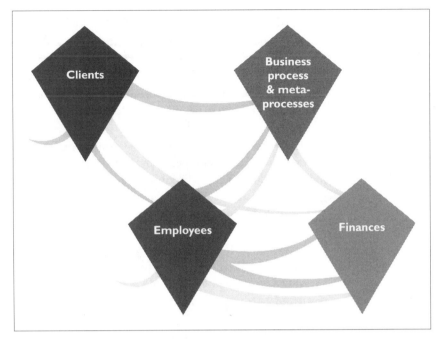

Figure 17.1
Knowledge fuels complex, ecological business interactions.

the company. Everyone is involved with and contributes to the new currency—knowledge, which is quite unlike the traditional paradigm wherein any element not involved in the most linear manner with generating cash is perceived as a "leech on the corporate purse." In fact, within a knowledge-based and forward-looking organization, intangible assets are considered quantitatively along with such tangible assets as cash, accounts receivable, computers and other amortizable equipment, facilities, and inventories.

17.4 Sample application: Tracing complexity and knowledge management through the proposal development process

Technically, a proposal is an offer prepared by a contractor (such as your company) to perform a service, supply a product, or a combination of both in response to a Request for Proposal (RFP) document by a procuring organization within government or private industry. Proposals are legal documents, often incorporated by reference into the final contract. That means that they must be factual and accurate—including information provided in résumés and project summaries, which summarize applicable contractual experience. In actuality, résumés and project summaries reflect a significant percentage of intellectual capital of a service organization. Relevant knowledge resident in the employees of the company as gained through education, training, and professional experience coupled with initiative, dedication, and innovation should be highlighted in each résumé. Similarly, to be effective, project summaries must reflect the organizational knowledge of the firm from technical, managerial, and geographic perspectives, to name but several.

Proposals are, first and foremost, knowledge products that include a host of marketing, technical, programmatic, institutional, pricing, and certification information. Through a choreographed process of knowledge generation, transfer, and congealment, a proposal is designed to sell both technical and managerial capabilities of a company to accomplish all required activities on time and at a reasonable cost. Your company's proposal document(s) is scored, literally, by the client's evaluators against formalized, specific standards in a process analogous to pattern recognition. A proposal is the tangible result of knowledge-building processes, supported in turn by hard and directed work and buoyed by a positive collective attitude within your company's proposal team of knowledge workers.

Crucial to proposal success both as a process and a product is the transfer of relevant marketing information to appropriate members of the proposal team. From there, that information must be interpreted, assigned value, and translated into knowledge that will ultimately be congealed into the actual proposal documents—into narrative, theme statements, graphics, captions, compliance matrices, tabs, and so forth. If marketing data and information do not find their way into the knowledge product called a proposal, then a company's proposal will not receive the maximal point score. Therefore, the marketing data are valueless. In effect, if intellectual capital measured in client, project, competitor, and political knowledge is not transferred to appropriate staff (knowledge workers) and converted efficiently into hardcopy or electronic proposal documents via established, articulated business processes, your company's proposal win ratios in terms of dollars (X dollars of contracts awarded for every $100 of contract value pursued) and number (X number of proposals won) will be suboptimal.

17.5 Summation

Evidence "suggests that the business world may be a complex system poised at the edge of chaos" [17]. If we grant that such is the case, then we who are an integral part of this world must leverage the complexity with which we are faced every day into optima (that is, business processes, business thinking, and the like) that lift us above the unpredictable knowledge landscape and ensure business sustainability. Implementing business processes that actively promote knowledge valuation, development, transfer, management, and congealment will help American companies maximize both intellectual and structural capital as we hurtle toward the twenty-first century.

We now turn our attention to a specific type of contracting with the federal government—the important SF254/255 solicitation process. These solicitations are presented in the Federal Acquisition Regulations (FAR) at Part 53.236-2, Architect-engineer services. SF254/255 documents are prepared for business opportunities with the federal government as well as states, municipalities, and quasi-governmental agencies such as port and transit authorities.

END NOTES

1. John Riordan and Lillian Hoddeson, *Crystal Fire: The Birth of the Information Age*, New York: W. W. Norton & Co., 1997, p. 10.

2. Information in this paragraph was drawn from Karl E. Svieby, "What Is Knowledge Management?," pp. 5–6. http://www.sveiby.com....owledgeManagement.html as well as Sacha Cohen, "Knowledge Management's Killer App," http://www.astd.org/magazine/current/cohen.htm and Laton McCartney, "Getting Smart About Knowledge Management," *Industry Week*, 14 May 1998, p. 32.

    See also Karl E. Sveiby, "Celemì's Intangible Assets Monitor," http://www.sveiby.com....Ass/CelemiMonitor.html. Celemì employs metrics such as efficiency (sales per customer, staff turnover), stability (repeat orders, growth in sales per administrative staff), and growth/renewal (average years of professional competence). Celemì is headquartered in Malmö, Sweden, and is dedicated to creating processes that help companies leverage the power of learning.

    Sacha Cohen notes that Hewlett-Packard employs a Web-based system called Connex to help identify subject experts in specific geographic locations. And Booz-Allen uses a best-practice application on its intranet, Knowledge On Line (KOL), an automated knowledge repository in which that firm's collective knowledge and expertise are captured, classified, and quantified.

3. Sutton, Michael H.,"Dangerous Ideas: A Sequel," Remarks delivered during the American Accounting Association 1997 Annual Meeting in Dallas, Texas, on 18 August 1997, p. 3; http://www.sec.gov/news/speeches/spch175.txt.

4. Quinn, James Brian, Philip Anderson, and Sydney Finkelstein, "Managing Professional Intellect: Making the Most of the Best," *Harvard Business Review*, Mar./Apr. 1996.

5. Quinn, Judy, "The Welch Way: General Electric CEO Jack Welch Brings Employee Empowerment to Light," *Incentive*, Sept. 1994, p. 50.

6. Hamilton, Martha A., "Managing the Company Mind: Firms Try New Ways to Tap Intangible Assets Such as Creativity, Knowledge," *Washington Post*, 18 Aug. 1996, p. H1.

7. Ibid., p. H5.

8. James, Geoffrey, "It's Time to Free Dilbert," *New York Times*, 1 September 1996, p. F-11.

9. Lissack, Michael R., "Complexity Metaphors and the Management of a Knowledge Based Enterprise: An Exploration of Discovery," http://www.lissack.com/writings/proposal.htm.

10. Stout, Donna M., "Focus on Systems: Reviews of Books on System Theory," http://www.wintu.edu/journals/wiuj0101-6.html, p. 2. (citing Margaret Wheatley).

11. Nasseri, Touraj, "Knowledge Leverage: The Ultimate Advantage," *Kognos: The E-Journal of Knowledge Issues*, Summer 1996, p. 4.

12. Ibid., p. 2.

13. Roos, Johan and Georg von Krogh, "What You See Depends on Who You Are," *Perspectives for Managers*, No. 7, Sept. 1995, p. 2.

14. Roos, Göran and Johan Roos, "Intellectual Performance: Exploring an Intellectual Capital System in Small Companies," 30 Oct., 1996, pp. 4, 9; http://www.imd.ch/fac/roos/paper_lr.html.

15. Nasseri, "Knowledge Leverage," p. 1.

16. Nasseri, "Knowledge Leverage," p. 2.

17. Phelan, Steven E., "From Chaos to Complexity in Strategic Planning," Presented at the 55th Annual Meeting of the Academy of Management, Vancouver, British Columbia, August 6–9, 1995, p. 7; http://comsp.com.latro...u.au/Papers/chaos.html.

Chapter 18

Planning and producing SF254/255 responses for architect-engineer services

18.1 SF254/255 and the FAR

TheFAR at Part 53.236-2, A-E services, describes a special type of solicitation that has *broad-based* applicability at the federal, state, municipality, and quasi-governmental levels such as port authorities and transit authorities. This type of solicitation is called a SF254 and SF255. Architectural and engineering firms of all sizes submit these documents routinely to establish their credentials with client organizations as diverse as the *U.S. Coast Guard* (USCG), *U.S. Army Corps of Engineers* (USACE), U.S. Navy, *U.S. Department of Agriculture* (USDA), the National Forest Service, and Harford County, Maryland. Presented in Figures 18.1 to 18.9

STANDARD FORM (SF) **254** Architect-Engineer and Related Services Questionnaire	1. Firm Name/Business Address:	2. Year Present Firm Established	3. Date Prepared:
		4. Specify type of ownership and check below, if applicable.	
1a. Submittal is for ☐ Parent Company ☐ Branch or Subsidiary Office		A. Small Business ☐	
		B. Small Disadvantaged Business ☐	
		C. Woman-owned Business ☐	

5. Name of Parent Company, if any:

5a. Former Parent Company Name(s), if any, and Year(s) Established:

6. Names of not more than Two Principals to Contact: Title/Telephone
1)
2)

7. Present Offices: City / State / Telephone / No. Personnel Each Office

7a. Total Personnel _____

8. Personnel by Discipline: (List each person only once, by primary function.)

___ Administrative	___ Electrical Engineers	___ Oceanographers
___ Architects	___ Estimators	___ Planners: Urban/Regional
___ Chemical Engineers	___ Geologists	___ Sanitary Engineers
___ Civil Engineers	___ Hydrologists	___ Soils Engineers
___ Construction Inspectors	___ Interior Designers	___ Specification Writers
___ Draftsmen	___ Landscape Architects	___ Structural Engineers
___ Ecologists	___ Mechanical Engineers	___ Surveyors
___ Economists	___ Mining Engineers	___ Transportation Engineers

9. Summary of Professional Services Fees
Received: (Insert index number)

Last 5 Years (most recent year first)

19____   19____   19____   19____   19____

Direct Federal contract work, including overseas _____
All other domestic work _____
All other foreign work* _____

*Firms interested in foreign work, but without such experience, check here: ☐

Ranges of Professional Services Fees
INDEX
1 Less than $100,000
2. $100,000 to $250,000
3. $250,000 to $500,000
4. $500,000 to $1 million
5. $1 million to $2 million
6. $2 million to $5 million
7. $5 million to $10 million
8. $10 million or greater

STANDARD FORM 254 PAGE 4 (REV. 11-92)

Figure 18.1 SF254.

10. Profile of Firm's Project Experience, Last 5 Years

Profile Code	Number of Projects	Total Gross Fees (in thousands)	Profile Code	Number of Projects	Total Gross Fees (in thousands)	Profile Code	Number of Projects	Total Gross Fees (in thousands)
1)			11)			21)		
2)			12)			22)		
3)			13)			23)		
4)			14)			24)		
5)			15)			25)		
6)			16)			26)		
7)			17)			27)		
8)			18)			28)		
9)			19)			29)		
10)			20)			30)		

11. Project Examples ... Last 5 Years

	Profile Code	"P," "C," "JV," or "IE"	Project Name and Location	Owner Name and Address	Cost of Work (in thousands)	Completion Date (Actual or Estimated)
1						
2						
3						
4						
5						
6						
7						

STANDARD FORM 254 PAGE 6 (REV. 11-92)

Figure 18.2 SF254 (continued).

20												
21												
22												
23												
24												
25												
26												
27												
28												
29												
30												

12. The foregoing is a statement of facts

Signature: _____ Typed Name and Title: _____ Date:

STANDARD FORM 254 PAGE 7 (REV. 11-92)

Figure 18.3 SF254 (signature page).

STANDARD FORM (SF) 254

Architect-Engineer and Related Services Questionnaires for Specific Project

1. Project Name/Location for which Firm is Filing:

2a. Commerce Business Daily Announcement Date, if any:

3b. Agency identification Number, if any:

3. Firm (or Joint-Venture) Name & Address

3a. Name, Title & Telephone Number of Principal in Contact

3b. Address of office to perform work, if different from item 3

4. Personnel by Discipline: (List each person only once, by primary function.) Enter proposed consultant personnel to be utilized on this project on line (A) and in-house personnel on line (B).

(A)___(B)___	Administrative	(A)___(B)___	Electrical Engineers	(A)___(B)___	Oceanographers
(A)___(B)___	Architects	(A)___(B)___	Estimators	(A)___(B)___	Planners: Urban/Regional
(A)___(B)___	Chemical Engineers	(A)___(B)___	Geologists	(A)___(B)___	Sanitary Engineers
(A)___(B)___	Civil Engineers	(A)___(B)___	Hydrologists	(A)___(B)___	Soils Engineers
(A)___(B)___	Construction Inspectors	(A)___(B)___	Interior Designers	(A)___(B)___	Specification Writers
(A)___(B)___	Draftsmen	(A)___(B)___	Landscape Architects	(A)___(B)___	Structural Engineers
(A)___(B)___	Ecologists	(A)___(B)___	Mechanical Engineers	(A)___(B)___	Surveyors
(A)___(B)___	Economists	(A)___(B)___	Mining Engineers	(A)___(B)___	Transportation Engineers
				(A)___(B)___	Total Personnel

5. If submittal is by JOINT-VENTURE list participating firms and outline specific areas of responsibility (including administrative, technical and financial) for each firm: (Attach SF 254 for each if not on file with Procuring Office)

5a. Has this Joint-Venture previously worked together? ☐ Yes ☐ No

STANDARD FORM 255 PAGE 3 (REV. 11-92)

Figure 18.4 SF255 (Blocks 1 to 5).

6. If respondent is not a joint-venture, list outside key Consultants/Associates anticipated for this project (Attach SF 254 for Consultants/Associates listed. If not already on file with the Contracting Office).

Name & Address	Specialty	Worked with Prime before (Yes or No)
1)		
2)		
3)		
4)		
5)		
6)		
7)		
8)		

STANDARD FORM 255 PAGE 4 (REV. 11-92)

Figure 18.5 SF255 (Block 6).

7. Brief resume of key persons, specialists, and individual consultants anticipated for this project.

a. Name & Title:

b. Project Assignment:

c. Name of Firm with which associated:

d. Years experience: With This Firm...... With Other Firms......

e. Education: Degree(s)/Year/Specialization

f. Active Registration: Year First Registered/Discipline

g. Other Experience and Qualifications relevant to the proposed project:

a. Name & Title:

b. Project Assignment:

c. Name of Firm with which associated:

d. Years experience: With This Firm...... With Other Firms......

e. Education: Degree(s)/Year/Specialization

f. Active Registration: Year First Registered/Discipline

g. Other Experience and Qualifications relevant to the proposed project:

STANDARD FORM 255 PAGE 8 (REV. 11-92)

Figure 18.6 SF255 (Block 7).

8. Work by firms or joint-venture members which best illustrates current qualifications relevant to this project (list not more than 10 projects).

a. Project Name & Location	b. Nature of Firm's Responsibility	c. Project Owner's Name & Address and Project Manager's Name & Phone Number	d. Completion Date (actual or estimated)	e. Estimated Cost (in Thousands)	
				Entire Project	Work For Which Firm Was/Is Responsible
(1)					
(2)					
(3)					
(4)					
(5)					
(6)					
(7)					
(8)					
(9)					
(10)					

STANDARD FORM 255 PAGE 9 (REV. 11-92)

Figure 18.7 SF255 (Block 8).

9. All work by firms or joint-venture members currently being performed directly for Federal agencies.

a. Project Name & Location	b. Nature of Firm's Responsibility	c. Agency (Responsible Office) Name and Address and Project Manager's Name & Phone Number	d. Percent Complete	e. Estimated Cost (in Thousands)	
				Entire Project	Work For Which Firm Is Responsible

STANDARD FORM 255 PAGE 10 (REV. 11-92)

Figure 18.8   SF255 (Block 9).

10. Use this space to provide any additional information or description of resources (including any computer design capabilities) supporting your firm's qualifications for the proposed project.

11. The foregoing is a statement of facts

Signature: _____ Typed Name and Title: _____ Date:

STANDARD FORM 255 PAGE 11 (REV. 11-92)

Figure 18.9 SF255 (Blocks 10 and 11).

are the forms and formats in which the government wants to receive the specific information regarding, for example, contractor personnel, office locations, and contract experience. These forms are available at the Web site http://www.gsa.gov/regions/r11/wpc/html/procurement/forms.htm. The SF254 may be used as a basis for selecting firms for discussions, or for screening firms preliminary to inviting submission of additional information.[1] The SF255 is a supplement to the SF254. Its purpose is to provide additional information regarding the qualifications of interested firms to undertake a specific federal A-E project.

18.2 Understanding the required structure of the response

SF 254/255[2] submittals are prepared in response to synopses for A-E services that appear in the *CBD*. Although generally prepared for Federal Government agencies such as the USACE, USCG, U.S. Army, or specific Air Force bases, SF254/255s or modified versions of them can be required submittals for states and municipalities as well. An example is the Form CEB 02, A-E and Related Services Questionnaire for Specific Project (City of Baltimore), which is analogous to the SF254/255. One major difference in contractor response to a government RFP versus an SF254/255 synopsis is that labor rates and other direct costs are included with a proposal but are *not* part of an SF254/255 response.

18.3 Overall strategy of response

The SF254/255 submittal consists essentially of a series of 11 blocks (SF255) plus the SF254. The content of the blocks is provided in Figure 18.10. As with response to RFPs, you should follow the guidance provided in the *CBD* synopsis very carefully, including references to Notes.

---

1 A note about Block 11 on the SF254, which requires "Project Examples, Last 5 Years." The letters "P," "C," "JV," and "IE" in the second column stand for "Prime Professional," "Consultant," "Joint Venture," and "Individual Experience," respectively. These designations are to be used to describe your company's role on each particular project listed in Block 11. For firms in existence less than five years, "Individual Experience" may be used rather than company experience.

2 SF255 is known formally as A-E and Related Services Questionnaire for Specific Project. SF254 is known formally as A-E and Related Services Questionnaire.

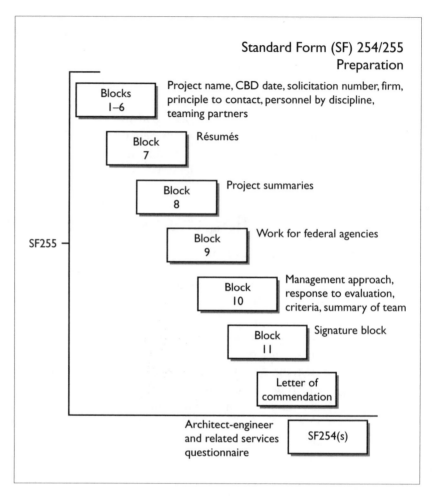

Figure 18.10
Overview of
SF254/255.

For example, the synopsis may ask that an offeror see Note 24.[3] Notes refer to explanatory and amplifying comments that appear at the back of the hardcopy version of the *CBD*. These notes are updated periodically. Look for specific requirements that will impact how you complete the

---

3   Note 24 states that "Architect-Engineer firms that meet the requirements described in this announcement are invited to submit: (1) a Standard Form 254, Architect-Engineer and Related Services Questionnaire, (2) a Standard Form 255, Architect-Engineer and Related Services questionnaire for Specific Project, when requested, and (3) any requested supplemental data to the procurement office shown. Firms having a current Standard Form 254 on file with the procurement office shown are not required to register this form." This note continues by explaining evaluation criteria and that responses submitted prior to the closing date of the announcement will be considered for selection, subject to certain limitations. Other notes stipulate such items as small and SDB requirements and proposal mailing and packaging instructions.

various blocks of the SF254/255 response. For example, the synopsis may state that all staff who work on the contract must be within 50 miles of a particular geographic location. This will affect how you complete Block 4, Personnel by Discipline. It will also affect Block 7, Personnel, as well as Block 10.

When strategizing your response to an SF254/255 opportunity, follow the steps as outlined in Figure 18.11, beginning with Block 7. During the pre-kickoff planning phase, reach a consensus regarding the personnel you will be proposing and exactly what position on the project organizational chart they will occupy. Particularly critical to decide upon are the program/project manager and the delivery/task order managers. Attempt to structure the project organizational chart in accordance with the government agency's organization and in compliance with the technical subareas named in the *CBD* synopsis. If, for example, the government agency is divided into three primary elements, then your company's project organization might reflect three similarly named elements as well. Your marketing intelligence should indicate the appropriate number of delivery/task order managers to propose. Does the particular government agency want streamlined contractor project management structures? In that case, one or two delivery/task order managers for each major organizational element might be appropriate. When selecting personnel for Block 7, keep in mind that these staff should have been integrally involved in the projects that you will select in Block 8. The government wants to be assured that the staff your company is proposing have actually worked on the contracts that you are presenting as similar technical/programmatic experience in Block 8.

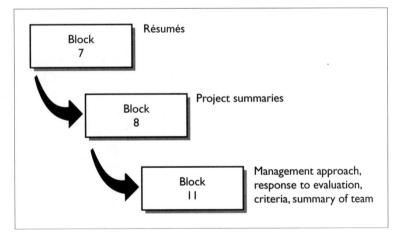

Figure 18.11
Strategizing to win
the SF254/255.

Once the project organizational chart has been constructed for Block 7, your next step is to identify 10 relevant projects for inclusion in Block 8. Most government agencies limit the number of projects to 10, although there are occasional exceptions.[4] When the applicable projects have been identified, your proposal manager can then turn his or her attention to Block 10. Most companies and individual proposal managers make the mistake of beginning their response to an SF254/255 with Block 10 because this is the technical heart of the document. However, it is extremely difficult to design a successful 254/255 response without first having made the pivotal decisions regarding Blocks 7 and 8. Remember that evaluators are reviewing 30 to 50 254/255s for even small procurements. They may spend only 30 min on any one proposal during their preliminary evaluation sequence. That is all the time your company will get to make a positive, lasting impression.

18.4 Build Block 7 first

To examine Block 7, Personnel, in more detail, résumés need to be placed into the special SF255 format as dictated by the FAR. Full-page résumés are generally required for the project/program manager, senior technical reviewer(s), QA director, and delivery/task order managers. Half-page résumés are used for other proposed technical staff. Avoid showing one person under two or more different technical areas, as this can be interpreted that your company does not have adequate technical depth to perform the work. The project organizational chart should appear at the beginning of Block 7.

You should consider including two text boxes in each résumé: one with the heading of "Basis of Team Selection" and the other with the heading of "Relevant Block 8 Projects." The former box should be a listing of three to five bulleted items. These bullets should highlight the relevant technical and programmatic accomplishments that provided the rationale for why the particular person was selected as a member of the proposed staff. Refer to the *CBD* synopsis frequently when preparing the "Basis of Team Selection" box. For example, for the proposed project manager or task managers, include statements that map directly to management-related evaluation criteria in the synopsis. Consider the following example.

CBD Synopsis: MANAGEMENT CAPABILITIES—ability to manage fluctuating workloads. Demonstrated ability to implement quality control procedures.

---

4   In a 1997 SF254/255 solicitation, Tinker A.F.B. in Oklahoma allowed up to 50 projects in Block 8!

Bullets for "Basis of Team Selection" Box: Highlight specific projects that the person has managed in which he or she applied QC procedures and in which he or she responded to changing levels of work.

- Applied effective quality control procedures on two $5M CPFF FAA contracts that resulted in accelerated design review processes.
- Met challenging, long-term systems engineering workload fluctuations while managing a $2.5M project for the Department of Transportation in New England.

The "Relevant Section" box should list the projects that appear in Block 8 for which the person has provided direct technical or programmatic support.

18.5 Block 8: Selling your project experience

Block 8 generally contains one critical figure along with the 10 project summaries in SF255 format. This figure is the Project Experience Matrix, which lists the 10 projects you have selected as key for this procurement on the *y*-axis and the relevant technical areas as enumerated in the *CBD* synopsis on the *x*-axis. The object of this figure is to place as many dots or bullets onto it as possible. You want to demonstrate maximal technical competence across the broad waterfront of technical areas and disciplines.

In selecting the 10 projects, you might apply the following criteria to determine their relevancy to the particular procurement.

- Same client (*USACE, Portland District*);
- Same agency (*USACE*);
- Same technical work (*landfill closure*);
- Same geographic location (*Pacific Northwest*);
- Same type of contract vehicle (*Indefinite Delivery/Indefinite Quantity* (ID/IQ)).

As with responding to RFPs, attempt to photodocument each project summary with relevant and appropriate imagery. Be sure that photos do not show health and safety violations or regulatory violations. And be certain to use photographs to which your company owns full rights so that you do not violate intellectual property laws. Stock photos can also be purchased for specific uses from photo houses in major cities nationwide or via the Internet.

You should consider including two text boxes in each project summary: "Relevance to the Requirements" and "Relevant Block 7 Résumés." For the "Relevance to the Requirements" box, list bulleted items that link your project directly to the technical and past performance evaluation criteria that appear in the *CBD* synopsis.

18.6 Block 10: Structure according to the evaluation criteria

Your company's response to the technical requirements and evaluation criteria found in the *CBD* synopsis is the foundation of Block 10. Ensure that each of the technical areas in the synopsis is addressed in your response. Use the verbiage of the particular synopsis to build the headings and subheadings of Block 10. In responses that are not page limited, Block 10 affords an opportunity to present additional project-specific experience to amplify and expand that which was provided in Block 8. There is one caveat, however. As with all proposal writing, keep the narrative in Block 10 crisp and focused on the requirements and how your company can meet them fully. A 20 page Block 10 is sufficient and preferable for most SF255 responses.

Block 10 also includes your company's project management plan. Link this narrative to the staff proposed in Block 7. Highlight the advantages of selecting your company at the conclusion of Block 10. Provide authenticated benefits of your technical and management approach to the client. In addition, Block 10 should include the specific rationales for why your company has teamed with other firms (if this is the case) in order to propose on this procurement.

18.7 Block 10 outlining

Let us suppose that the USCG, your client, has released a synopsis in the *CBD* for miscellaneous environmental services. The weighted evaluation criteria are as follows: (a) technical excellence; (b) prior experience and past performance; (c) management capabilities; (d) personnel qualifications; and (e) computer capabilities. How should you outline Block 10?

Build the Block 10 outline around these stated Evaluation Criteria, making sure to include subheadings for each one of the points stated under a given Evaluation Criterion. For example, if the management capabilities criterion reads as follows:

(c) Management capabilities (23%) as demonstrated by internal quality control and quality assurance procedures used to ensure the technical accuracy and the coordination of disciplines, the structure of the firm as it relates to the overall approach to project management, capability to perform work in house, and ability to manage fluctuating workload.

You might consider outlining this section of Block 10 as shown in Table 18.1.

10.3	Management Capabilities (Evaluation Criterion c)
	10.3.1 Overview
	10.3.2 Management of Projects of Similar Size and Scope
	• Cost-effective strategies
	• Time-efficient strategies
	10.3.3 Structure of the Firm
	10.3.4 Project Execution Strategies
	10.3.4.1 Project Initiation and Kickoff
	10.3.4.2 Internal QC/QA Procedures
	• QC/QA for Deliverables
	10.3.4.3 Ability to Manage Fluctuating Workloads
	10.3.4.4 Cost and Schedule Control Approaches
	10.3.4.5 Performance Measurement and Control Systems
	10.3.5 In-House Capabilities
	10.3.6 Project Leadership
	• Project Manager's Role and Span of Authority
	• USCG-Company Interfaces
	• Monthly Reporting
	10.3.7 Subcontractor Management

Table 18.1
Block 10:
Management
Section Outline

18.8 Other Blocks

Block 1, the project name, as well as Blocks 2a and 2b, should be taken directly from the *CBD* synopsis. Block 3, the firm name & address, is your headquarters location, whereas Block 3b should indicate your office location that will actually oversee the project. Block 4, personnel, contains two columns, A and B. Column A should include subcontractor personnel when your company is the prime. (Some federal agencies want *all* subcontractor personnel to be listed; others want only the number representing those subcontractor staff who will actually be assigned to support the project.) Column B would then include the appropriate numbers of staff from your company. Block 6 is a listing of subcontractor companies and their area of speciality for the proposal at hand. The listing should be in descending order based upon the percentage of the contract that each subcontractor will receive. If subcontractor A is to receive 8% of the work and subcontractor B is to receive 5%, subcontractor A is listed first.

Block 9 is a listing of federal contract work currently being performed. Your company's contracts department should be responsible for maintaining the accuracy and completeness of this information.

Block 11 must be signed by a senior manager who is authorized to represent your company. Your company's SF254 must also be signed by an authorized senior manager. Build in adequate time to secure these signatures as well as a signature on the transmittal letter.

Your ongoing information management activity should include collecting and archiving letters of commendation from the government and private-sector clients for use in your proposals and responses to SF254/255s. Consider creating an electronic subdirectory of high-impact client quotations that will help to authenticate your marketing documents. Ensure that you have your clients' permission to quote from their letters of commendation.

Consider building and maintaining an electronic subdirectory of high-impact client quotations to authenticate your proposals and SF254/255s.

18.9 Subcontractor participation

During the pre-kickoff planning phase of the SF254/255 process, ensure that written teaming agreements are in place with all of your subcontractors. Call on them during the writing phase to provide résumés for Block 7, project summaries for Block 8 (with photos), a company overview and technical information for Block 10, a current SF254 with signature, any appropriate letters of commendation, and a letter of commitment from the president of the subcontracting firm to your company's president or senior vice president. (See Table 18.2 for a comprehensive list of materials required and guidelines for submittal.) Include résumés and project summaries from your subcontractors in accordance with the percentage of the

A. Software Application Requirements

1. Submit all text in *Microsoft Word 97*.

2. Submit all line art graphics as *CorelDRAW!* files or Windows Metafiles.

3. Any photos, submit as .tif files or as 35mm slides or prints.

4. All text and graphics provided in electronic form MUST be provided on diskette or CD-ROM, not via e-mail.

5. No graphics from the Internet. We need 200 dpi resolution photos/images.

B. Formatting Requirements

1. Use as few formatting commands (e.g., **bold**, <u>underline</u>, *italic*, centering, indenting, etc.) as possible in creating your text. Make text flush left, and separate paragraphs with [Hard Returns].

C. Required Materials from Each Teaming Partner

1. Signed, original letters of commitment on company letterhead.

2. Signed, fully executed SF254s.

3. Tailored résumés, including all relevant professional certifications and registrations along with associated dates and states.

4. Tailored project summaries.

5. Overview of company history and relevant technical, programmatic, and contractual capabilities.

6. Statement of financial strength and solvency.

7. Camera-ready company logos.

8. Project-specific photos and identifying captions.

9. Relevant client references, along with full title, name, address, telephone, FAX, and e-mail.

10. Recent ACASS (Architect/Engineer Contract Administration Support System) performance ratings.

11. Office locations map.

12. Total number of projects performed for the client organization.

13. Mileage from nearest office locations to client HQ.

14. Specific numbers of personnel by discipline in the relevant geographic area for Column A of Block 4 of the SF255.

15. Relevant client commendation letters/awards.

Table 18.2
Comprehensive
Guidelines for
Teaming Partners

contract that is being subcontracted. In effect, if only 15% of the contract will be performed by the subcontractors, then one or two project summaries should be from the subcontractors.

Your subcontractors can also be used in a review capacity for the SF254/255s. As with responses to RFPs, your SF254/255s should also undergo internal reviews and quality checking throughout the response lifecycle. SF254/255s have specific due date and times, just like formal RFPs. Ensure that your company receives a time- and date-stamped receipt for your submittal.

18.10 Building teaming agreements

Teaming agreements between your company and other organizations are established for several important reasons. Many times, *CBD* synopses for SF254/255s (and RFPs as well) stipulate that a certain percentage of the contract must be performed, for example, by SDBs, WOBs, and *historically black colleges and universities* (HBCUs). Depending on its size, your company may need to enter into teaming agreements with these organizations to fulfill these stipulated contractual requirements. It is beneficial to build a database of potential teaming partners that fall into these categories and have been prequalified with the particular client agency, if necessary. Your company can and should also consult Pro-*Net* (http://pro-net.sba.gov/pro-net/search.html), a Web-based procurement information source for and about more than 171,000 small, disadvantaged, 8(a), and women-owned businesses. Pro-*Net* is a search engine for contracting officers, a marketing tool for small firms, and a link to procurement opportunities and other important information. Pro-*Net* is free to federal and state government agencies as well as prime and other contractors seeking small business contractors, subcontractors, and/or partnership opportunities. Pro-*Net* is open to all small firms seeking federal, state, and private contracts. Businesses profiled on the Pro-*Net* system can be searched by SIC codes, key words, location, quality certifications, business type, ownership race and gender, electronic data interchange (EDI) capability, etc. Your government client's contracting office can provide you with the names and points of contact of prequalified SDBs, WOBs, and HBCUs. This is another reason to visit your client or potential client early in the marketing lifecycle.

Another common reason for entering into a teaming agreement is that each company will benefit from the other's technical expertise, contractual experience, programmatic strength, geographic proximity to the government site, and number and diversity of professional staff. For example, perhaps your company offers many important and relevant environmental services, but the *CBD* synopsis for a specific USCG project also requires support in the areas of unexploded ordnance and noise reduction. Under these circumstances, teaming is essential in order to meet the USCG's technical needs completely.

Teaming agreements are formal documents that define the specific association of companies and other entities, and which offer the basis for legal action if they are violated. In actual practice, they are not infallible but do articulate the basis of mutual interaction for the purpose of acquiring business. Agreements may be exclusive, or nonexclusive, depending upon the procurement. In effect, a given company may elect to join several teams to heighten its chances of winning a portion of the contract work. When your company serves as the prime contractor, your goal will be to secure *exclusive* teaming agreements with your subcontractors. This helps to eliminate the possibility of company-sensitive or procurement-sensitive information leaking to your competitors.[5]

The language of teaming agreements varies depending upon the procurement, the exact nature of the working relationship, and the business culture of the companies involved. Articles that might be built into a teaming agreement include (1) the particular *purpose* of the agreement, (2) the *relationship* of the parties involved, (3) *scope of services* to be performed upon contract award, (4) *costs* incurred, (5) the process for *executing contracts,* (6) *termination* procedures, (7) *notices* (i.e., points of contact in each party), (8) nondisclosure of *proprietary information,* and (10) grounds for and limits of any *liability.* Presented in Figure 18.12 are topics that your company may want to consider including in your teaming agreements when you serve as the prime contractor. It is advisable to include a specific list of proposal-related information and materials (as presented previously in Table 18.2) as an attachment to your formal teaming agreement. Too often, proposal-related requirements and the specific process of information exchange are poorly defined and result in production challenges and delays throughout the proposal response life cycle. These topics include those for SF254/255s as well as RFPs and international procurements.

It is strongly suggested that your company's legal counsel review the terms and conditions of all teaming agreements into which you enter.

---

5   I have had the experience of coordinating a major IRS proposal effort that involved a team of companies, some of which were important members of competitor proposal teams. The logistics of conducting effective strategy meetings and comprehensive review sessions were quite challenging.

Potential Teaming Agreement Topics

1. Mutual support during the entire proposal and post-proposal life cycle, including negotiations.

   Timely submittal of all proposal materials.

   Availability of qualified staff to assist in developing the proposal or SF254/255 as well as in conducting discussions and negotiations with the government.

2. Prime contractor's right to include additional subcontractors on the team.

3. Each party will bear all costs, risk, and liabilities incurred. Prime will be responsible for the layout, printing, binding, and delivery costs of the proposal.

4. Prime will have sole right to decide the form and content of all documents submitted to the government.

5. Prime will make every reasonable effort to subcontract to the sub that portion of the work stipulated in a Statement of Work (attached to the teaming agreement).

6. Any news releases, public announcements, advertisement, or publicity released by either party concerning the agreement, any proposals, or any resulting contracts or subcontracts will be subject to the prior approval of the other party.

7. Non-disclosure and protection of proprietary data and information.

   Each party will hold the other party's information confidential, unless such information becomes part of the public domain or unless subject to lawful demand, e.g., subpoena.

8. Duration/term of the agreement.

   Termination of the agreement.

9. Non-compete clause.

10. Percentage of the contract to be allocated to the subcontractor.

11. Interfaces with the government.

12. Patentable inventions and software developed pursuant to the work performed as a result of the agreement.

13. Non-proselytizing clause.

    Neither party shall solicit for employment any employee(s) of the other party to work on the project contemplated in the agreement.

14. Condition/mechanisms for amending, modifying, or extending the agreement.

15. Communication between the two parties.

16. Authorized signatures from both parties, with witness.

17. Date of execution.

18. Agreement shall be enforced and interpreted under applicable state laws.

Figure 18.12
Potential topics for teaming agreements.

Epilogue

Thinking to win small business competitive proposals

The magic is in the people. And so are the results.
—*The PeopleWise Organization*, Kepner-Tregoe, Inc.

S MALL FIRMS ARE IN the enviable position of being able to respond rapidly to changing business environments. As a key builder of your company's business future, you should maximize your agility as a small contracting firm and establish as soon as possible in your corporate history appropriate internal and external business development and knowledge management processes and patterns of thinking and behavior that will facilitate winning proposals in the federal, private-sector, and international marketplace. Genuine teamwork is a critical element of success in the contracting arena. Proposalmanship is a capability that needs to be cultivated throughout the levels of your organization in order to fully harness the talent and energy resident there. Many companies struggle because only a few of their staff are trained and experienced in proposal management, design, and development. And often, development and knowledge management are ill-defined and poorly implemented.

Ongoing, positive communication throughout your company is a second measure of your successful business culture. That *multidirectional* communication extends among the acquisition team and the proposal team, business development staff and technical staff, business development staff and your client, proposal manager and senior management, proposal manager and proposal writers, and proposal manager and publications staff.

Management support must necessarily assume many different forms. But it must be manifested clearly in order to be fully effective. That support should be present in ensuring that efforts to build appropriate résumé and project summary files on your company's staff and contractual history are met with complete and timely support throughout the technical and programmatic ranks. Management support should take the shape of assisting the proposal manager in enforcing the proposal milestone schedule and of empowering the proposal manager to meet the challenges of a given proposal effort. Ensure qualified leadership with the authority to do the job. And management involvement and support also lie in committing the resources—such as human, financial, equipment, and floor space—to make every proposal your company elects to pursue a superior client-centered sales document. Support for an apprentice system for proposal managers would be well-targeted energy.

As a leader, you are able to infuse an ethos of rational and formal planning into your company's business development and proposal development infrastructure. Expend the time and effort early in your firm's history to develop a mission statement and strategic plan. Generate and have all of your management team follow written, albeit revisable, business development and knowledge management protocol and processes. This applies particularly to bid/no bid decision making. Too many times, small and large businesses dilute their collective resources in pursuing marketing opportunities that do not support their primary lines of business. Planning also extends to developing the proposal kickoff package prior to conducting the formal kickoff meeting. Up-front planning, analysis, and decision making yield significant dividends downstream in the proposal process.

And finally, recognition and reward. People need to know in a variety of ways that they are meeting your company's business and proposal development expectations. They need to be recognized when they contribute in any one of several ways to a proposal victory. Senior management should develop and implement a definitive recognition and incentive plan that is communicated clearly to everyone in your company. Success should be noted in a big way. Those people in your firm who are *thinking to win* should be recognized and should receive tangible benefits. Remember that your *people* and their knowledge are the most important ingredients in your company's success in establishing a recognizable *brand* of excellence in the marketplace.

Appendix A

One agency's response
to acquisition streamlining

The Federal Aviation Administration's (FAA) new Acquisition Management System (AMS)

Since 1994, processes associated with the acquisition of goods and services within the federal arena have undergone significant changes. DoD and civilian agencies of the Federal Government have responded to legislation, executive directives, and mandates/guidelines to streamline their acquisition processes. The Federal Aviation Administration (FAA) developed and implemented an Acquisition Management System (AMS) that emphasizes preference for commercial and nondevelopmental solutions to mission requirements; strives to provide small businesses with attainable opportunities to participate as contractors and subcontractors; and delegates source selection responsibility, authority, and accountability to integrated product teams (see Figure A.1).

| FAST HOME | HELP | WHAT'S NEW | COMMENTS | SEARCH |

POLICY

AUTHORITY FOR THE FEDERAL AVIATION ADMINISTRATION ACQUISITION MANAGEMENT SYSTEM--APRIL 1, 1996

INTRODUCTION

I am inherently and expressly authorized to acquire goods, services, and property needed to carry out my aviation safety duties and powers. All of the Federal Aviation Administration's (FAA) acquisitions are in furtherance of these responsibilities. On October 31, 1995, Congress passed an act, *Making Appropriations for the Department of Transportation and Related Agencies, for the Fiscal Year Ending September 30, 1996, and for Other Purposes* (The *1996 DOT Appropriations Act*). On November 15, 1995, the President signed this bill into law (Public Law 104-50). In Section 348 of this law, Congress directed me to develop and implement a new acquisition management system that addresses the unique needs of the agency. At a minimum, this system is to provide for more timely and cost-effective acquisitions. By signing this document, I am making effective FAA's new acquisition management system.

STATUTORY EXEMPTIONS

Under Section 348, I was instructed by Congress to develop and implement a new acquisition management system for FAA "notwithstanding provisions of Federal acquisition law." Congress added that the following provisions of acquisition law "shall not apply" to this new acquisition management system:

1. Title III of the Federal Property and Administrative Services Act of 1949 (41 U.S.C. 252-266);

2. Office of Federal Procurement Policy Act (41 U.S.C. 401 et seq.);

3. Federal Acquisition Streamlining Act of 1994 (Public Law 103-355);

4. Small Business Act (15 U.S.C. 631 et seq.), except that all reasonable opportunities to be awarded contracts shall be provided to small business concerns and small business concerns owned and controlled by socially and economically disadvantaged individuals;

5. Competition in Contracting Act;

6. Subchapter V of Chapter 35 of Title 31, relating to the procurement protest system;

7. Brooks Automatic Data Processing Act (40 U.S.C. 759); and

http://fast.faa.gov/archive/v199/ams/authority.htm   4/5/99

Figure A.1 FAA Web page and contents.

8. Federal Acquisition Regulation and any laws not listed in (1) through (7) above, providing authority to promulgate regulations in the Federal Acquisition Regulation.

Although the combination of these provisions in Section 348 exempts the new acquisition management system from all acquisition laws, FAA has the discretion to adopt the substance of portions of acquisition law into its system as FAA deems appropriate. Unless stated specifically otherwise in this document or in legislation subsequently enacted, no acquisition statute or regulation shall apply to FAA acquisitions. The parties will, however, remain bound to the terms of any contract existing on this date unless the contract is modified by agreement of the parties or in accordance with existing contract terms.

LEGAL EFFECT OF THIS DOCUMENT

This document brings FAA's new acquisition system into effect and establishes the policies, guiding principles, and internal procedures for FAA's new acquisition system. Nothing in this document creates or conveys any substantive rights.

MODIFICATION OF THIS SYSTEM

FAA reserves the right to modify, add to, or delete any portion of this acquisition management system, either in whole or in part, as deemed appropriate by the Administrator or his designee. In addition to continuous improvement feedback, three years after implementation there will be an independent assessment of the acquisition management system and changes will be made, as necessary.

PENDING CASES

Unless the parties agree otherwise, all acquisition litigation timely filed and pending before forums of competent jurisdiction on or before the effective date, April 1, 1996, of this new acquisition management system may remain under the jurisdiction of that tribunal in accordance with the applicable contract or solicitation provision.

David Hinson

David R. Hinson
April 1, 1996

PREVIOUS SECTION | **AMS-HOME** | NEXT SECTION

http://fast.faa.gov/archive/v199/ams/authority.htm   4/5/99

Figure A.1   (continued)

| FAST HOME | HELP | WHAT'S NEW | COMMENTS | SEARCH |

POLICY

1.3 PRECEDENCE

The Acquisition Management System defines all acquisition and procurement policy within the FAA. It replaces existing policy pertaining to the elements of acquisition management, including functional disciplines such as logistics support, test and evaluation, human factors, configuration management, contracting, and transition to operational use. Policy for the functional disciplines is contained in Section 2.9, while related guidance (e.g., instructions, best practices, lessons learned, other job-related aids) are in the FAA Acquisition Management System Toolset - FAST (see Section 1.7). FAA's policy regarding legal participation is defined elsewhere.

The Acquisition Management System supersedes the Major Acquisition Policies and Procedures of the Department of Transportation, and all other acquisition and procurement statutes and regulations including the Federal Acquisition Regulation. Contracts awarded prior to April 1, 1996, remain under the Federal Acquisition Regulation until bilateral modification brings them under the Acquisition Management System. The FAA will continue to follow certain other statutes identified in the guidance section of FAST.

PREVIOUS SECTION | AMS-HOME | NEXT SECTION

http://fast.faa.gov/archive/v199/ams/ams1-3.htm   4/5/99

Figure A.1   (continued)

FAST HOME	HELP	WHAT'S NEW	COMMENTS	SEARCH

POLICY

1.8 KEY FEATURES

The following are key features of the Acquisition Management System:

Lifecycle Acquisition Management Policy

- Establishes a lifecycle partnership between users and providers so final products and services are what users/customers want and need.

- Creates a seamless lifecycle acquisition management process that extends from mission analysis to product disposal.

- Explores advanced technology opportunities and non-traditional operational concepts in full partnership between providers and users/customers.

- Provides a framework for evolutionary product development so the upgrade of complex systems can be faster and cheaper.

- Stresses preference for commercial and nondevelopmental solutions to mission needs.

- Streamlines policy so effort and resources are focused on products.

- Establishes a rigorous configuration control process for improving the Acquisition Management System continuously.

- Places resource decisionmaking at the Corporate level and program decisionmaking with Integrated Product Teams to increase the pace of doing business and stabilize program execution.

- Establishes a strong capability for mission analysis that looks forward in time to identify and prioritize needs before they become operational problems.

- Establishes a strong capability for

http://fast.faa.gov/archive/v199/ams/ams1-8.htm   4/5/99

Figure A.1   (continued)

investment analysis that ensures rigorous and impartial treatment of alternative strategies for satisfying mission need, while also achieving "buy-in" from the users who must live with the solution and from the providers who deliver it.

- Unifies Acquisition Management System processes with agency planning, programming, and budgeting; the NAS Architecture; and long-range strategic planning.

Procurement Policy

- Establishes competition for products and services among two or more sources as the preferred method of source selection.

- Strives to provide small businesses with attainable and reasonable opportunities to participate as contractors and subcontractors.

- Enables tailoring of processes and guidance to meet the goals of each requirement.

- Encourages industry participation in the development of requirements and solutions throughout the lifecycle acquisition management process.

- Establishes lists of qualified vendors for products and services based on their capabilities and past performance.

- Eliminates the requirement for formal solicitation and allows screening to narrow offerors to only those likely to receive an award based on capabilities and past performance.

- Delegates source selection responsibility, authority, and accountability to the Integrated Product Team.

- Resolves protests and contract disputes at the agency level through the FAA Dispute Resolution System.

http://fast.faa.gov/archive/v199/ams/ams1-8.htm   4/5/99

Figure A.1   (continued)

| FAST HOME | HELP | WHAT'S NEW | COMMENTS | SEARCH |

POLICY

Section 1: Acquisition Management System Overview

1.1 Purpose

The Federal Aviation Administration (FAA) Acquisition Management System (AMS) establishes policy and guidance for all aspects of the acquisition lifecycle from the determination of mission needs to the procurement and lifecycle management of products and services that satisfy those needs. It is intended to simplify, integrate, and unify the elements of lifecycle acquisition management into an efficient and effective system that increases the quality, reduces the time, and decreases the cost of delivering needed services to its customers.

Figure A.1   (continued)

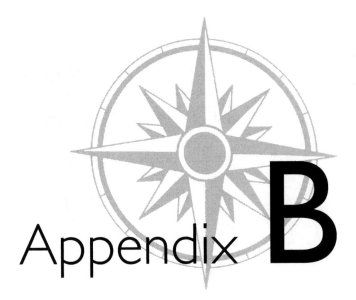

Appendix B

Sample Proposal Kickoff Package

A WELL-PREPARED Proposal Kickoff Package—which includes important components such as a milestone schedule, annotated outline, page allocations, and key person contact list—is a valuable tool for proposal managers to maximize the value of bringing staff together for the formal Kickoff Meeting. The Kickoff Package also serves as a helpful, easy-to-follow reference tool for proposal writers, contract staff, teaming partners, production staff, and senior management throughout the proposal response lifecycle.

KICKOFF PACKAGE
GROUNDWATER TREATMENT SYSTEM
for
Any County, Virginia

Prepared by:

Acquisition Team
XYZ, Inc.

Proposal #:
Department Code:

PROPRIETARY INFORMATION

Agenda

1. Proposal Summary

2. Schedule for Technical and Business Management Volumes

3. Volume Responsibilities

4. Evaluation Factors for Award

5. Technical Outline and Writing Assignments

   a. Themes and Critical Issues

   b. Page Allocations

   c. Proposal Writing Guidelines

6. Business Management Writing Assignments

7. Action Items

8. Items for Management Approval

9. Telephone List

10. Proposal Security

Proposal Summary Sheet

Title:	Groundwater Treatment System
Solicitation Number:	
Client:	Any County, Virginia
1. Proposal Due Date and Time:	Technical proposals and bids are due by 5 P.M. on Monday, in Virginia
2. Client Name (Point of Contact): Telephone #:	Contract Specialist
3. Period of Performance:	25 weeks + 1 month O&M
4. Copies to Client:	3 copies of the proposal and 1 copy of the bid [p. 8][1]

---

1   Page number references are from the RFP.

Proposal Team
Procurement Name

Proposal Manager:

Proposal Team Members:

    Technical:

    Program Management:

    Cost:

    Legal:

    QA:

Evaluation Factors for Award
Procurement Name

The evaluation factors are listed below:

(1)	Project Organization	15
(2)	Project Execution	25
(3)	Schedule	15
(4)	Qualifications and Experience	30
(5)	Health and Safety	5
(6)	SB, SDB, or WOB	5
(7)	Financial Resources	5
Total		100

PROPOSAL MILESTONE SCHEDULE

Sunday	Monday	Tuesday	Wednesday	Thursday	Friday	Saturday
			December 28 Meet to discuss the contract scope and proposal team membership.	29 Develop Kickoff Package that includes a milestone schedule and outline.	30 Develop Kickoff Package that includes a milestone schedule and outline.	31
January 1	2	3	4 Develop Kickoff Package that includes a milestone schedule and outline.	5	6 9 A.M.: Meet to review Kickoff Package. 11 A.M.: Proposal Kickoff Meeting.	7 Proposal writing.
8 Proposal writing.	9 Proposal writing.	10 Proposal writing. Proposal Manager to contact all writers to assess status. Writers to provide input to him as sections are complete.	11 Proposal writing.	12 Proposal writing.	13 Proposal writing. Proposal Manager to contact all writers to assess status. Writers to provide input to him as sections are complete.	14

Sunday	Monday	Tuesday	Wednesday	Thursday	Friday	Saturday
22 Address Red Team comments.	23 Address Red Team comments.	22 Address Red Team comments. Proposal Manager to provide hardcopy MASTER markup document to Production Services by 4 P.M. Document freezes. No additional changes accepted.	25 Final production.	26 Final production.	27 Final production. QC, phtotcopying, and assembly.	28
29	30 Proposal due in Virginia.					

Technical Proposal Outline

Prepared by Proposal Manager

EXECUTIVE SUMMARY

1. PROJECT ORGANIZATION   10 pages

   1.1   Project Management (include an organizational chart)

   1.2   Key Personnel Résumés

   Project Manager

   Construction Superintendent

   Health and Safety Representative

   Quality Control Manager

   Field Engineering Staff

2. PROJECT EXECUTION   20 pages

   2.1   Construction Operations Work Plan

      2.1.1   Civil Works and Building Installation

         Understanding the Work

         Approach to Ensuring Quality

      2.1.2   Equipment Purchasing

         Understanding the Work

         Approach to Ensuring Quality

      2.1.3   Mechanical Work

         Understanding the Work

Approach to Ensuring Quality

        2.1.3.1   Piping

        2.1.3.2   Electrical Conduit

        2.1.3.3   Termination Control and Field Instruments

     2.1.4   Functional Testing and Startup

            Understanding the Work

            Approach to Ensuring Quality

  2.2   Community Relations Plan

  2.3 Subcontractors

    2.3.1   Scope Items and Approximate Dollar Amount

    2.3.2   Use of SB, SDB, or WOB

3. SCHEDULE   5 pages

   3.1   Capacity to Accomplish Work in Required Timeframe (includes a Critical Path Method diagrams that shows critical equipment purchases)

    3.1.1   Staff Time Commitments

    3.1.2   Equipment Commitments

    3.1.3   Strategies for Resolving Competing Demands

4. QUALIFICATIONS AND EXPERIENCE   15 pages

   4.1   Institutional Overview
         Number of years in business
         Superfund and hazardous waste site work experience, by number of years and specific role
         Annual Experience Modification Rating (EMR)

OSHA violations
Construction Equipment

4.2 Staffing Profile (by trade)

4.3 Case Histories (minimum of 3)
References, including name, position, and phone
Company's role and responsibility
Contract amount
Date completed/anticipated completion date
Percentage of contract performed by workforce

5. HEALTH AND SAFETY   10 pages

5.1 Site Safety Plan

5.2 Site Safety Officer (name, qualifications, and résumé)

5.3 Corporate Health and Safety Program

6. BIDDER STATUS   1 page

7. FINANCIAL RESOURCES   5 pages

7.1 Overview of Financial Stability

7.2 Summary of Gross Billings (for past 10 years)

7.3 Bonding Capacity

7.4 History of Work Performance

COMPLIANCE MATRIX

TOTAL PAGE COUNT   65 pages

Action Item
Procurement Name

#	Person Responsible	Assignment	Date Assigned	Date Due	Status
1		Select appropriate résumés.	1/6/99	1/10/99	CLOSED
2		Select appropriate project summaries.	1/6/99	1/10/99	OPEN
3		Complete the bid portion of the proposal.	1/6/99	1/26/99	OPEN

Phone List
Procurement Name

Name:
  Work#:
  Home#:
  FAX#:
    Days on vacation, out of office, or otherwise unavailable:

Name:
  Work#:
  Home#:
  FAX#:
    Days on vacation, out of office, or otherwise unavailable:

Name:
  Work#:
  Home#:
  FAX#:
    Days on vacation, out of office, or otherwise unavailable:

Name:
  Work#:
  Home#:
  FAX#:
    Days on vacation, out of office, or otherwise unavailable:

Name:
  Work#:
  Home#:
  FAX#:
    Days on vacation, out of office, or otherwise unavailable:

Appendix C

Template to capture
important résumé information

C RITICAL TO THE SUCCESS of your organization is the ongoing
capture of information about your technical and programmatic
staff's experience and knowledge base. Appendix C provides
a sample template for infor- mation collection for an Information Technology (IT) firm. Having this information readily accessible in electronic
form can make the proposal process easier and more cost-effective.

Workforce Questionnaire for an IT Firm

Employee ID No:

Division:

Current job title:

Software experience:

Ada
C
C++
COBOL
DOS
Easipace
FORTRAN-77
FTP
IDL
IHS
IMSL
ISPF
Lotus 1–2–3
Motif
Natural
Pascal
PL/1
Quattropro
Statistical Analytical System (SAS)
TCP/IP
Tex
TSO
UNIX
VMS
Windows
WordPerfect
X

Hardware Experience:

Cray Y-MP supercomputers
Cyber supercomputers
DEC PDP mini-computers

DEC VAX mini-computers
DEC workstations (including VMS, Ultrix, ans Alpha)
Ethernet
HP workstations
IBM workstations
IBM mainframes
IBM-compatible PC microcomputers
Macintosh
SGI workstations
Sun workstations (Sparcstations)
VAXcluster
X-terminals

Database experience:

Adabase
dBASE
Oracle
Paradox
R:Base
SQL

Experience supporting or working for government organizations:

Department of Agriculture (USDA)
Department of Defense (DoD)

> Army
> DISA
> DLA
> Navy

>> Marine Corps
>> NAVAIR
>> NAVSEA
>> Office of Naval Research (ONR)
>> SPAWAR

> Office of the Secretary of Defense (OSD)

Department of Energy (DOE)
Department of Justice (DOJ)
Department of Transportation (DOT)
General Services Administration (GSA)
NASA/HQ
Treasury

> IRS
> FMS

Technical areas of expertise:

CASE tools
Communication systems
Computer programming
Computer networking
Configuration management (CM)
Database development and management
Data systems modeling
Data analysis and planning
Data quality engineering
Documentation
Electronic Commerce/Electronic Data Interchange (EC/EDI)
4th generation languages
Graphical User Interface (GUI)
Information Resources Management (IRM)
Information engineering
Information systems design and development
Integration and Test (I&T)
IV & V
Large-scale data systems
Logistics engineering
Numerical analysis
Requirements analysis
Risk assessment
Software Quality Assurance (SQA)
Software engineering
Software systems development
Systems analysis
Systems engineering
Systems integration
Telecommunications
Total Quality Management
Training
User support

Project/program management or technical leadership experience:

Maximum number of staff supervised:

Publications: Yes

Knowledge of and experience with military standards: Yes

Security clearance:

Appendix D

Marketing information and intelligence sources: federal, international, and private sector

THERE ARE THREE major information sources for federal competitive procurements: (1) publicly available information, (2) client-derived information, and (3) community-derived information. The latter refers to friends, competitors, seminars, subcontractors, vendors, and published scientific and/or conference reports.

D.1 Sources of federal marketing leads and information

Acquisiton forecasts

Public Law 100-656, the Business Opportunity Development Reform Act of 1988, amended the Small Business Act to place new emphasis on acquisition planning. The law requires agencies to annually compile and make available one-year projections of contracting opportunities that small and small disadvantaged companies can perform. The NASA Acquisition Forecast, for example, includes projections of all anticipated contract actions above $25,000 that small and small disadvantaged businesses can perform under direct contracts with the government or as part of a subcontract arrangement. NASA's Acquisition Forecast is divided into sections for Headquarters in Washington, D.C., and for each Center (Goddard, Marshall, Ames, Lewis, etc.). Retain these forecasts in your company's proposal data center or library for three to five years in order to track contracts as they become available for recompetition.

Aerospace Daily (International Standard Serial Number (ISSN) 0193-4546/90)

1156 15th Street, N.W.
Washington, DC 20005
(202) 383-2378
A publication of the Aviation Week Group of McGraw-Hill, Inc.
Air Force Electronic Commerce/Electronic Data Interchange (EC/EDI) Newsletter, Internet: http://infosphere.safb.afmil/

Air Force Small Business Bulletin Board

This service allows small companies to identify Air Force activities. It provides descriptions of what the Air Force buys along with points of contact, addresses, and telephone numbers. Small businesses may access the Bulletin Board Service by calling 1-800-638-9636.

Aviation Week & Space Technology

P.O. Box 503
Hightstown, NJ 08520-9899

Baltimore Business Journal (newspaper)

American City Business Journals, Inc. (Baltimore)
117 Water Street

Baltimore, MD 21202
(410) 576-1161
Editor, Marsie Freaney
Circ.: 13,500

Business Credit (journal)

National Association of Credit Management
8815 Centre Park Drive, Suite 200
Columbia, MD 21045-2158

Commerce Business Daily (CBD)

An absolutely essential daily list of government procurement invitations, contract awards, sales of surplus supply, and overseas business opportunities published by the U.S. Department of Commerce. For federal agencies, the *CBD* publishes synopses of proposed contract actions that exceed $25,000 in value. The Reader's Guide is published in every Monday edition (except when federal holidays occur on a Monday, in which case the Guide will appear in the next day's issue) and includes the *CBD*'s numbered notes as well as other information. If a note or notes are referenced in a *CBD* announcement in which your company is interested, it is important to review and comply with the instructions contained in those notes. Announcement of proposed contract actions and contract awards in the *CBD* is governed by FAR 5.

The *CBD* is available by subscription via the U.S. Mail; on line via such services as DIALOG (Knight-Ridder), Files 194 (historical) and 195 (current 3 months); or through electronic subscription services. In the case of U.S. Mail, first-class subscriptions are recommended over second class to ensure timely delivery.

With DIALOG, searches can be done at will and tailored to your specific requirements. You also have the capacity to search the *CBD* for announcements that have appeared several weeks, months, or years ago.

Service firms will perform specific *CBD* searches (for example, keyword searches of particular technical topics, geographic locations, and agencies) for your company and then transmit the search results to you electronically on a daily or weekly basis.

A word of caution, however, about using the *CBD*. Announcement of the upcoming release of an RFP should definitely not be the time to begin a proposal response if the announcement is the first time your company learned of the opportunity and/or the client.

Competitors' newsletters and annual reports

These can be obtained by contacting public relations departments. You may need an outsider such as a college student or consultant to do this.

Contract Management (journal)

National Contract Management Association
1912 Woodford Road
Vienna, VA 22182-3728
Editor, Terry Hoskins

Contractual exposure

When your company's personnel are on the client's site to perform existing work, they can certainly ask questions.

Defense Logistics Agency (DLA)

"An Identification of Commodities and Services Purchased by the Defense Logistics Agency." Cameron Station Alexandria, VA 22304-6100

Defense Technical Information Center (DTIC)

Defense Logistics Agency (DLA)
Office of User Services and Marketing
Attn: DTIC-BLU
Building #5, Cameron Station
Alexandria, VA 22304-6145
(703) 274-6434
http://www.dtic.dla.mil
DTIC is the central point within the DoD for acquiring, storing, retrieving, and disseminating scientific and technical information to support the management and conduct of DoD research, development, engineering, and studies programs. DTIC services are available to DoD and its contractors, and to other U.S. Government organizations and their contractors.

Department of Defense Electronic Commerce Information Center

1-800-334-3414
Provides information on government-certified VANS and other technical information related to the initiative to migrate toward electronic commerce as stipulated by the *Federal Acquisition Streamlining Act* (FASA) of 1994.

Department of Defense Telephone Directory

Issued three times a year, the directory contains numbers for the Army, Navy, and Air Force.

DIALOG Information Services, Inc. (Knight-Ridder)

DIALOG Information Services, Inc.
3460 Hillview Avenue
Palo Alto, CA 94304
1-800-334-2564

The online, modem-accessed service called DIALOG is available for use in gathering marketing intelligence. DIALOG can be utilized to query any of 350+ business and commercial, defense, aerospace, academic, career placement, Federal Acquisition Regulations (FAR)—File 665, Federal Research in Progress (FEDRIP)—Files 265 and 266, and other databases. The cost for this service is determined by access time, specific database being used, and method of downloading the data. Searches can be performed, for example, on public or private companies, specific market segments, general subject areas, government policy and research forecasts, competitor acquisitions and contract award history, and general fiscal trends.

Electronic Government

5204 Woodleaf Court
Centerville, VA 20120-4100
(703) 502-1033
FAX: (703) 968-7826

Quarterly trade publication designed to cover the people and technologies that are working to bring the government online.

Federal Business Council

(a nonprofit association)
10810 Guilford Road
Suite 105
Annapolis Junction, MD 20701
(301) 206-2940; 1-800-878-2940
FAX: (301) 470-6313

The Federal Business Council sponsors a variety of business seminars to give your company procurement direction, contacts, and specific knowledge regarding the federal marketplace.

Federal Computer Week (ISSN 0893-052X)

FCW
P.O. Box 602
Winchester, MA 01890-9948
(617) 729-4200

Federal Information & News Dispatch, Inc. (FIND)
http://www.find-inc.com.
(202) 544-4800

The Federal Marketplace Home Page (copyrighted by Wood River Technologies, Inc.)
The Federal Marketplace provides assistance to companies in marketing and selling products and services to the federal government.
www.fedmarket.com
Federal Data ContractSearch: ContractSearch is a detailed Oracle-powered database of virtually every federal agency contract more than $25,000. Data are updated quarterly, and the annual price for unlimited use is $600 (1997).
Federal Data TeamingSearch: TeamingSearch is a service that provides access to an Oracle-powered database with roughly 80,000 companies that do business with, or have done business with, the federal government. This comprehensive resource can be used to locate teaming partners, subcontractors, and candidates for merger and acquisition. TeamingSearch provides contact information, along with information on what kind of federal work the company does, where the company is performing the work, and for which agencies. Data are updated quarterly, and the annual price for unlimited use is $250 (1997).
Procurement Assistance Jumpstation: Lists federal procurement Web sites, department and agency Web pages, business resources, *Commerce Business Daily*, active federal contracts, laws and regulations, technology transfer, EDI, SBIR, MIL SPECS, federal laboratories, proposal writing, state and local procurement, international procurement, and trade links.

Federal Register
Daily newspaper that informs of proposed, interim, or final regulations issued by the government. It is in the *Federal Register* that the FAR receives its official publication. Changes to the FAR are published as *Federal Acquisition Circulars* (FACs). Careful attention to the *Federal Register* will indicate program changes and new priorities.

Federal Yellow Book 1997 (ISSN 0145-6202)
Leadership Directories, Inc.
1301 Pennsylvania Avenue, N.W.
Suite 925
Washington, DC 20004
(202) 347-7757
Internet: http://www.info.gov/info/html/fed_yellow_pgs.htm

The *Federal Yellow Book* is an organizational directory of the departments and agencies of the Executive Branch of the Federal Government. It provides organizational position, addresses, and telephone numbers of more than 30,000 senior people.

Freedom of Information requests

Freedom of Information Act (5 U.S.C. § 552). Consider asking for only one item per written request to avoid delay if a particular item undergoes further legal review or is denied.

govcon™ (Web site)

http://www.govcon.com
Lists business opportunities in the CBD, FACNET, and the SBIR program, as well as proposal resources.

Government Information Quarterly (ISSN 0740-624X)

JAI Press Inc.
Greenwich, CT

Government Solutions: a free resource for identifying and capturing new business.

http://www.govsolutions.com
Government Solutions
8000 Towers Crescent Drive
Suite 1350
Vienna, VA 22182
(703) 847-3601

Minority business associations (for example, Minority Business Association of Northern Virginia)

The Nash & Cibinic Report

Keiser Publications
2828 Pennsylvania Avenue
Suite 309
Washington, DC 20007
(202) 337-1000
A monthly report of analysis and advice on U.S. Government contracts prepared by Professors Ralph C. Nash, Jr., and John Cibinic, Jr., of The George Washington University.

National Association of Small Disadvantaged Businesses
Mr. Hank Wilfong, President
Los Angeles, CA

National Coalition of Minority Businesses
Washington, DC

National Federation of 8(a) Companies
Ms. Babielyn Trabbic, President
Arlington, VA

National Small Business United
1155 15th Street, NW, Suite 710
Washington, DC 20005
(202) 293-8830

National Technical Information Service (NTIS)
Sills Building
5285 Port Royal Road
Springfield, VA 22161
(703) 487-4650
NTIS is a part of the U.S. Department of Commerce and is an excellent source for U.S. and foreign government-sponsored *research and development* (R&D) results, business information, and engineering solutions. Complete technical reports for most government-sponsored R&D and engineering activities are provided through NTIS. The Service also makes available U.S. Army and Air Force Manuals and Regulations as well as a wide variety of software, data files, and databases, and information on federal laboratory inventions and technologies.
For copies of the *NTIS Products & Services Catalog*, call or write to NTIS and request catalog number PR-827.

Procurement Automated Source System (PASS)
The U.S. Small Business Administration's Office of Procurement and Technical Assistance maintains PASS, a computer-based inventory and referral system that includes capability profiles of small businesses interested in Federal Government procurement opportunities. Having your company listed in PASS could potentially lead to a contract or subcontract.
PASS contains the profiles of over 200,000 small businesses that are interested in competing for federal procurements. Company profiles are online in the fields of R&D, manufacturing, construction, services, SIC, Federal

Supply Codes/Commodity/Product Service Code(s), and Quality Assurance. There is no cost associated with listing your company on PASS.

Resources for U.S. Government Contractors

U.S. Government Printing Office
P.O. Box 371954
Pittsburgh, PA 15250-7954

Small Business Success

National magazine that includes articles on successfully operating a business, international trade, marketing, and trends for the 1990s. It also includes a resource directory. Partners for publication include the SBA, SCORE, and Pacific Bell. Available from your local SBA office.

SBA's Office of Women's Business Ownership

(202) 205-6673

Space Business News

1401 Wilson Blvd.
Suite 900
Arlington, VA 22209
(703) 528-1244
(800) 424-2908
Space Business News provides business intelligence on the space industry and NASA programs.

Trade association meetings

Trade association meetings are announced in trade journals and newsletters as well as on the Internet.

U.S. General Services Administration (GSA)

Federal Procurement Data Center (FPDC): An "FPDS[1] Federal Procurement Report" is published annually and is free of charge. A Special Report can be customized for your company at any time for a fee. Send a letter directed to "Special Reports" at the address below, outlining the search parameters your company wants to use. For example, perhaps your company's Mission Statement and Strategic Plan have identified the FAA as one of the company's primary federal clients. You

---

1   FPDS is an acronym that stands for Federal Procurement Data System, which was designed under the guidance of the *Office of Federal Procurement Policy* (OFPP).

may then wish to focus a Special Report on FAA support services contracts.

An "FPDS Federal Procurement Report" contains information on products and services purchased by the Federal Government, arranged in the following manner: (I) Total Federal View, (II) Geographic Views, and (III) Agency Views. Section I includes federal procurement information by SIC as well as the top 100 federal contracting firms, ranked by contract dollars. Section II presents federal procurement spending within each state, subdivided by such elements as the type of product or service and type of business. Finally, Section III offers a detailed agency-by-agency listing of procurement activities, subdivided by such parameters as type of contract, product or service, type of business/contractor, and competitive or noncompetitive. Because the report contains statistics on the procurements of 60 federal agencies, your company can utilize this information to build information on market size. From this, you can begin to establish target values for market share.

U.S. General Services Administration

Automated External Information Division
Federal Procurement Data Center
ROB, Room 5652
Washington, DC 20407
(202) 401-1529

Washington Business Journal (ISSN 0737-3147)

American City Business Publications, Inc. (Arlington)
2000 14th Street North
Suite 500
Arlington, VA 22201
(703) 875-2200

Washington Technology (ISSN 1058-9163)

Technews, Inc.
8500 Leesburg Pike
Suite 7500
Vienna, VA 22182-2412
(703) 848-2800
WT Online: http://www.wtonline.com/wtonline
Washington Technology is published semimonthly. A supplemental publication is *Washington Technology's 8(a) and Small Business Report*, which appears twice per year. This publication offers news and information important to minority-owned businesses, including how to establish

relationships with federal agencies, 8(a) program upgrades, and financing opportunities for 8(a) companies.

D.2 Sources of international marketing leads and information

From the U.S. federal government

- U.S. Agency for International Development (U.S. AID); Washington, D.C.
- U.S. Asian Environmental Partnership (U.S. AEP); Washington, D.C.
- U.S. Environmental Protection Agency (U.S. EPA); Washington, D.C.
- U.S. Trade and Development Agency (U.S. TDA); Washington, D.C.
- Office of Environmental Technologies Exports (U.S. Department of Commerce); Washington, D.C. (202) 482-3889

From state-level organizations (example for Maryland)

- Maryland State—Development of International Trade; Baltimore, Maryland
- World Trade Center Institute; Baltimore, Maryland
- Maryland Sister State Program; Baltimore, Maryland

From other organizations and media

- Multilateral Banks: World Bank, Asian Development Bank (ADB), and Inter-American Development Bank (IDB); Washington, D.C.
- Border Environmental Cooperation Commission (BECC)
- *Business in China* (1995), electronic book available through Industry Canada (http://info.ic.gc.ca/ic-data/industry/china/eng/indexe.html).
- Canadian China Business Association
- *Commerce Business Daily* (*CBD*)
- *The China Business Review*
- *China Daily;* published in the Peoples' Republic of China, distributed from New York, New York

- *Diario Official*; Mexico City, Mexico
- World Bank Monthly Operational Summary; Washington, D.C.
- Asian Development Bank *Business Opportunities Bulletin* published by the ADB Information Office; Manila, Philippines
- Internet notices from *U.S. Asian Environmental Partnership* (U.S. AEP); Washington, D.C.
- *U.S. Trade and Development Agency* (U.S. TDA); published biweekly; Washington, D.C.
- *Bu$iness Mexico* (magazine); American Chamber/Mexico; (800) 227-4909.
- *The Economist*, London
- The *Economist Intelligence Unit* (EIU), London

D.3 Sources of U.S. private-sector marketing leads and information

El Digest

Environmental Information, Ltd.
4801 West 81st Street
Suite 119
Minneapolis, MN 55437
(612) 831-2473
FAX (612) 831-6550

The *EI Digest* research service provides strategic information on the business, policy, and technology of industrial and hazardous waste. It assists firms to identify market opportunities and to develop new services that meet client requirements. *EI Digest* is a monthly collection of new reports. Environmental Information Limited's mission is to be a leading provider of quality research and strategic analysis on the business, regulation, policies, and technology of environmental protection.

The PEC Report

Industrial Information Resources, Inc.
Columbia Centre
11011 Richmond Avenue
4th Floor, Houston, TX 77042
(713) 783-5147

This report can be FAXed to your company on a weekly basis.

The Green Book Report

The Green Book, Inc.
Corporate Place
100 Burtt Road
Andover, MA 01810
1-800-527-2204
This report is published on a weekly basis.

Appendix E

Glossary of proposal-related terms

A-76: Office of Management and Budget (OMB) Circular A-76 was first issued more than 30 years ago. Revised in 1967, 1979, and 1983, this Circular sets forth federal policy for determining whether recurring commercial activities associated with conducting the government's business will be performed by federal employees or private contractors. Recent revisions to the A-76 *Supplemental Handbook* were designed to enhance federal performance through competition and choice, seek the most cost-effective means of obtaining commercial products and support services, and provide new administrative flexibility in agency decisions to convert to or from in-house, contract, or Interservice Support Agreement (ISSA) performance. OMB circular A-76 is a tool created by the federal government to maximize the effectiveness of the services for which it contracts while minimizing their cost. These guidelines are used to determine the feasibility of contracting out (i.e., "outsourcing") services performed by federal employees. The scope of A-76 is often referred to as "managed competition." The three pieces of legislation that affect outsourcing in general and the methodology of outsourcing specifically are Clinger-Cohen/ITMRA, the Government Performance Results Act (GPRA), and the Federal Acquisition Streamlining Act (FASA). Together these statutes require agencies to make determinations regarding outsourcing and to give guidance on measuring performance. In summary, Circular A-76 requires that a federal agency consider outsourcing before the agency commits to using government personnel for either undertaking a new requirement or expanding an existing commercial activity. A new requirement is a newly established need for a commercial product or service. An expansion is the modernization, replacement, upgrading, or enlargement of an in-house commercial activity or capability. An expansion involves a 30 percent increase in the total capital investment or a 30 percent increase in the annual personnel programs and functions to identify which areas are the best candidates for privatization.

Action title　Figure caption that conveys a sales message. Captions should emphasize themes. *Example:* "Well-defined Interfaces at the Project Manager/COTR Level Will Ensure Mutual Understanding of Changing Requirements."

Armed Services Procurement Act (1948)　ASPA, 10 U.S.C. § 2301 et seq. Along with FPASA, this Act covered most areas of government procurement.

BAFO　Best and final offer. Response to client request after submission of initial proposal. BAFOs are not required, therefore your company's initial proposal should contain the *best* terms from a cost or price and technical standpoint (see FAR 52.215.16).

Base Year　The first year of a given contract.

Best value　Best value is a process used in competitive negotiated contracting to select the most advantageous offer by evaluating and comparing factors in addition to cost or price. Best value allows offerors flexibility in selection of their best proposal strategy through trade-offs that may be made between the cost and noncost evaluation factors. It should result in an award that will give the government the greatest, or best value, for its money. Executive Order 12931 issued on October 13, 1994, directs executive agencies to "place more emphasis on past performance and promote best value rather than simply low cost in selecting sources for supplies and services" (Section 1, Item d). Best value has emerged as the centerpiece of acquisition reform.

Bidders conference   See *Preproposal Conference.*

Blackout   Cessation of information exchange on the part of the client once the RFP is announced in the *CBD* or is actually released.

Black Team   Low profile, highly confidential contractor activity designed to assess the competition.

BOE   Basis of estimate. The basic purpose of the BOE is to *sell* your company's technical approach, staffing, and cost to the client. The BOE is a *nontechnical* communication mechanism to present an argument in favor of expending a stated quantity of resources to accomplish a given task. It is documentation of how an estimate was developed and includes your company's assumptions that impact cost, a description of the methodology used in developing the estimate, and justification for using a particular methodology. BOEs must reflect an internal consistency among the Technical, Management, and Cost Volumes of your proposal. And BOEs must be traceable to task descriptions, the SOW, and the WBS. BOEs are most appropriate when the government has not specified hours or any quantitative *level of effort* (LOE).

  Instead of merely stating that Task A will require five staff-months of Labor Category 1 personnel and eight staff-months of Labor Category 4 personnel, a BOE-based approach might build on the following.

  Task A requires five staff-months of Labor Category 1 personnel and eight staff-months of Labor Category 4 personnel because on a similar task performed in 1995 under our XYZ-11 contract with Marshall Space Flight Center, this ratio of senior to junior labor proved to be extremely effective in bringing the work to a successful, in-budget conclusion. Such a labor mix will meet all task requirements as discussed in our Technical Approach section of this proposal.

  BOEs may be substantiated by historical or current corporate contractual or commercial work, trade-off studies (particularly in the case of design-oriented procurements), vendor quotes, purchase orders, etc. Show your calculations, as you would on a math test, and document your references, as you would in a technical or scientific paper. Your client's source selection board needs quantifiable data to evaluate your proposal. *Make sure that you would buy your products or services based on the data you have provided.* (Requirements for submission of BOEs are established by FAR 15.804-6 Table 15-2.)

Bogey   "Ballpark" dollar amount used for preliminary costing.

Boilerplate   Frequently used, "standard" materials that go into proposals, such as phase-in plans, configuration management plans, quality assurance plans, project descriptions/summaries, professional compensation plan, make-or-buy policy, etc. All boilerplate, however, should be tailored and customized for the specific proposal. As Tom Hewitt, chief executive of Federal Sources, Inc., has noted, "Boilerplate Loses!" Federal Sources is a research firm based in McLean, Virginia, which tracks the procurement industry.

Brooks Act (ADPE Contracting) (1965)   40 U.S.C. § 759. This Act (Public Law 92-582) grants exclusive purchasing authority for automatic data processing equipment (ADPE) to the Administrator of the General Services Administration (GSA), subject to the fiscal control of the Office of Management and Budget (OMB). This statute is implemented by the Federal Information Resources and Management Regulation (FIRMR), 41 CFR 201-2.001 and Part 201-23, which supplements the FAR. (Cibinic and Nash, *Formation of Government Contracts*, p. 15.)

Capture plan (or strategy)   Formalized, documented sequence of activities that includes the formation of an Acquisition Team, the generation of a call plan, and actual proposal design and development. A capture plan is directed to the acquisition of a specific business target. Prepared early in the proposal lifecycle, the capture plan should include a call plan to define and orchestrate frequent contact with the client at all levels—from senior executive management *to* business development staff *to* all appropriate technical ranks. Specifically, the call plan is designed to gather intelligence about the client, the particular procurement, and the competition. Information collected must then be analyzed by members of the acquisition team, and eventually factored into your company's proposal where appropriate.

Competition in Contracting Act (1984)   CICA, P.L. 98-369, July 18, 1984. This Act provides for increased use of competitive procedures in contracting for procurement. CICA modified both FPASA (1949) and ASPA (1947). (Cibinic and Nash, *Formation of Government Contracts*, pp. 13–14, 287.)

Competitive range   Proposals deemed within the competitive range are technically sufficient and exhibit a reasonable relationship between cost/price and the product or service to be procured. Proposals with a reasonable chance of being selected for contract award are deemed within the competitive range.

DD Form 254   Department of Defense Contract Security Classification Specification.

Defense Contract Administration Service (DCAS)   Comprised of nine regional offices which administer most of the contracts for the DoD and many civilian agencies. Ensures that contractors perform their contracts successfully and on time.

Defense Contract Audit Agency (DCAA)   Performs contract audit functions required by the DoD and many civilian agencies. Determines the suitability of contractor accounting systems for government work, evaluates the validity of proposed costs, and also reviews the efficiency of contractor operations. Negotiated solicitations contain a clause that authorizes the government Contracting Officer (CO) or his representative to examine the contractor's records under circumstances prescribed in FAR 52.215-2. The DCAA auditor is generally the CO's representative. (See McVay, *Proposals That Win Federal Contracts*, p. 16.)

Desktop publishing (DTP)   Desktop publishing refers to a method by which text and graphic materials of publishable quality are produced under *author control* (as opposed to outside printing company control) at relatively low cost. DTP can introduce higher quality, more professional documentation in your company's publication group. Desktop publishing software applications include CorelVentura and Aldus PageMaker.

Discriminators   Authenticated, conclusive statements that demonstrate clearly the differences between your company and the competition, and demonstrate the benefits of those differences to your client. Also known as "competitive themes."

Electronic commerce   Electronic commerce (EC) is the paperless exchange of business information, using electronic data interchange (EDI), electronic mail, electronic bulletin board systems (BBS), electronic fund transfer (EFT), and other related technologies.

Electronic data interchange   Electronic data interchange (EDI), a major part of electronic commerce (EC), is the computer-to-computer exchange of business data in a standardized format.

Electronic source selection (ESS)   ESS uses standard office software applications as well as hardware to produce a "paperless environment" for source selections and to streamline RFP preparation. ESS has been used in major DoD programs such as Joint Strike Fighter, Landing Platform Dock, and Global Broadcast System.

FACNET   The Federal Acquisition Streamlining Act (FASA) of 1994 established the Federal Acquisition Computer Network (FACNET) to evolve the federal government's acquisition process from one driven by paperwork into an expedited process based upon electronic data interchange (EDI).

Federal Acquisition Regulations System   Established for the codification and publication of uniform policies and procedures for acquisition by all executive agencies. The Federal Acquisition Regulations System consists of the FAR, which is the primary document, and agency acquisition regulations that implement or supplement the FAR. The FAR System is articulated in Title 48 of the CFR. The FAR is organized into Subchapters a–h, Parts (of which there are 53), Subparts, Sections, and Subsections. Example: 25.108-2 Part 25, Subpart 1, Section 08, and Subsection 2. The FAR is published by the GSA FAR Secretariat and maintained by the DoD, the General Services Administration (GSA), and NASA under several statutory authorities of those agencies. Any critical understanding of the FAR must include the protest decisions of the Comptroller General and the General Services Board of Contract Appeals (GSBCA).

In addition to hardcopy, the FAR is available on diskette from private vendors to facilitate searches. DIALOG Information Services, Inc., maintains a *Federal Acquisitions Regulations* database (File 665) as part of their on-line service. File 665 provides full text of the FAR and its supplements and is updated with amendments contained in the Federal Acquisition Circulars (FACs), Defense Acquisition Circulars (DACs), and other sources. Approximately 25 specific government agency regulations are included in this file, as well as the Defense Contract Audit Agency Manual (DCAAM), DoD FAR Supplement (DFAR), and Federal Information Resource Management Regulations (FIRMR). Call the GPO at (202) 512-1800 to place a subscription.

Federal Property and Administrative Services Act (1949)   FPASA, 41 U.S.C. § 252. This Act covered most areas of government procurement along with the Armed Services Procurement Act (ASPA).

FOIA   Enacted in 1966, the Freedom of Information Act (5 U.S.C. 552, as amended by the Electronic Freedon of Information Act Ammendments of 1996) provides that select information is to be made available to the public either by publication in the *Federal Register*, providing an opportunity to read or copy records, or providing copies of records.

Foldout page   An 11-in by 17-in double page, usually graphical in nature. Certain RFPs limit the number of foldout pages in a proposal. Foldouts are important space-savers for page-limited proposals.

Gantt Chart   A bar chart named after Henry Gantt. Sometimes referred to as project time lines, Gantt Charts are commonly used scheduling charts. Gantt Charts ordinarily have a list of tasks down the left side. A bar or line on the Gantt Chart shows the date when each task begins and ends. Gantt Charts are useful for envisioning an entire project over time.

Ghosting   Lines of argumentation used in proposal text to discredit or play down the competitor's strong points.

Government Performance and Results Act (GPRA)   GPRA (P.L. 103-62), enacted in 19993, legislates that federal agencies are to be more responsive and accountable to the public/customers relative to achieving program results. GPRA requires that agencies develop comprehensive strategic, business, and performance plans documenting the organizational objectives, goals, strategies, and measures for determining results. The organizational plans and measures for achieving the goals are key elements in determining an agency's need for services.

Hypermedia   Richly formatted documents that contain a variety of information types, such as textual, image, movie, and audio.

Incumbent   The contractor or contractor team that is currently performing on a contract that is approaching recompetition.

Matrix management   Management configuration in which staff are "assigned" to support a program or project manager outside their own line organizations or groups.

Micropurchase   A subcategory of simplified acquisitions that refers to purchases of less than $25,000.

Minority-owned business   A business that is at least 51% owned by one or more individuals classified by the U.S. government as socially and economically disadvantaged and whose management and daily business operations are controlled by one or more such individuals. Socially or economically disadvantaged individuals, by government classification, include Black Americans, Hispanic Americans, Native Americans (includes Eskimos, Aleuts, and native Hawaiians), Asian-Pacific Americans, and Asian-Indian Americans.

Modular   A proposal volume format in which each section strictly adheres to a page allocation that contains facing text and art pages, usually one page of each per section.

Modular contracting   Type of contracting that provides for the delivery, implementation, and testing of a workable information technology (IT) system or solution in discrete increments, or *modules*. Modular contracting is one of many approaches that can be used by federal agencies to acquire major IT systems. It may be achieved by a single procurement, or multiple procurements, but is intended to ensure that the government is not obligated to purchase more than one module at a time. Modular contracting is intended to balance the government's need for fast access to rapidly changing technology and incentivized contractor performance, with stability in program management, contract performance, and risk management.

To help improve federal agencies' acquisition and management of major IT systems, Congress passed, and the president signed, the Information Technology Management Reform Act (ITMRA, P.L. 104-106, February 1996; also known as the Clinger/Cohen Act). Section 5202 of this law directs federal agencies to use modular contracting, to the maximum extent practicable, for the acquisition of major IT systems. Following the ITMRA, the president issued Executive Order No. 13011, which instructs agencies to apply modular contracting "where appropriate" and "to the maximum extent practicable."

Note that one of great enablers of modular contracting is the ready availability of task order, multiagency, and Government-wide Agency Contracts (GWACs). With these tools, program managers have access to many major IT service and product providers through these contracts. In some cases, task orders can be in place within 60 days of the initial request. This allows program managers to compete modular tasks quickly and easily.

Modular contracting is one of the federal government's strategies that moves the acquisition strategy from a traditional "grand design" to a more manageable incremental approach. A significant benefit offered by modular contracting is its potential to refocus acquisition management attention and strategies to a more responsive and realistic "systems development acquisition model." (See also http://www.itpolicy.gsa.gov/mks/whitepr/modawpex.htm.)

Option years   Additional years available on a given contract if the government elects to exercise its option to renew that contract.

Oral presentation   An oral examination in which an offeror submits information bearing on its capability to a panel of government source selection officials. This information may include, for example, a description of, and justification for, the offeror's performance policies, processes, and plans. The oral presentation is distinct from the offer (i.e., the proposal) in that it is not itself an offer or part of an offer, and does not become a part of any resultant contract. The source selection panel uses the oral information to determine the offeror's understanding of the prospective work and, thus, its capability to perform successfully.

Out years   The final years exercised on a given contract.

Performance Based Service Contracting (PBSC)   PBSC is the process of contracting for services by using mission-related, outcome-oriented statements of work (Performance Work Statements) and quality assurance performance measures. PBSC focuses on the desired outcome and its quality measures rather than on the "how" of providing the required services. A Performance Work Statement (PWS) is the basic document used in PBSC that describes the specific requirements the contractor must meet in performance of the contract in terms of "output" and a "measurable standard" for the output.

PERT Chart   PERT, or Program Evaluation and Review Technique, is a "flow diagram showing the sequence of ... various activities and their interdependence in terms of completion dates. In a sense it is a road map to a destination to show the intermediate points and distances between them. Activities are sequenced from left to right." (See Krathwohl, *How to Prepare a Research Proposal*, p. 75.) Developed initially by the U.S. Navy, Lockheed, and Booz-Allen and Hamilton, PERT charts are used iin project management and project planning activities.

Portrait/landscape pages   *Portrait* refers to pages of a document oriented such that the vertical dimension is 11 in and the horizontal dimension is 8½ in. (This page is in portrait format.) *Landscape* refers to pages oriented such that the vertical dimension is 8½ in and the horizontal dimension is 11 in. Landscape pages are often used for full-page graphics or tables.

Procurement Technical Assistance Centers   Procurement Technical Assistance Centers (PTACs) are government and privately funded organizations dedicated to helping small companies do business with the federal government.

Profit center   The smallest organizationally independent segment of a company charged by that company's management with profit and loss (P&L) responsibilities.

SBIR   Small Business Innovation Research program. To receive SBIR information as soon as it becomes available, write to:

U.S. Small Business Administration
Office of Innovation Research & Technology
409 Third Street, SW
Washington, DC 20416
Tel. (202) 205-6450

The Small Business High Technology Institute (SBHTI) publishes a very limited printed collection of winning SBIR proposals that includes the text of 15 Phase I and 2 Phase II proposals. Contact SBHTI at:

Small Business High Technology Institute
346 West Georgia
Phoenix, AZ 85016
Tel. (602) 277-6603
FAX (602) 279-7175

Set aside   A kind or class of procurement reserved for contenders who fit a certain category, for example, business size, region, or minority status (see FAR 19.501).

SF   Standard Forms (SFs) and optional forms may be obtained from the Superintendent of Documents, Government Printing Office (GPO), Washington, DC 20402, or from the prescribing agency (see FAR 53.107).

SF 33   Solicitation offer and award. SF 33 is prescribed for use in soliciting bids for supplies or services and for awarding contracts that result from bids.

SF 129   Standard Form 129 is the Solicitation Mailing List Application. The list is used for government planning purposes to match direct procurement opportunities with companies that may be able to provide the specific products or services required, as specified in FAR 14.205-1(d). This form is now being replaced by the EDI 838 contractor registration process.

SF 254/255   SF 254 is known formally as Architect-Engineer and Related Services Questionnaire. SF 255 is known formally as Architect-Engineer and Related Services Questionnaire for Specific Project.

SF 1411   Contract Pricing Cover Sheet.

Simplified Acquisition Procedures (SAPs)   Streamlined techniques and guiding principles designed to reduce the administrative burden of awarding the lower dollar value procurements that account for the vast majority of DoD acquisition. They allow for informal quoting and competition procedures, encourage accepting oral quotes versus written quotations, prefer comparing quoted prices versus conducting negotiations, and provide streamlined clauses to support the award document. Saving money, improving opportunity and efficiency, and avoiding administrative burden are at the core of the SAP program. The Federal Acquisition Streamlining Act (FASA) and the Federal Acquisition Reform Act (FARA) were integral in the creation of simplified acquisitions and the simplified acquisition threshold (SAT) of $100,000.

Small business   A business that is independently owned and operated and is not dominant in its field; a business concern meeting government size standards for its own particular industry type. A small business in the manufacturing field normally employs, with its affiliates, not more than 500 persons. In the services arena, small businesses are usually defined as businesses that generate less than $5 million in gross annual receipts.

Small Business Act (1953)   This Act established the U.S. Small Business Administration. Public Law 95-507, enacted in October 1978, made major revisions to strengthen the Small Business Act.

Small Business Administration (SBA)   The federal government agency whose function is to aid, counsel, provide financial assistance to, and protect the interests of the small business community (from McVay, *Proposals That Win Federal Contracts*, p. 324). The U.S. SBA has the responsibility of making certain that small business obtains a fair share of Government contracts and subcontracts. Their mission is articulated in the Small Business Act of 1953, which established that organization. The SBA's Office of

Business Initiatives serves as the small business liaison to the expanding arena of EC/ED. FAX: (202) 205-7416; http://www.sba.gov.

Sole source acquisition   A contract for the purchase of supplies or services that is entered into or proposed to be entered into by the contracting officer of an agency after soliciting and negotiating with only one source (vendor, contractor). This is done on the grounds that only that one source is capable of satisfying the government's requirements.

Standard Industrial Classification (SIC)   The SIC was developed for use in the classification of business establishments by type of activity in which they are engaged. The SIC is intended to cover the entire field of economic activities. The U.S. Census Bureau has announced that the U.S. SIC system is being replaced by the North American Industry Classification System (NAICS). NAICS was developed jointly by the United States, Canada, and Mexico to provide comparable statistics about business activity across North America's three NAFTA trading partners. NAICS will make it easier to compare North American business data with the International Standard Industrial Classification System (ISIC), developed and maintained by the United Nations. The *NAICS Manual* is available through the National Technical Information Service (NTIS) at 1-800-553-6847 or on the Web at http://www.ntis.gov/yellowbk/1nty205.htm.

Stet   Literally, "let it stand"; a proofreader's notation meaning that the indicated change should not be made.

Strawman RFP or proposal   Mock, simulated, or preliminary RFP or proposal.

Technical leveling   The elimination of differences between competitors by repeated discussions and exchange of information. It is inappropriate for the government to assist offerors by suggesting modifications that improve their proposals (FAR 15.610(d)(1)). (Cibinic and Nash, *Formation of Government Contracts*, pp. 529, 625.)

Technical transfusion   The transfer of one competitor's ideas to another competitor during the course of discussions with the government. (Cibinic and Nash, *Formation of Government Contracts*, pp. 529, 622.)

Themes   Authenticated, substantiated statements that clearly articulate why the client should select your company over your competition.

Examples: common themes, unique themes, competitive themes. Supported claims; platforms; sound arguments.

Tiger Team   Specialized working group generally assigned to track one particular aspect of a proposal effort, such as cost strategy.

Walsh-Healey Act (1936)   "41 U.S.C. §§ 35-45, requires that, for contracts exceeding $10,000, the contractor be a 'manufacturer of' or a 'regular dealer in the materials, supplies, articles, or equipment to be manufactured or used in the performance of a contract.' (Cibinic and Nash, *Formation of Government Contracts*, p. 226.)

War room   A facility or portion of a facility dedicated to proposal operations, planning, writing, and review. The walls of a war room are often used to display proposal (and RFP) sections for review and comment. Seeing specific proposal sections in the context of other sections aids in overall continuity and flow, as well as helps eliminate unnecessary redundancy. A war room should be a highly secure area because of the open display of proposal materials.

Woman-owned business (WOB)   A business that is at least 51% owned, controlled, and operated by a woman or women. "Control" is defined as exercising the power to make policy decisions, and "operate" is defined as actively involved in day-to-day management. WOBs now constitute 32 percent of all U.S. small businesses, according to the Census Bureau. That number will expand to 40 percent by 2000. According to a 1995 study by the National Foundation for Women Business Owners and Dun and Bradstreet, there are now 7.7 million women-owned firms in the United States that provide jobs for 15.5 million people.

Selected list of acronyms and abbreviations

ACA	after contract award
ACASS	architect/engineer contract administration support system
ACEC	American Consulting Engineers Council
ACH	automated clearing house
ACO	administrative contracting officer
ACOP	Acquisition Center Business Opportunity (CECOM)
ACRI	acquisition cost reduction initiatives

ACWP	actual cost of work performed
ADA	Americans with Disalilities Act
ADB	Asian Development Bank; African Development Bank
ADP	automated (or automatic) data processing
ADPE	automatic data processing equipment
ADR	administrative dispute resolution; alternative dispute resolution
A-E	architect-engineer
A/E/C	Architect/Engineer/Construction
AEP	Asian Environmental Partnership
AF	Air Force
AFAA	Air Force Audit Agency
AFARS	Army Federal Acquisition Regulation Supplement
AFB	Air Force base
AFCEE	Air Force Center for Environmental Excellence
AFFARS	Air Force Federal Acquisition Regulation Supplement
AFI	Air Force Instruction (e.g., AFI 32-4002, "Hazardous Materials Emergency Planning and Response Compliance")
AFMC	Air Force Material Command
AFR	Air Force Regulation
AID	Agency for International Development (320 21st Street, N.W., Washington, D.C., 20523; phone: (202) 647-9620)
AISB	American Institute for Small Business (Sarasota, Florida; Http://www.smallbizz.org/aisb/)

ALC additional labor categories

ALN asynchronous learning network

AMC Army Material Command

AMD Acquisition Management Directorate (DCA)

AMSDL acquisition management source data list

ANSI American National Standards Institute

AOL America Online

APEC Asian-Pacific Economic Cooperation

APMP Association of Proposal Management Professionals (Internet: http://www.apmp.org/)

APR agency procurement request

AR Army regulation; acquisition reform

ARNet Acquisition Reform Network

ARO after receipt of order

ARDEC Armament Research, Development and Engineering Center

ASBA American Small Business Association (Washington, D.C.; http://www.asba.net)

ASBCA Armed Services Board of Contract Appeals

ASBCD Association of Small Business Development Centers

ASC Accredited Standards Committee (ANSI)

ASPA Armed Services Procurement Act (1948, 10 U.S.C. § 2301 et seq.)

ASPM Armed Services Pricing Manual

ASPR	Armed Services Procurement Regulation
ASSIST	Acquisition Streamlining and Standardization Information System
ATR	assistant technical representative
B&P	bid and proposal
BA	basic agreement
BAA	broad agency announcement
BABC	British-American Business Council
BAFO	best and final offer
BARFO	best and revised final offer
BBS	bulletin board system
BCA	Board of Contract Appeals
BCWP	budgeted cost for work performed
BCWS	budgeted cost for work scheduled
BDG	business development group
BECC	Border Environmental Cooperation Commission
BI	SBA's Office of Business Initiatives
BIC	SBA Business Information Center. BICs provide small businesses with access to high-technology hardware, software, and telecommunications. Each BIC offers electronic bulletin boards, computer databases, on-line information exchange, periodicals and brochures, videotapes, reference materials, interactive media, and counseling. Call 1-800-8-ASK-SBA for the location nearest your office or visit http://www.sba.gov/bi/bics on the Web.
B/L	bill of lading

BLS   Bureau of Labor Statistics, U.S. Department of Labor

BOA   basic ordering agreement

BOD   bid opening date; the final date that a bid must be received by the appropriate government office.

BOE   basis of estimate

BOM   bill of materials

BOTP   British Overseas Trade Board

BPA   blanket purchase agreement; this agreement between the government and a vendor gives the government the option to purchase goods or services from the vendor when needed on an on-call basis.

BPCR   breakout procurement center representative

BPO   blanket purchase order

BPR   business process reengineering

BSO   Business Support Centers (Japan)

BT   British Telecom

BVS   best value selection

CAA   Council Civilian Agency Acquisition Council

CAAS   contract for advisory and assistance services

CACES   computer-aided cost estimation system

CACO   corporate administrative contracting officer

CAD   computer-aided design

CAGE   Commercial and Government Entity (Code). Contact the Defense Logistics Service Center in Battle Creek, Michigan at 1-888-352-9333 (toll free) to begin the process of applying for a CAGE code. This five-character

CAGE code is required for doing business with the DoD, but not with civil federal agencies. The codes are used by the DoD for record-keeping purposes.

CAIV   cost as an independent variable

CANDI   commercial and nondevelopmental item

CAR   Commerce Acquisition Regulation

CAS   cost accounting standard (enumerated at FAR 30); contract administration services

CASB   Cost Accounting Standards Board

CBD   *Commerce Business Daily*

CBSC   Canada Business Service Centre

CCASS   construction contract appraisal support system

CCF   Contract Cases Federal

CCH   Commerce Clearing House

CCI   consolidated contracting initiative (management process that emphasizes developing, using, and sharing contract resources to meet government agency objectives)

CCN   Cooperating Country Nationals

CCP   current cost or pricing

CCR   Central Contractor Registry (Columbus, Ohio, and Ogden, Utah; an electronic database of government vendor firms that is maintained by the DoD); Central Consultancy Register (Brussels, Belgium)

CDA   Contract Disputes Act

CDL   contract data list

CDRD   contract data requirements document

CDRL	contract data requirements list; contractor establishment code
CEB	cost element breakdown
CEC	Center for Electronic Commerce; contract establishment code
CECOM	U.S. Army's Communications-Electronics Command
CEI	contract end item
CENDI	Commerce, Energy, NASA, and Defense Information
CEO	chief executive officer
CER	cost estimating relationship
CERN	Conseil European pour la Recherche Nucleaire; European Organization for Nuclear Research. The originators of the HTTP and HTML concepts.
CEU	continuing education unit
CFE	contractor furnished equipment
CFP	customer furnished property
CFR	*Code of Federal Regulations* (e.g., Title 41, Public Contracts)
CGLI	comprehensive general liability insurance
CI	configuration item; contractor inquiry
CICA	Competition in Contracting Act of 1984 (P.L. 98-369, 98 Stat. 1175)
CID	commercial item description
CIF	cost, insurance, freight
CIO	chief information officer
CIS	capability/interest survey; Commonwealth of Independent States (*former* Soviet Union)

CITI	Centre for Information Technology Innovation
CKO	chief knowledge officer
CLIN	contractor line item
CLS	contractor logistics support
CM	configuration management
CMM	Capability Maturity Model (Software Engineering Institute)
CO	contracting officer
COB	close of business
COC	certificate of competency (issued by SBA)
COCO	contractor owned, contractor operated; Chief of the Contracting Office (DOT)
COD	cash on delivery
CODSIA	Council of Defense and Space Industries Associates
COI	conflict of interest
COMET	Committee for Middle East Trade (part of the British Overseas Trade Board)
COMSEC	communications security
CONUS	continental United States
CORDUS	Community R&D Information Service (Europe)
COTR	contracting officer's technical representative
COTS	commercial off-the-shelf
CPAF	cost plus award fee

CPARS	contractor performance assessment reporting system
CPD	comptroller's procurement decisions
CPFF	cost plus fixed fee
CPI	Consumer price index
CPIF	cost plus incentive fee
CPM	critical path method
CPMD	Contractor Performance Measurement Organization
CPPF	cost plus percentage fee
CPRS	contractor purchasing system review
CPU	central processing unit (computer platform)
CR	clarification request
CRA	continuing resolution authority
CRADA	cooperative research and development agreement
C/SCSC	cost/schedule control system criteria
CTW	Contracts and Tenders Worldwide (White Plains, New York, and London)
CWBS	contract work breakdown structure
CY	calendar year
D&B	Dun & Bradstreet
D&F	determinations and findings
DABBS	Defense Communications Agency's Acquisition Bulletin Board System
DACON	Data on Consulting Firms (automated World Bank and Asian Development Bank systems)

DAPS	Defense Automated Printing Service (Philadelphia)
DAR	Defense Acquisition Regulation
DARO	days after receipt of order
DARPA	Defense Advanced Research Projects Agency
DBA	Davis-Bacon Act (1931)
DBE	Disadvantaged business enterprise
DCA	delegation of contracting authority
DCAA	Defense Contract Audit Agency
DCAAM	Defense Contract Audit Agency Manual
DCADS	Defense Contract Action Data System
DCAS	Defense Contract Administration Service, Defense Logistics Agency
DCI	data collection instrument
DCMC	Defense Contract Management Command
DCSC	Defense Construction Supply Center
DESC	Defense Electronics Supply Center
DFAR	Defense Federal Acquisition Regulations (also called DAR)
DFARS	Defense Federal Acquisition Regulation Supplement
DFAS	Defense Finance Accounting Service
DGSC	Defense General Supply Center
DHHS	Department of Health and Human Services
DICON	Data on Individual Consultants (automated Asian Development Bank system)

DID	data item description
DIRMM	DOT's IRM Manual
DISA	Defense Information Systems Agency; Data Interchange Standards Association
DISC	Defense Industrial Supply Center
DLA	Defense Logistics Agency; Direct Labor Analysis
DLIS	Defense Logistics Information Service
DLSC	Defense Logistics Services Center (now called DLIS)
DM	division manager
DN	deficiency notice
D.O.	delivery order
DOC	Department of Commerce
DoD	Department of Defense
DODD	Department of Defense Directive (e.g., DODD 5000.1)
DODI	Department of Defense Instruction (e.g., DODI 5000.2)
DODISS	Department of Defense Index of Specifications and Standards
DODSSP	Department of Defense Single Stock Point for Specifications and Standards
DOE	Department of Energy
DOI	department operating instruction
DOJ	Department of Justice
DOS	Department of State

DOSAR	Department of State Acquisition Regulation
DOSC	delivery order selection criteria
DOT	Department of Transportation
DPA	Delegation of Procurement Authority (given by GSA)
DPAS	Defense Priorities and Acquisition System
DR	deficiency report
DRD	data requirements document
DRFP	draft request for proposal
DRL	data requirements list
DROLS	Defense Research Development, Test, and Evaluation Online System
DSARC	Defense Systems Acquisition Review Council
DSP	Defense Standardization Program
DTC	design to cost
DT&E	development, test, and evaluation
DTIC	Defense Technical Information Center
DTP	desktop publishing
DUAP	Dual-Use Applications Program (DoD)
DUNS	Data Universal Numbering System; to apply for a DUNS number, call Dun & Bradstreet at 1-800-333-0505.
EAC	estimate at completion (cost)
EBRD	European Bank for Reconstruction and Development
EBS	electronic bid set

EC	electronic commerce; *also* European Community; European Commission
ECAPMO	Electronic Commerce Acquisition Program Management Office
ECIC	Electronic Commerce Information Center
ECP	engineering change proposal
ECPN	electronic commerce processing node
ECPO	Electronic Commerce Program Office; a multi-agency group assembled under the coleadership of the General Services Administration and the Department of Defense. Its purpose is to develop a federal strategy to implement EC/EDI for the federal acquisition programs (eca.pmo@gsa.gov).
ECRC	Electronic Commerce Resource Center. (The mission of the ECRC Program is to promote awareness and implementation of electronic commerce and related technologies into the U.S. integrated civil-military industrial base. See http://www.ecrc.ctc.com/ecrcpr.html).
ECU	European currency unit
EDF	European Development Fund
EDI	electronic data interchange
EDIFACT	EDI for Administration, Commerce and Transport (Europe and Asia)
EEA	European Economic Area (Norway)
EELV	evolved expendable launch vehicle
EEO	Equal Employment Opportunity
EFT	electronic funds transfer
EIB	European Investment Bank
EIC	Euro Information Centre(s)
EICN	Euro Info Centre Network

EIU   Economist Intelligence Unit (London)

EO   Executive Order

EOI   expression of interest

E&MD   engineering and manufacturing development

EPA   Environmental Protection Agency

EPAAR   Environmental Protection Agency Acquisition Regulations

EPIC   Electronic Processes Initiatives Committee

EPIN   European Procurement Information Network

EPLS   excluded parties list (excluded from participating in federal procurement programs)

EPS   equipment performance specification; electronic publishing system; electronic proposal submission

EPTF   Electronic Procurement Task Force (APMP)

ESS   electronic source selection

EU   European Union

EVMS   earned value management system

EWCP   export working capital program (provides short-term, transaction-specific financing for certain loans)

F3I   form, fit, function and interface

FAA   Federal Aviation Administration

FAC   Federal Acquisition Circular

FACNet   Federal Acquisition Computer Network

FAQ   frequently asked questions

FAR	Federal Acquisition Regulation
FARA	Federal Acquisition Reform Act (included in the 1996 Defense Authorization Act, Clinger-Cohen Act)
FAS	free along side
FASA	Federal Acquisition Streamlining Act (October 1994) (Public Law 103-355)
FAT	factory acceptance test; first article testing. When DoD buys certain goods, they may perform extensive tests on the first item delivered.
FCCM	facilities capital cost of money
FCIA	Foreign Credit Insurance Association
FCR	federal contracts report
FDI	foreign direct investment
FDO	(Government) Fee Determination Official (determines amount of award fee on a contract based upon contractor performance)
FEDRIP	Federal Research in Progress (database)
FED-STD	federal standard
FEMA	Federal Emergency Management Agency
FESMCC	Federal EDI Standards Management Coordinating Committee
FFP	firm fixed price
FFRDC	federally funded research and development center
FIND	Federal Information & News Dispatch, Inc.
FIP	Federal Information Processing
FIPS	Federal Information Processing Standards (Department of Commerce)

FIRMR   Federal Information Resources Management Regulation (41 CFR Chapter 201). Government-wide regulations that govern the purchase of computer goods and services.

FLC   Federal Laboratory Consortium

FLSA   Fair Labor Standards Act

FMS   foreign military sales

FMSS   Financial Management Systems Software

FMV   fair market value

FOB   free on board

FOIA   Freedom of Information Act

FOIR   Freedom of Information Request

FPASA   Federal Property and Administrative Services Act (1949, 41 U.S.C. § 251 et seq.)

FPDC   Federal Procurement Data Center

FPDS   Federal Procurement Data System

FP-EPA   fixed price with economic price adjustment

FPI   fixed price incentive

FPIF   fixed price incentive fee

FPLE   fixed price level of effort

FPLH   fixed price labor hour

FPMR   Federal Property Management Regulations

FPO   for position only (used in reference to graphics or photos)

FPR   Federal Procurement Regulations; fixed price with redetermination

FR   Federal Register

FSC   Federal Supply Classification (or Class) (associated with SAACONS Vendor Information Program)

FSN   Federal Stock Number; a code used to identify documents sold by the U.S. Government Printing Office, Superintendent of Documents

FSS   Federal Supply Service. An organization within the U.S. General Services Administration.

FTE   full-time equivalent

FTP   file transfer protocol (a method of transferring files to and from remote locations)

FTR   federal travel regulation

FY   fiscal year

FYDP   five-year defense plan or program; future years defense plan

G-2   intelligence (information on competitors, contracts, etc.)

G&A   general and administrative

GAAP   generally accepted accounting principles

GAAS   generally accepted auditing standards

GAO   General Accounting Office

GATT   General Agreement on Tariffs and Trade

GBL   government bill of lading

GDP   gross domestic product

GEF   Global Environmental Fund (World Bank)

GEMS   Australian Government Electronic Marketplace Service

GFE	government furnished equipment
GFF	government furnished facilities
GFM	government furnished material
GFP	government furnished property
.GIF	graphics interchange format
GIS	geographic information system
GMSS	guaranteed maximum shared saving
GOCO	government owned, contractor operated
Gopher	text-based information distribution system developed at the University of Minnesota
GPO	Government Printing Office
GPRA	Government Performance Results Act (Public Law 103-62, 1993). This Act holds federal agencies accountable for achieving performance results.
GRA&I	Government Reports Announcements and Index
GSA	General Services Administration
GSBCA	General Services Board of Contract Appeals
GSD	Government Supplies Department (Hong Kong, China)
GSFC	Goddard Space Flight Center
GST	goods & services tax (Canada)
GTN	Global Technology Network
GWAC	government-wide acquisition contract
HBCU	historically black colleges and universities

HCA	head of contracting activity
HHS	Health and Human Services
HIA	high impact agencies
HMSO	Her Majesty's Supply Office (Great Britain)
HOA	head of the operating administration
H&S	health and safety
HS	Harmonized system (international trade)
HTML	Hypertext Markup Language (coding mechanism used to author Web pages; HTML is a subset of SGML)
HTRW	hazardous, toxic, and radioactive waste
HTTP	HyperText Transport Protocol (the protocol, or planned method of exchanging text, graphic images, sound, video, and other multimedia files used by the WWW servers)
HUBZone	historically underutilized business zone
IAPSO	Inter-Agency Procurement Services Office (United Nations)
IAW	in accordance with
IBRD	International Bank for Reconstruction and Development (part of the World Bank Group)
ICB	international competitive bidding
ICE	independent cost estimate
ICR	intelligent character recognition (scanning technology)
ICSID	International Centre for Settlement of Investment Disputes (part of the World Bank Group)
IDA	International Development Agency (part of the World Bank Group)

IDB	Interamerican Development Bank
IDC	indefinite delivery contract
IDE	Institute of Developing Economies (Japan)
ID/IQ	indefinite delivery/indefinite quantity (contract type, examples of which are the Air Force's Desktop V and the Army's Small Multiuser Computer Contracts)
IE	individual experience (SF254, Block 11)
IEI	invitation for expression of interest
IESC	International Executive Service Corps
IEW	Intelligence and electronic warfare
IFB	invitation for bid
IFC	International Finance Corporation (part of the World Bank Group)
IG	inspector general
IGCE	independent government cost estimate
ILA	integrated logistics assessment
ILS	integrated logistics support
IMAS	International Marketing Assistance Service (SUNY-Albany program)
IMF	International Monetary Fund (World Bank)
IOC	initial operating capability
IOT&E	initial operational test and evaluation
IPD	integrated logistics assessment
IPPA	Integrated Public Procurement Association (Europe)

IPPD	integrated product and process development
IQC	indefinite quantity contract
IRAD	independent research and development
IR&D	internal (independent) research and development
IRM	information resources management
IRS	Internal Revenue Service
ISBC	International Small Business Consortium (Norman, Oklahoma; http://www.isbc.com/isbc/)
ISDN	Integrated services digital network (128Kbps)
ISIC	International Standard Industrial Classification system
ISP	Internet service provider (current top commercial ISPs are AOL, the Microsoft Network, and Prodigy)
ISSA	Interservice Support Agreement
ISSAA	Information Systems Selection and Acquisition Agency (U.S. Army)
ISSN	International Standard Serial Number
I&T	integration and test
IT	information technology
ITAR	International Traffic in Arms Regulations
ITMRA	Information Technology Management Reform Act (1996)
ITOP	Information Technology Omnibus Procurement (a DOT-sponsored National Partnership for Reinventing Government experiment to provide streamlined information technology services and products to DOT and other federal agencies)
ITT	invitation(s) to tender

IV&V	independent verification and validation
J&A	justification and approval
JDAM	Joint Direct Attack Munition
JEPCO	Joint Electronic Commerce Program Office
JETRO	Japanese Government Procurement Database (Japan External Trade Organization)
JN	job number
JOFOC	justification for other than full and open competition
JPEG	Joint Photographic Expert Group (a method of storing an image in digital format)
JSC	Johnson Space Center
JTR	joint travel regulations
JV	joint venture
JWOD Act	Javits-Wagner-O'Day Act. The Javits-Wagner-O'Day (JWOD) Program creates employment and training opportunities for people who are blind or have other severe disabilities and, whenever possible, prepares them for competitive jobs. Under the JWOD Program, government employees are required to buy selected supplies and services from nonprofit agencies employing such persons. As a result, federal customers obtain quality products and services at reasonable prices, while JWOD employees are able to lead more productive, independent lives. The Committee for Purchase From People Who Are Blind or Severely Disabled is the federal agency that administers the JWOD Program and maintains a Procurement List of mandatory source items. Two national organizations, NIB (National Industries for the Blind) and NISH (serving people with a range of disabilities), have been designated to provide technical and financial support to more than 550 nonprofit agencies participating in the JWOD Program. The Federal Acquisition Streamlining Act of 1994 continues the legal requirement to buy JWOD items.
Kbps	Kilobits per second

KISS	"Keep it short and simple"
KME	knowledge management environment
KSC	Kennedy Space Center
LAN	local area network
LCC	lifecycle costs
LCOTR	lead COTR
L-H	labor-hour
LIB	limited international bidding
LLNL	Lawrence Livermore National Laboratory
LOB	line of business
LOE	level of effort
LOI	letter of interest (Canadian equivalent of RFI); also letter of intent
LOSP	Liaison Outreach and Services Program (DOT)
LPO	local project overhead
LPTA	lowest price, technically acceptable
LS	lump sum
LSA	labor surplus area; a federal program to set aside certain contracts to businesses located in areas with high unemployment
LTOP	lease-to-ownership program
LWOP	lease with option to purchase
MAC	multiple award contracts
MAN	metropolitan area network

MBDA	Minority Business Development Agency (U.S. Department of Commerce)
MBDC	Minority Business Development Center (DOT)
MBE	minority business enterprise
Mbps	megabits per second
MBRC	Minority Business Resource Center
MCTL	Military Critical Technology List
MDA	Multilateral Development Agencies
MED	Minority/female/disabled-owned business; minority enterprise development
MEGA	Minority Enterprise Growth Assistance
MEO	most efficient organization
MERX	Internet-based national electronic tendering service (Canada)
MI	Minority institutions
M&IE	meals and incidental expenses
MIGA	Multilateral Investment Guarantee Agency (part of the World Bank Group; established in 1988)
MIL-HBK	Military Handbook
MILSPEC	Military Specification
MIL-STD	Military Standard
MIS	management information system
MMO	Materials Management Office
MNC	multinational corporation

MOA	memorandum of agreement
MOU	memorandum of understanding
MPC	most probable cost
MPCG	most probable cost to the government
MPP	Mentor–Protégé Program (DoD)
MPT	modular proposal technique
MR	modification request
MRO	maintenance, repair, and operating supplies
MSFC	Marshall Space Flight Center
MYP	multiyear procurement
NABDC	Native American Business Development Center
NAF	nonappropriated fund
NAFTA	North American Free Trade Agreement
NAICS	North American Industry Classification System (NAFTA, 1997)
NAIS	NASA Acquisition Internet Service
NAPM	National Association of Purchasing Management
NAPS	Navy Acquisition Procedures Supplement
NASA	National Aeronautics and Space Administration
NASAPR	NASA Procurement Regulation
NAVSEA	Naval Sea Systems Command
NAVSUP	Naval Supply Systems Command

NAWBO	National Association of Women Business Owners
NBA	National Business Association
NCB	national competitive bidding
NCMA	National Contract Management Association
NCMB	National Coalition of Minority Businesses
NCSA	National Center for Supercomputing Applications (at the University of Illinois Urbana-Champaign)
NDI	non-developmental items
N.E.C.	Not Elsewhere Classified (related to SIC codes)
NECO	Navy's Electronic Commerce On-Line
NEP	Network Entry Point (EC/EDI; two NEPs in Ohio and Utah)
NFS	NASA FAR Supplement
NFWBO	National Foundation for Women Business Owners (Silver Spring, Maryland; http://www.nfwbo.org/nfwbo/nfwboabt.htm)
NHB	NASA Handbook
NIB	National Industries for the Blind
NIC	Newly industrializing countries
NICRA	Negotiated Indirect Cost Rate Agreement
NIGP	National Institute of Governmental Purchasing, Inc. (Canada)
NIS	New Independent States (of the former Soviet Union)
NIST	National Institute of Standards and Technology
NLT	no later than

NMI	NASA Management Instruction
NMSO	National Master Standing Offers (Canada)
NOAA	National Oceanic and Atmospheric Administration
NPR	National Partnership for Reinventing Government (formerly National Performance Review)
NRA	NASA Research Announcement
NRC	Nuclear Regulatory Commission
NSF	National Science Foundation
NSN	national stock number; a unique number assigned by the General Services Administration that catalogs a wide range of items by commodity, group, and class.
NSNA	no stock number assigned
NSP	not separately priced
NTE	not to exceed
NTIS	National Technical Information Service
NTP	notice to proceed
o/a	on or about
O&M	operations and maintenance
OA	office automation
OBS	organizational breakdown structure; Open Bidding Service (Canada)
OCI	organizational conflict of interest
OCR	optical character recognition (scanning technology)
ODC	other direct cost

OEM	original equipment manufacturer
OFCC	Office of Federal Contract Compliance
OFPP	Office of Federal Procurement Policy
OGAS	other government agency system(s)
O/H	overhead
OICC	officer in charge of construction
OIRM	Office of Information Resources Management
OJ	*Official Journal of the European Communities*
OMB	Office of Management and Budget
OPBA	Ontario Public Buyers Association, Inc. (Canada)
OPM	Office of Personnel Management
OPW	Office of Public Works (Ireland)
O&S	operations and support
OSD	Office of the Secretary of Defense
OSDBUs	Offices of Small and Disadvantaged Business Utilization (Small Business Administration; these offices provide procurement assistance to small, minority-8(a), and woman-owned businesses)
OSHA	Occupational Safety and Health Administration (established in 1970)
OSTA	Optical Storage Technology Association (Santa Barbara, California)
OTA	Office of Technology Assessment
OT&E	operational test and evaluation
OTS	off-the-shelf

P&L	profit and loss
P&S	product & service code
P³I	pre-planned product improvement
PAD	Project Approval Document (part of NASA acquisition process)
PALT	procurement administrative lead time
PAR	proposal analysis report
PARC	principal assistant responsible for contracting
PASS	procurement automated source system (now defunct; replaced by Pro-*Net*)
PAT	Process Action Team (DoD)
PBBE	performance-based business environment
PBC	performance-based contracting
PBO	performance-based organization
PBSC	performance-based service contracting
PC	personal computer
PCO	procuring (or procurement or principal) contracting officer
PCR	procurement center representative
PDF	personnel data form; (.PDF) portable document format
PEA	Procurement Executives Association
PEAG	proposal evaluation analysis group
PEDS	program element descriptive summary
PEN	program element number

PEP	project execution plan
PERT	program evaluation and review techniques (diagrams)
PI	principal investigator
PID	procurement initiation document (German procurement law)
PIIN	procurement instrument identification number
PIN	pre-invitation notice; a summary of a solicitation package sent to prospective bidders, who may then request the entire solicitation package
PIO	procurement information online
PIP	procurement improvement plan
PL	public law
PLI	professional liability insurance
PM	project manager, program manager
PMMP	Pilot Mentor–Protégé Program
P/N	part number
PO	purchase order
POA	plan of action
POC	point of contact
POG	paperless order generator
PPBS	planning, programming, and budgeting system
PPI	proposal preparation instruction; pixels per inch; past performance information
PPM	principal period of maintenance

PPT	performance–price–trade-off (U.S. Air Force evaluation strategy)
PQ^{2}D	price quantity quality delivery
PQS	personnel qualification sheet
PR	purchase request
PRDA	program R & D announcements
PROCNET	Procurement Network (ARDEC)
PSA	Pre-Solicitation Announcement (for SBIR and STTR opportunities, published quarterly by SBA); professional services agreement
PSC	product service code
PSD	private sector development
PSPQ	potential supplier profile questionnaire
PTAC	Procurement Technical Assistance Center
PWBS	program/project work breakdown structure
PWS	performance work statement
QA	quality assurance
QASP	quality assurance surveillance plan
QBL	qualified bidders list
QC	quality control
QCBS	quality and cost-based selection
QML	qualified manufacturers list
QPL	qualified products list
QSL	qualified suppliers list

QVL	qualified vendors list
RAFV	risk adjusted functional value
RAM	random access memory; reliability and maintainability
RAM-D	reliability, availability, maintainability or durability
R&D	research and development
RDT&E	research, development, test and evaluation
RFA	request for application (grants)
RFC	request for comment
RFI	request for information
RFO	request for offer
RFP	request for proposal
RFQ	request for quotation (normally prepared on Standard Form 18, FAR 53.301-18. Unlike bids or proposals following an IFB or RFP, quotations submitted in response to an RFQ are not considered to be offers. Standing alone, a quotation cannot be accepted by the government to form a contract for goods or services. (Cibinic and Nash, *Formation of Government Contracts*, p. 158.) *Also* Request for Qualifications
RFTP	request for technical proposals
RIP	raster image processing
RM	risk management
RM&S	reliability, maintainability and supportability
ROI	return on investment
ROICC	resident officer in charge of construction
ROM	rough order of magnitude (cost)

ROS   return on sales

RTN   routing and transfer number (electronic payments)

SA   supplemental agreement

SAACONS   Standard Army Automated Contracting System

SADBUS   small and disadvantaged business utilization specialist

SAME   Society of American Military Engineers

SAP   Simplified Acquisition Procedures (FAR Part 13)

SAT   Simplified Acquisition Threshold

SBA   Small Business Administration

SBANC   Small Business Advancement National Center (University of Central Arkansas; http://wwwsbanet.uca.edu/)

SBD   Standard Bidding Documents (e.g., the World Bank)

SBDC   Small Business Development Center (There are 56 SBDCs.)

SBE   small business enterprise

SBHTI   Small Business High Technology Institute (Phoenix, Arizona)

SBIR   Small Business Innovation Research program; $916M in awards to small businesses in FY96. In 1992, Congress reauthorized the SBIR Program to continue through 30 September 2000. Under this program, federal agencies with large research and development (R&D) budgets must direct designated amounts of their R&D contracts to small businesses. Small businesses must be American-owned and independently operated, for-profit, principal researcher employed by the business, and have no more than 500 employees. Federal agencies that participate in the SBIR Program include the DoD, USDA, DOC, DOE, DHHS, DOT, EPA, NASA, NRC, NSF, and the Department of Education. These agencies designate R&D topics and accept proposals for a three-phase program. Phase I (feasibility studies) has awards up to $100K for six months, and Phase II (full-scale research) has awards up to $750K for up to two years.

Phase III (commercialization) is the period during which the Phase II innovation moves from the laboratory into the marketplace. No SBIR funds support this phase. The small business must find the funding in the private sector or non-SBIR federal agency. The U.S. SBA serves as the coordinating agency and information link for SBIR. Examples of successful SBIR proposals include thin-film optical filters for intense laser light, membrane-based process for debittering citrus juice, and neonatal/infant/fetal pump-oxygenator system. Since the inception of the SBIR Program in 1983, nearly $4 billion in competitive federal R&D awards have been made to qualified small business concerns under the program. Only Phase I winners may submit Phase II proposals.

SBLO	small business liaison officer
SBSA	small business set aside
SCORE	Service Corps of Retired Executives (SBA program with more than 400 offices)
SDB	small and disadvantaged business
SDR	Special Drawing Rights (an international currency unit set up by the International Monetary Fund)
SDRL	subcontract data requirements list
SEB	Source Evaluation Board
SEC	Securities and Exchange Commission
SEDB	Socially and Economically Disadvantaged Business
SEI	Software Engineering Institute (Carnegie Mellon)
SEMP	system engineering management plan
SF	standard form
SFAS	Statement of Financial Accounting Standard
SGML	Standard generalized markup language

SIC	Standard Industrial Classification
SIMAP	Système d'information pour les Marchés Public (European Commission)
SITC	Standard international trade classification (devised by the United Nations to classify commodities used in international trade)
SLIN	subcontractor line item
SMC	Space and Missile Systems Center (Air Force)
SMDC	U.S. Army's Space and Missile Defense Command
SME	small and medium-sized enterprises (European Commission)
SML	solicitation mailing list
SOC	statement of capability
SOHO	small office, home office
Sol.	solicitation
SON	statement of need (equivalent of SOW)
SOO	statement of objectives
SOQ	statement of qualifications
SOW	statement of work
SP&BDG	Strategic Planning and Business Development Group
SPE	senior procurement executive
SPI	single process initiative
SSA	source selection authority
SSAC	Source Selection Advisory Council
SSCASS	Service & Supply Contract Appraisal Support System

SSEB	Source Selection Evaluation Board
SSO	source selection official
SSQAAP	Small Supplier Quality Assurance Assistance Program
STOP	sequential thematic organization of proposals (technique developed at Hughes Ground Systems)
STR	senior technical review; senior technical representative
STRICOM	U.S. Army Simulation Training and Instrumentation Command
STTR	Small Business Technology Transfer Pilot Program
SUNY	State University of New York
SWOT	Strengths, Weaknesses, Opportunities, and Threats
Ts & Cs	terms and conditions
TAC	Transportation Acquisition Circular
TAM	Transportation Acquisition Manual
TAR	Transportation Acquisition Regulation
TASBI	Transatlantic Small Business Initiative (US and European Union (EU))
TBD	to be determined
TBN	to be negotiated
TBP	to be proposed
TCN	Third County Nationals
TCO	termination contracting officer; total cost of ownership
TCP/IP	Transmission Control Protocol/Internet Protocol; a set of rules that establish the method with which data are transmitted over the Internet between computers

TD	task description
TDA	U.S. Trade and Development Agency
TDI	technical data interchange
TDP	technical data packages (U.S. Army)
TDY	temporary duty
T&E	test and evaluation
TED	Tenders Electronic Daily
TEP	total evaluated price
TEV	total evaluated value
TIFF	Tag Image File Format (a file format used to store image files)
TILO	Technical Industrial Liaison Office (Army)
TIN	Taxpayer Identification Number; call the IRS at 1-800-829-1040 to obtain a TIN
TINA	Truth in Negotiations Act (P.L. 87-653 (1962); extended in 1984)
T&M	time and materials
TMIS	technical management information system
T.O.	task order
TOA	total obligation authority
TOPS	Total Operating Paperless System (CECOM)
TORFP	task order RFP
TPCR	traditional procurement center representatives
TPIN	trading partner identification number

TQM	total quality management
TRCO	technical representative of the contracting officer
U2	unclassified/unlimited
UCA	undefinitized contract action
UCC	uniform commercial code
UCF	uniform contract format
UCLINS	uniform contract line item numbering system
UN	United Nations
UNCITRL	United Nations Commission on International Trade Law
UNBD	UN Development Business
UNDP	United Nations Development Programme
URL	Uniform Resource Locator (path or pointer for locating files and sites on the World Wide Web)
URN	unique reference number
USACE	United States Army Corps of Engineers
USAFE	United States Air Force-Europe
USAREU	United States Army-Europe
USC	United States Code
USCG	United States Coast Guard
USDA	United States Department of Agriculture
USEAC	U.S. Export Assistance Program (SBA is a key partner)
USEPA	United States Environmental Protection Agency

USTDA   United States Trade and Development Agency

VA   Department of Veterans Affairs

VAAR   Veterans Administration Acquisition Regulation

VAN   value-added network

VAS   value-added service

VE   value engineering

VECP   value engineering change proposal

VEP   value engineering proposal

VP   vice president

VUSME   Virtual University for Small and Medium Sized Enterprises; not-for-profit joint venture of business organizations and educational institutions promoting an understanding of how Internet technology can be used to start and grow profitable businesses (http://www.vusme.org/)

W2W   Work to Welfare Program

WAIS   Wide Area Information Server (a database)

WBE   woman-owned business enterprise

WBS   Work breakdown structure

WinPET   Windows proposal evaluation tool

WISE   Web-integrated solicitation elements

WMDVBE   Women, Minority and Disabled Veteran Business Enterprise

WNET   Women's Network for Entrepreneurial Training

WOB   woman-owned business

WOBREP   women-owned business representative (federal agencies such as the Department of Agriculture, Commerce, Defense, Education, and Veterans' Affairs have WOBREPs)

WTO   World Trade Organisation

WTO GPA   World Trade Organisation Agreement on Government Procurement

WWW   World Wide Web (allows for the presentation and linkage of information dispersed across the Internet in an easily accessible way; WWW = W3 = The Web, a distributed HyperText-based information system conceived at CERN to provide its user community with an easy way to access global information). The Internet II is projected to be 100 times faster than the Internet. Nearly 100 U.S. colleges and universities are participating in the upgrade process. The Internet is an international computer network of networks that connect government, academic, and business institutions.

Selected bibliography

"A Brief History of the National Performance Review," February 1997, http://www.npr.gov/lib...pers/bkgrd/brief.html.

Abramson, Neil R., and Janet X. Ai, "You Get What You Expect in China," *Business Quarterly*, Winter 1996, pp. 37–44.

"Acquisition Reform: Obstacles to Implementing the Federal Acquisition Computer Network." Letter Report, GAO/NSIAD-97-26, January 3, 1997.

Anderson, Tania, "8(a) Companies Learn the Secret of Partnering," *Washington Technology*, Vol. 10 Oct. 1996, p. 22.

"Annual Report on the State of Small Business," The White House to the Congress of the United States, 6 June 1996.

Anthes, Gary H., "Procurement Horror Stories Draw Feds' Scrutiny," *Computerworld*, 21 Feb. 1994.

Behr, Peter, "Just Say Know," *Washington Post*, 1 April 1996.

Beveridge, James M., and Edward J. Velton, *Positioning to Win: Planning and Executing the Superior Proposal*, Radnor, PA: Chilton Book Company, 1982.

Blankenhorn, Dana, "11 Lessons from Intranet Webmasters," *Netguide*, Oct. 1996, pp. 82–90.

Bowman, Joel P., and Bernadine P. Branchaw, *How to Write Proposals That Produce*, Phoenix, AZ: Oryx Press, 1992.

Brockmeier, Dave, "Help Shape Federal Acquisition Regulations," *Business Credit*, Vol. 97, March 1995, pp. 40–42.

Brusaw, C. T., G. J. Alred, and W. E. Oliu, *Handbook of Technical Writing*, New York: St. Martin's Press, 1976.

Caudron, Shari, "Spreading Out the Carrots," *Industry Week*, 19 May 1997, pp. 20, 22, 24.

The Chicago Manual of Style, thirteenth edition, Chicago and London: The University of Chicago Press, 1982.

Cibinc, John, and Ralph C. Nash, *Formation of Government Contracts*, second edition, Washington, DC: The George Washington University, 1986.

Clemons, Erid K., and Bruce W. Weber, "Using Information Technology to Manage Customer Relationships: Lessons for Marketing in Diverse Industries," *Proceedings of 26th Hawaii International Conference on System Sciences*, January 1993.

Clipsham, Neil B., "Appraising Performance Appraisal Systems," The Military Engineer, No. 591, April–May 1998, pp. 31–32.

Cohen Sacha, "Knowledge Management's Killer App," http://www.astd.org/magazine/current/cohen.htm (American Society for Training and Development).

"Community KM Glossary," http://knowledgecreators.com/km/kes/glossary.htm.

"Complying with the Anti-Kickback Act," *Developments in Government Contract Law*, No. 10, Sept. 1990 (Ropes & Gray 1001 Pennsylvania Avenue, N.W. Suite 1200 South Washington, DC 20004).

"Congress Says Transportation Has Best Strategic Performance Plans," *Reinvention Express,* 26 June 1998, Vol. 4, No. 9; http://www.npr.gov.

Cranston, H. S., and Eric G. Flamholtz, "The Problems of Success," *Management Decision*, Vol. 26, Sept. 1988, p. 17.

Cremmins, Edward T., *The Art of Abstracting*, Philadelphia, PA: ISI Press, 1982.

Davenport, Thomas H., and Laurence Prusak, *Working Knowledge: How Organizations Manage What They Know,* Boston, Harvard Business School Press, 1997.

Davenport, Tom, and Larry Prusak, "Know What You Know," http://www.brint.com/km/davenbport/cio/know.htm.

Davidson, Paul, "U.S. Air Force SMC Innovations in Electronic Procurement," *APMP Perspective*, Vol. I, July/Aug. 1996, pp. 7, 12, 13, 15.

Day, Robert A., *How to Write and Publish a Scientific Paper*, Philadelphia, PA: ISI Press, 1979.

December, John, and Neil Randall, *The World Wide Web Unleashed*, second edition, Indianapolis: Samsnet, 1995.

Delphos, William A., *Inside the World Bank Group: The Practical Guide for International Business Executives,* Washington, DC: Delphos International, 1997.

Desktop Guide to Basic Contracting Terms, second edition, Vienna, VA: National Contract Management Association, 1990, 1989.

"Disadvantaged Firms to Get Edge in Bids for Federal Contracts," *Baltimore Sun,* 25 June 1998, p. 2A.

"DOD Acquisition Chief Pushes COTS/Dual-Use, Wants More Suppliers," *Technology Transfer Week*, Vol. 30, April 1996.

"DOD's Airborne Recon Plan Pushes Sensors, Image Recognition," *Sensor Business News,* 8 May 1996.

Doebler, Paul D., "Who's the Boss?: Sorting Out Staffing and Workflow Issues," *EP&P*, April 1988, pp. 21–26.

Do-It-Yourself Proposal Plan, TRW Space & Defense Proposal Operations, Rev. 3, February 1989.

Donnelly, John, "Preston Touts Acquisition Reform Results," *Defense Week*, Vol. 7, Aug. 1995.

Douglas, Susan P., C. Samuel Craig, and Warren J. Keegan, "Approaches to Assessing International Marketing Opportunities for Small- and Medium-Sized Companies," *Columbia Journal of World Business*, Fall 1982, pp. 26–32.

Dreifus, Claudia, "Present Shock," *The New York Times Magazine*, 11 June 1995, p. 48.

Edvinsson, Leif, and Michael S. Malone, *Intellectual Capital: Realizing Your Company's True Value by Finding its Hidden Brainpower*, New York: HarperBusiness, 1997.

Edwards, Vernon J., *Competitive Proposals Contracting*, Falls Church, VA: Educational Services Institute, 1989 (in association with The George Washington University School of Government and Business Management).

Edwards, Vernon J., *Federal Contracting Basics*, Falls Church, VA: Educational Services Institute, 1990 (in association with The George Washington University School of Government and Business Management).

Edwards, Vernon J., "Oral Presentations: New Development and Challenges for Proposal Managers," *The Executive Summary: The Journal of the APMP's National Capital Chapter*, November 1996, pp. 1, 3, 4, 5.

Elbert, Bruce, and Bobby Martyna, *Client/Server Computing*, Boston, MA: Artech House, 1994.

Erramilli, M. Krishna, and C. P. Rao, "Choice of Foreign Market Entry Modes by Service Firms: Role of Market Knowledge," *Management International Review*, Vol. 30, 1990, pp. 135–150.

Farrell, Michael, "Many Firms Want to Do Business Internationally But Lack the Wherewithal," *Capital District Business Review* (Albany, N.Y.), Vol. 22, 4 Sept. 1995, p. 15.

Federal Acquisition Regulation (as of May 1, 1990), Chicago, IL: Commerce Clearing House, Inc., 1990.

"Federal Acquisition Streamlining Act Enacted," *Business Credit*, Vol. 97, March 1995, p. 6.

Flagler, Carolynn, "The ANY Aspects of International Proposals," *Contract Management*, March 1995, pp. 11–19.

Fox, Harold W., "Strategic Superiorities of Small Size," *Advanced Management Journal*, Vol. 51, 1986, p. 14.

Frank, C., "Uncle Sam Helps Business Get High Tech to Market," *Missouri Tech Net News*, 4 November 1997; http://www.umr.edu/~tscsbdc/sbirsumm.html.

Freed, Richard C., Shervin Freed, and Joe Romano, *Writing Winning Business Proposals: Your Guide to Landing the Client, Making the Sale, Persuading the Boss*, New York: McGraw-Hill, 1995.

Frey, Robert S., "Effective Small Business Response Strategies to Federal Government Competitive Procurements," *The Journal of Business and Management*, Summer 1997, Vol. 4, No. 1, pp. 40–74.

Gibson, Rowan, ed., *Rethinking the Future: Rethinking Business, Principles, Competition, Control, Leadership, Markets and the World*, London: Nicholas Brealey Publishing, 1997.

Goldberg, Mim, "Listen to Me," *Selling Power*, July/August 1998, pp. 58–59.

Goretsky, M. Edward, "When to Bid for Government Contracts," *Industrial Marketing Management*, Vol. 16, Feb. 1987, p. 25.

Gouillart, Francis J., and Frederick D. Sturdivant, "Spend a Day in the Life of Your Customers," *Harvard Business Review*, Jan./Feb. 1994.

Graham, Charles E., "Electronic Proposal Submission Basic Concepts," *NCURA Newsletter*, Feb./March 1996.

Graham, John R., "Getting to the Top, And Staying There," The New Daily Record, (Baltimore, MD), 26 July 1997, pp. 7A–7B.

Grundstein, Michel, "Companies & Executives in Knowledge Management," http://www.brint.com/km/cko.htm.

Guide to Doing Business in All 50 States for A/E/P and Environmental Consulting Firms; Natick, MA: Zweig White and Associates, 1998.

A Guide for Preparing BOEs (Basis of Estimates), TRW Space & Defense Sector, 1989.

Hackeman, Calvin L., "Best Value Procurements: Hitting the Moving Target," McLean, VA: Grant Thornton, 1993 (a White Paper prepared in connection with Grant Thornton's Government Contractor Industry Roundtables).

Hall, Dane, "Electronic Proposal Evaluation: How One U.S. Government Agency Does It," *The Executive Summary: The Journal of the APMP's National Capital Area Chapter*, June 1996, pp. 6, 8–9.

Hamilton, Martha A., "Managing the Company Mind: Firms Try New Ways to Tap Intangible Assets Such as Creativity, Knowledge," *Washington Post*, 18 Aug. 1996, pp. H1, H5.

Hanson, Doug, "Electronic Commerce and NASA's Scientific and Engineering Workstation Procurement (SEWP)," http://www.sewp.nasa.gov/edidoc/hanson9711.html.

Harrfeld, Heather, "Reinventing for Results," *Federal Computer Week*, 21 September 1998.

Hartman, Curtis, and Steven Pearlstein, "The Joy of Working," *INC.*, Nov. 1987, pp. 61–71.

Heiman, Stephen E., and Diane Sanchez, *The New Strategic Selling*®, *Revised and Updated for the 21st Century*, New York: Warner Books, 1998.

Helgeson, Donald V., *Engineer's and Manager's Guide to Winning Proposals*, Boston, MA: Artech House, 1994.

Henry, Shannon, "The 8(a) Dating Game: Primes and Small Business Find Each Other Through Hard Work and Happenstance," *Washington Technology*, Vol. 12, Sept. 1996, p. 24.

Hesselbein, Frances, Marshall Goldsmith, and Richard Beckhard, eds., *The Organization of the Future*, San Francisco: Jossey-Bass Publishers, 1997.

Hevenor, Keith, "Signs of Life," *Electronic Publishing*, 1 Aug. 1996, p. 4.

Hevenor, Keith, "Storage Product Buyers Guide," *Electronic Publishing*, 1 Aug. 1996, p. 10.

Hewitt, Tom. "Preparing Winning Proposals and Bids," VA: Federal Sources, Inc.

Hill, James W., ed., *How to Create and Present Successful Government Proposals: Techniques for Today's Tough Ecomomy*, IEEE, 1993.

History at NASA [NASA HHR-50], Washington, DC: NASA Headquarters, 1986.

Hoft, Nancy L., *International Technical Communication: How to Export Information Abou High Technology*, New York: John Wiley & Sons, 1995.

Holtz, Herman, and Terry Schmidt, *The Winning Proposal: How to Write It*, New York: McGraw-Hill, 1981.

"How the Pros See the Global Business Scene," *Rochester Business Journal*, Vol. 12, 21 June 1996, p. 10.

Howard, James S., and John Emery, "Strategic Planning Keeps You Ahead of the Pack," *D & B Reports*, Vol. 33, March/April 1985, p. 18.

Hoyt, Brad, "What Is KM?," *Knowledge Management News,* 1998; http://www.kmnews.com/Editorial/whatkm.htm.

"IBM, U.S. Chamber of Commerce Announce Results of New Study on U.S. Small Business and Technology," http://www.uschamber.org/news/sb980601a.htm.

Ingram, Thomas N., Thomas R. Day, and George H. Lucas, "Dealing with Global Intermediaries: Guidelines for Sales Managers," *Journal of Global Marketing*, Vol. 5, 1992.

Ink, Dwight, "Does Reinventing Government Have an Achilles Heel?" *The Public Manager: The New Bureaucrat*, Vol. 24, Winter 1995, p. 27.

Introduction to the Federal Acquisition Regulation Training Course, Vienna, VA: Management Concepts, Inc.

Jain, Subash C., *International Marketing Management*, Boston, MA: Kent Publishing Company, 1993.

James, Geoffrey, "It's Time to Free Dilbert," *New York Times*, 1 Sept. 1996, p. F-11.

Jennings, Robert W., *Make It Big in the $100 Billion Outsourcing Contracting Industry*, Westminster, CO: Westfield Press, 1997.

Jones, Stephen, "Navigating the Federal Bureaucracy Labyrinth," *Computerworld*, 14 Nov. 1988, p. 145.

Joss, Molly W., "Authoring Alchemy: Ingredients for Brewing Up a Multimedia Masterpiece," *Desktop Publishers*, Jan. 1996, pp. 56–65.

Kasser, Joe, *Applying Total Quality Management to Systems Engineering*, Boston, MA: Artech House, 1995.

Kelleher, Kevin, "Feds, State Go On Line with Contracts," *San Francisco Business Times*, Vol. 9, April 1995, p. 3.

Keninitz, Donald, "The Government Contractor's Dictionary: A Guide to Terms You Should Know," http://www.kcilink.com/govcon/contractor/gcterms.html.

Kelman, Steven, "GWACs. Time to Put Competition Back in Task Order Contracts."

Kniseley, Sina Fusco, "Point-and-Click Government," *Washington Technology*, 11 July 1996, pp. 44–45.

Knudsen, Dag, "The Proposal Game: Two Strikes and You're Out," *American Consulting Engineer*, May/June 1998, p.13.

Krathwohl, David R., *How to Prepare a Research Proposal: Guidelines for Funding and Dissertations in the Social and Behavioral Sciences*, third edition, Syracuse, NY: Syracuse University Press, 1988.

LaFlash, Judson, "The Target Proposal Seminar," Woodland Hills, CA: Government Marketing Consultants, June 1981.

Leibfried, Kate H. J., and Joe Oebbecke, "Benchmarking: Gaining a New Perspective on How You Are Doing," *Enterprise Integration Services Supplement*, Oct. 1994, pp. 8–9.

Lev, Baruch, "Accounting Needs New Standards for Capitalizing Intangibles," *ASAP: Forbes Supplement on the Information Age*, 7 April 1997; http://www.forbes.com/asap/97/0407/034.htm.

Lissack, Michael R., "Complexity Metaphors and the Management of a Knowledge Based Enterprise: An Exploration of Discovery," http://www.lissack.com/writings/proposal.htm.

Loring, Roy, and Harold Kerzner, *Proposal Preparation and Management Handbook*, New York, NY: Van Nostrand Reinhold Co., 1982.

Maher, Michael C., "The DOD COTS Directive—What About Radiation Hardness?" *Defense Electronics*, Vol. 26, Oct. 1994, pp. 29–33.

Malone, Thomas W., and John K. Rockart, "Computers, Networks, and the Corporation," Scientific American, September 1991, pp. 128–36.

Mariotti, John, "Nursery-Rhyme Management," *Industry Week,* 5 May 1997, p. 19.

Markie, Tracy, and Mark Tapscott, "COTS: The Key to a Tougher Military Marketplace," *Defense Electronics*, Vol. 25, Nov. 1993, p. 29–33.

Marshall, Colin, "Competing on Customer Service: An Interview with British Airways' Sir Colin Marshall," *Harvard Business Review*, Nov./ Dec. 1995.

Martin, James A., "Team Work: 10 Steps for Managing the Changing Roles in Your Desktop Publishing Work Group," *Publish!*, Dec. 1988, pp. 38–43.

McCarthy, Shawn P., "OMB Documents Pave the Way to Explore Electronic Commerce," *Government Computer News,* 23 March 1998.

McCartney, Laton, "Getting Smart About Knowledge Management," Industry Week, 4 May 1998, pp. 30, 32, 36–37.

McConnaghy, Kevin V., and Robert J. Whitehead, "RFPs: Winning Proposals," *Environmental Lab*, May/June 1995, pp. 14–19.

McCubbins, Tipton F., "Three Legal Traps for Small Businesses Engaged in International Commerce," *Journal of Small Business Management*, Vol. 32, July 1994, p. 95–103.

McVay, Barry L., *Proposals That Win Federal Contracts: How to Plan, Price, Write & Negotiate to Get Your Fair Share of Government Business*, Woodbridge, VA: Panoptic Enterprises, 1989 (Panoptic Federal Contracting Series).

McVey, Thomas W., "The Proposal Specialist as Change Agent," *APMP Perspective*, May/June 1997, pp. 1, 3, 13.

Meador, R., *Guidelines for Preparing Proposals: A Manual on How to Organize Winning Proposals for Grants, Venture Capital, R&D Projects, and Other Proposals,* Chelsea, MI: Lewis Publishers, 1985.

Merry, Uri, "Postings #1–7 in Complexity & Management," Posting #4, "Knowledge Management," http://pw2.netcom.com/~nmerry/post1.htm#post4.

Michaelson, Herbert B., *How to Write and Publish Engineering Papers and Reports*, Philadelphia, PA: ISI Press, 1982.

"Minds at Work: How Much Brainpower Are We Really Using?: A Research Report," Princeton, NJ: Kepner-Tregoe, 1997.

Miner, Jeremy T., and Lynn E. Miner, *A Guide to Proposal Planning and Writing*, http://www.orzypress.com/miner.htm.

Miner, Lynn E., and Jeremy T. Miner, *Proposal Planning and Writing*, Second ed., Phoenix, AZ: Orzy Press, 1998.

"The Modular Proposal Technique" (Volume I: A Review of Its Need, Origin, & Benefits; Volume II: The Step-By-Step Guide to Implementation; Volume III: Proposal Preparation Checklist, Categorical/Topical Outline Examples, Storyboards & Finished Modules), Van Nuys, CA: Litton Systems, Inc.

Mulhern, Charlotte, "Round 'Em Up," *Entrepreneur*, August 1998, Vol. 28, No. 8, pp. 117–22.

Munro, Neil, "8(a) Program Survives Republican Attacks," *Washington Technology's 8(a) and Small Business Report*, 12 Sept. 1996, pp. S-6, S-8.

Munro, Neil, "Clinton Set-Aside Plan Becomes Election-Year Pawn," *Washington Technology*, 11 April 1996, pp. 5, 93.

Murray, Bill, "Looking Beyond Graduation," *Washington Technology's 8(a) and Small Business Report*, 7 March 1996, pp. S-4, S-6, and S-8.

NASA Contractor Financial Management Reporting System: A Family of Reports, Washington, DC: National Aeronautics and Space Administration (c. 1986).

"NASA Mentor-Protégé Program Award Presented to SAIC," *SAIC Magazine*, Fall 1996; http://www.saic.com/publications/.

Nash, Ralph C., and John Cibinic, *Administration of Government Contracts*, Washington, DC: George Washington University Government Contracts, 1985.

Nash, Ralph C., and John Cibinic, *Federal Procurement Law*, 2 vols., Washington, DC: George Washington University Government Contracts, 1977, 1980 (Contract Formation Series).

Nasseri, Touraj, "Knowledge Leverage: The Ultimate Advantage," *Kognos: The E-Journal of Knowledge Issues,* Summer 1996; http://magi.com/~godbout/kbase/kognos11.htm.

Newton, Fred J., "Restoring Public Confidence in Government Contractors," *Management Accounting,* Vol. 67, June 1986, p. 51.

O'Guin, Michael, "Competitive Intelligence and Superior Business Performance: A Strategic Benchmarking Study," *Competitive Intelligence Review,* Vol. 5, 1994, pp. 4–12.

Osborne, David, and Ted Gaebler, *Reinventing Government: How the Entrepreneurial Spirit Is Transforming the Public Sector,* Reading, MA: Addison-Wesley, 1992.

"Outstanding Information Technology," General Services Administration, February 1998, http://www.itmweb.com/essay528.htm.

Paine, Lynn Sharp, "Managing for Organizational Integrity," *Harvard Business Review,* April/May 1994.

Peterson, Ralph R., "Plotting a Safe Passage to the Millennium," http://bstconsulta.../whatshot/ch2masce.htm.

Phelan, Steven E., "From Chaos to Complexity in Strategic Planning," Presented at the 55th Annual Meeting of the Academy of Management, Vancouver, British Columbia, August 6–9, 1995; http://comsp.com.latro...u.au/Papers/chaos.html.

Pine, B. Joseph, *Mass Customization: The New Frontier in Business Competition News,* Boston: Harvard Business School Press, 1993.

Piper, Thomas S., "A Corporate Strategic Plan for General Sciences Corporation," Spring 1989 (unpublished).

Porter, Kent, "Usage of the Passive Voice," *Technical Communication,* First Quarter 1991, pp. 87–88.

Power, Kevin, "Agencies Opt Out of FACNET," *Government Computer News,* 10 February 1997.

Preparing Winning Proposals and Bids, U.S. Professional Development Institute (USPDI), Washington, DC, Sept. 16–17, 1987.

"Presenting Your Proposal Orally and loving it!" TRW, Inc., 1979.

"President Signs Law Establishing First Performance-Based Organization," 7 October 1998; http://www.npr.gov/library/announc/100798.html.

Proposal Preparation Manual, 2 vols., Covina, CA: Procurement Associates, Inc., 1989 (revised).

Prusak, Laurence, ed., Knowledge in Organizations, Boston: Butterworth-Heinemann, 1997.

Quinn, James Brian, Philip Anderson, and Sydney Finkelstein, "Managing Professional Intellect: Making the Most of the Best," *Harvard Business Review*, March/April 1996.

Quinn, Judy, "The Welch Way: General Electric CEO Jack Welch Brings Employee Empowerment to Light," *Incentive*, Sept. 1994, pp. 50–52, 54, 56.

Ramsey, L. A., and P. D. Hale, *Winning Federal Grants: A Guide to the Government's Grantmaking Process*, Alexandria, VA: Capital Publications, 1996.

Ray, Dana, "Filling Your Funnel: Six Steps to Effective Prospecting and Customer Retention," *Selling Power*, July/August 1998, pp. 44–45.

"Reinventing the Business of Government: An Interview with Change Catalyst David Osborne," *Harvard Business Review*, May/June 1994.

Reis, Al, and Jack Trout, *The 22 Immutable Laws of Marketing: Violate Them at Your Own Risk*, NY: HarperBusiness, 1994.

Ricks, David A., *Blunders in International Business*, Cambridge, MA: Blackwell Publishers, 1993.

Rifkin, Jeremy, *The End of Work: The Decline of the Global Labor Force and the Dawn of the Post-Market Era*, New York: G.P. Putnam and Sons, 1995.

Riordan, John, and Lillian Hoddeson, *Crystal Fire: The Birth of the Information Age*, New York: W. W. Norton & Co., 1997

Roberts, J. B., The Art of Winning Contracts: Proposal Development for Government Contractors, J. Melvin Storn Co., 1996.

Roos, Johan, "Intellectual Performance: Exploring and Intellectual Capital System in Small Companies," 30 October 1996, pp. 4, 9; http://www.imd.ch/fac/roos/paper_lr.html.

Roos, Johan, and Georg von Krogh, "What You See Depends on Who You Are," *Perspectives for Managers,* September 1995, No. 7.

Ruggles, Rudy L., *Knowledge Management Tools,* Boston: Butterworth-Heinemann, 1997.

Root, Franklin R., *Entry Strategies for International Markets*, New York: Lexington Books, 1994.

Root, Franklin R., *Foreign Market Entry Strategies*, New York: AMACOM, 1987.

Sabo, Bill, "Orals: Your Worst Nightmare or Your Greatest Competitive Advantage?" *APMP Perspective*, Sept./Oct. 1996, pp. 1, 11.

Safford, Dan, *Proposals: On Target, On Time,* Washington, DC: American Consulting Engineers Council, 1997.

Scarborough, Norman M., and Thomas W. Zimmerer "Strategic Planning for Small Business," *Business*, Vol. 37, April/June 1987, p. 11.

"Selected Viewgraphs from Judson LaFlash Seminars on Government Marketing and Proposals," Government Marketing Consultants, July 1980.

Selling to the Military, DOD 4205.1-M, Washington, DC: Department of Defense.

Schillaci, William C., "A Management Approach to Placing Articles in Engineering Trade Journals," *Journal of Management in Engineering*, Sept./Oct. 1995, pp. 17–20.

Shuman, Jeffrey C., and John A. Seeger, "The Theory and Practice of Strategic Planning in Smaller Rapid Growth Firms," *American Journal of Small Business*, Vol. 11, 1986, p. 7.

Slaughter, Jeff, "New Way of Doing Business with Uncle Sam: Electronic Commerce/Electronic Data Interchange," *Mississippi Business Journal* (Jackson, MS), Vol. 18, Feb. 1996, p. 10.

Soat, Douglas M., *Managing Engineers and Technical Employees: How to Attract, Motivate, and Retain Excellent People*, Boston, MA: Artech House, 1996.

Solomon, Robert, "International Effects of the Euro," Brookings Institution Policy Brief #42, Washington, D.C.: The Brookings Institution, 1999.

Source Evaluation and Selection, Vienna, VA: Management Concepts, Inc.

Source Selection: Greatest Value Approach, Document #KMP-92-5-P, Washington, DC: U.S. General Services Administration, May 1995.

Standard Industrial Classification Manual, Executive Office of the President, Office of Management and Budget, 1987.

Statements of Work Handbook [NHB 5600.2], Washington, DC: Government Printing Office, Feb. 1975.

Steiner, Richard, *Total Proposal Building: An Expert System Dedicated to One Result: Winning Grants & Contracts from Government, Corporations & Foundations*, second edition, Albany, NY: Trestleetree Publications, 1988.

Stewart, Rodney D., and Ann L. Stewart, *Proposal Preparation*, New York: John Wiley & Sons, 1984.

Stewart, Thomas A., *Intellectual Capital: The New Wealth of Organizations*, New York: Doubleday, 1997.

Stewart, Thomas A., "Mapping Corporate Brainpower," Fortune, 1995; http://www.pathfinder...51030/leadingedge.html.

Stout, Donna M., "Focus on Systems: Reviews of Books on System Theory," http://www.wintu.edu/journals/wiuj0101-6.html, p. 2, (citing Margaret Wheatley).

Sutton, Michael H., "Dangerous Ideas: A Sequel," Remarks delivered during the American Accounting Association 1997 Annual Meeting in Dallas, Texas, August 18, 1997, p. 3; http://www.sec.gov/news/speeches/spch175.txt.

Sveiby, Karl E., "Celemì's Intangible Assets Monitor," http://www.sveiby.com...Ass/CelemiMonitor.html.

Sveiby, Karl E., "What Is Knowledge Management," http:www.sveiby.com...owledgeManagement.html.

Tapscott, Mark, "COTS Regulation Reforms Sought by Industry Coalition," *Defense Electronics*, Vol. 25, Oct. 1993, pp. 8–11.

Ten Steps to a Successful BOE: Basis of Estimate Estimator's Checklist, TRW Space & Defense Sector, 1987.

Tepper, Ron, *How to Write Winning Proposals for Your Company or Client*, New York: John Wiley & Sons, 1990, 1989.

Terpstra, Vern, *International Marketing*, fourth edition, New York: The Dryden Press, 1987.

"Test Your Global Mindset," *Industry Week*, 2 November 1998, p. 12.

"Thriving on Order" (an interview of Steve Bostic), *INC.*, Dec. 1989, pp. 47–62.

Thompson, Arthur A., and A. J. Strickland, *Strategy Formulation and Implementation: Tasks of the General Manager*, fourth edition, Homewood, IL and Boston, MA BPI/IRWIN, 1989.

Tittel, Ed, and James Michael Stewart, "The Intranet Means Business," *NetGuide Magazine*, July 1996, pp. 121–127.

Toffler, Alvin, and Heidi Toffler, *Creating a New Civilization: The Politics of the Third Wave*, Atlanta, GA: Turner Publishing, 1995, 1994.

Training and Validation in Basis of Estimates, TRW Space & Defense Sector, 1987 [vugraphs].

Twitchell, James B., *AdcultUSA: The Triumph of Advertising in American Culture*, New York: Columbia University Press, 1996.

Unisys Defense Systems Proposal Development Digest, McLean, VA: Unisys, 1989.

Ursey, Nancy J., *Insider's Guide to SF254/255 Preparation*, Second ed., Natick, MA: Zweig White & Associates, 1996.

U.S. Small Business Administration, "Lockheed Martin Division, Nova Group Earn SBA's Top Awards for Subcontracting," Press Release Number 98-43, 4 June 1998; http://www.sba.gov/news/current98-43.html.

U.S. Small Business Administration, "BSA Streamlines 8(a) Contracting," Press Release Number 98-24, 4 June 1998; http://www.sba.gov/news/current98-24.html.

U.S. Small Business Administration, "Vice President and SBA Administrator Announce Pact with Big Three Automakers," Press Release Number 98-09, 19 February 1998; http://www.sba.gov/news/current98-09.html.

U.S. Small Business Administration, "Vice President Announces Plan to Help Women-Owned Firms Win Federal Contracts," Press Release Number 98-67, 31 July 1998; http://www.sba.gov/news/current98-67.html.

U.S. Small Business Administration, "Vice President Gore Praises SBA for Expanding Economic Opportunity: Congratulates Agency on 45th Birthday," Press Release Number 98-68, 30 July 1998; http://www.sba.gov/news.

Verespej, Michael A., "Drucker Sours on Terms," *Industry Week,* 6 April 1998, p.16.

Verespej, Michael A., "The Old Workforce Won't Work," *Industry Week,* 21 September 1998, pp. 53, 54, 58, 60, 62.

Verespej, Michael A., "Only the CEO Can Make Employees King," *Industry Week,* 16 November 1998, p. 22.

Vivian, Kaye, *Winning Proposals: A Step-by-Step Guide to the Proposal Process,* Jersey City, NJ: American Institute of Certified Public Accountants, 1993.

Wall, Richard J., and Carolyn M. Jones, "Navigating the Rugged Terrain of Government Contracts," *Internal Auditor,* April 1995, pp. 32–36.

Warburton, Charles, "Electronic Commerce," *The Military Engineer,* October–November 1997, No. 587, pp. 63–64.

Weil, B. H., I. Zarember, and H. Owen, "Technical-Abstracting Fundamentals," *Journal of Chemical Documentation,* Vol. 3, 1963, pp. 132–136.

Whitely, Richard, and Diane Hessan, *Customer-Centered Growth: Five Proven Strategies for Building Competitive Advantage,* Reading, MA: Addison-Wesley, 1996.

Whitman, Marina, and Rosabeth Moss Kanter, "A Third Way?: Globalization and the Social Role of the American Firm," *Washington Quarterly,* Spring 1999.

Wodaski, Ron, "Planning and Buying for Multimedia: Effective Multimedia for the Small Office," *Technique: How-To Guide to Business Communication,* Oct. 1995, pp. 16–25.

Words Into Type, third edition, Englewood Cliffs, NJ: Prentice-Hall, 1974.

The World Bank Annual Report, Washington, D.C.: The World Bank, 1994.

Worthington, Margaret, "The Ever Changing Definition of Allowable Costs," *Government Accountants Journal*, Vol. 35, Spring 1986, p. 52.

Writing Winning Proposals, Shipley Associates, 1988.

Yukins, Christopher R., "Relaxed Rules Won't Fix the FAA's Problems," *Washington Technology*, 11 April 1996, pp. 8, 12.

Zimmerman, Jan, *Doing Business With the Government Using EDI: A Guide for Small Businesses*, Van Nostrand Reinhold, 1996.

Zuckman, Saul, "Is There Life After 8(a)?," *The Columbia (Md.) Business Monthly*, Oct. 1994, p. 37.

About the author

Robert S. Frey has performed proposal management; end-to-end proposal development, design, and production; and business planning and development for 13 years. That successful experience is coupled with two decades of writing and publication-related activities. He has conceptualized and established effective proposal development infrastructures for five federal contracting firms in Virginia and Maryland. Four of those firms were small businesses operating under the Small Business Administration's 8(a) program. In addition, Mr. Frey has provided end-to-end proposal development, design, and production support for 500+ government, private-sector, and international proposals. He has coordinated all phases of up to 12 scientific and engineering proposals concurrently.

Of particular note is that Mr. Frey has hands-on experience with most facets of proposal development. In the process of establishing and implementing proposal infrastructures for various companies, he has actually performed the myriad tasks associated with responding to government and private-sector requests for proposal (RFPs), including graphics conceptualization, electronic text editing, computer troubleshooting, and photocopying along with original writing and review. During various periods, he was the only full-time staff person in a business development capacity. Bottom line—he has developed a very thorough understanding of both the "big picture" and the day-to-day, hour-to-hour logistical, organizational, and interpersonal activities associated with successful proposalmanship.

Mr. Frey also develops and presents innovative proposal training curricula, having delivered interactive seminars nationwide to businesses and academe. These seminars have included a 2-day short course at the University of California at Los Angeles (UCLA) in 1999 and a workshop at Gallaudet University in Washington, D.C., in 1998, sponsored by the Deaf and Hard of Hearing Entrepreneurship Council. He possesses a detailed understanding of marketing intelligence resources as well as collection and analysis techniques. Mr. Frey participates in business strategizing and planning for numerous federal, civilian, and defense agencies under a broad umbrella of contract vehicles, such as Government-wide Agency Contracts (GWACs), ID/IQs, cost-reimbursable, firm-fixed price (FFP), and time and materials (T&M).

Federal agencies for which he has prepared proposals include the U.S. Army Corps of Engineers (USACE), Air Force Center for Environmental Excellence (AFCEE), NASA, National Oceanic and Atmospheric Administration (NOAA), Environmental Protection Agency (EPA), Department of Transportation (DOT), National Institutes of Health (NIH), Department of Veterans Affairs (VA), Department of the Treasury, the Air Force Space Command, Defense Logistics Agency (DLA), and the Defense Information Systems Agency (DISA).

Mr. Frey is an active member in the nationwide Association of Proposal Management Professionals (APMP, http://www.apmp.org), serving as the assistant managing editor and member of the Editorial Review Board for the new *APMP Professional Journal: Journal of the Association of Proposal Management Professionals.* He was selected to deliver a presentation entitled "Managing Critical Proposal Information Effectively," at the APMP Annual Conference held in Colorado Springs, Colorado, in April 1998. In addition, Mr. Frey had an article entitled "Effective Small Business Response Strategies to Federal Government Competitive Procurements" published in *The Journal of Business and Management* (Western Decision Sciences Institute/California State University, Dominguez Hills) that appeared in the Summer 1997 issue.

Successful Proposal Strategies for Small Businesses is the fourth book that Mr. Frey has written since 1985. He holds a bachelor of science degree (*cum laude*) in biology and a master of arts degree (with highest honors) in history. Currently, Mr. Frey serves as director of Knowledge Management and Proposal Development for RS Information Systems (http://www.rsis.com), a highly successful 8(a) information technology consulting firm headquartered in McLean, Virginia.

You can contact Mr. Frey at BRIDGES23@AOL.com, or by calling (410) 329-3055 (Maryland).

Index

A-76, 381
Acquisition, 90
    federal process, 87–113
    forecasts, 49
    sole source, 390
Acquisition teams
    formation of, 165
    functions, 165–66
    kickoff meeting participation, 169
    makeup of, 165
Action title, 382
Advanced planning group. *See* Business
    development groups (BDGs)
Advertising, 65–69
    baseline recognition, 66
    brand name recognition, 65
    Business Wire, 66, 67
    consistency, 68
    news releases, 66

PR, 65, 67
    publications, 68
    special issue publications, 66
    trade shows, 65
    Web pages, 68
    See also Marketing
American National Standards Institute (ANSI)
    X12 standards, 101–2
Arabnet, 221–22
Architect-Engineer Contract Administration
    Support Services (ACASS), 273
Armed Services Procurement Act (1948)
    (ASPA), 382
Asian Development Bank (ADB), 200
Asia-Pacific Economic Cooperation
    (APEC), 219–20
Australian Trade Commission Online
    (Austrade), 220–21

Base Year, 382
Basis of estimate (BOE), 383
Benchmarking, 27, 307
Best and final offer (BAFO), 16, 127,
    139–40, 382
Best-in-class, 27
Best value determination, 69, 78, 382
Bid and proposal (B&P), 3, 176, 263,
    264, 265, 266
Bidders conference. *See* Preproposal Conference
Bid/no bid decision-making process
   as common failure point, 129
   company consensus, 129
   contract desirability, 128
   resource requirements, 128
   winning probability, 127
Blackout, 383
Black Team review, 139, 383
Block 7, 327, 328–29
Block 8, 329–30
Block 10, 330–31
Blue Team review, 137
Bogeys, 161, 383
Boilerplate, 72, 289–90, 384
Bottom line prices, 161
Bridge Grant program, 11
British-American Business Council
    (BABC), 224–25
Brooks Act (ADPE Contracting) (1965), 384
Bullet drafts, 133–35, 167
Bulletin board systems (BBSs), 46
Business development groups (BDGs), 13
   de-scoped, 17
   functional charter activities, 14
   LOBs, 15
   management recommendations, 14–15
   organization of, 16
Business process reengineering (BPR), 259, 305
Business Wire, 66, 67

Call plans, 51, 52–57
   buying organizations, 52
   client contact reports, 54–55
   client current performance, 52
   client information interviewing tool, 57
   client plans, 53
   company performance, 54

competition, 54
competitor's performance, 53
incumbent personnel, 53
information necessary for, 52–54
key personnel, 52
LOE contracts, 55
pre-proposal interview form, 55–56
relevant documents/articles, 53
SEB, 53
subcontractors, 54
Capture plans, 51–52, 384
Central Consultancy Register (CCR), 199–200
Central Contractor Registry (CCR), 97
Clients, 24–25
   contact reports, 54–55
   external, 24
   information interviewing tool, 57
   internal, 24
Code of Federal Regulations (CFR), 88, 111
Commerce Business Daily (CBD), 38, 46, 369
Commercial and Government Entity (CAGE)
    code registration, 107
Commercial-off-the-shelf (COTS)
    procurement, 60–61, 141
Company organization
   BAFO, 16
   BDGs, 13–14, 16, 17
   centralized business development and
    control, 16
   corporate image, 17
   decentralized business activities, 16
   intelligence, 16
   LOBs, 15
   management choices, 14–15
   sales duties, 15–16
   time considerations, 17
   See also Knowledge management
Competition in Contracting Act (1984)
    (CICA), 384
Competitive negotiation, 89
Competitive proposals, 142
   8(a) program, 7, 8, 25–26
   attitude changing, 7
   B&P, 3
   BDGs, 13–14
   business acquisition process, 8
   client knowledge, 3

clients, 24–25
company organization, 13–17
development, 3
FAR, 6
inadequacy of, 3
informal organizations, 7
Internet sites, 7
IR&D, 3
knowledge management, 19–25
LOBs, 15
maximizing company strengths, 9–10
MED, 7
mission planning, 17–19
organizational empowerment, 7–8
planning, 3, 4
presentations, importance of, 6
PSA, 11
revenue pipeline, 4, 5
SBA, 7
SBIR, 10–13
small business constraints, 8–9
Small Business Week, 7
strategic planning processes, 7, 17–19
STTR, 10, 13
successful bidding factors, 5–6
transfer ratios, 9
winning, 3
win ratios, 4–5
Competitive range, 384
Compliance matrix, 149–50
Compliant submittal, 76
Confederation of German Trade Fair and
Exhibition Industries
(AUMA), 217–18
Construction Contractor Appraisal Support
System (CCASS), 297
Continuous process improvement (CPI), 305
Contract Data Requirements Lists (CDRLs), 78
Contracting officer (CO), 49, 256
Contracting officer's technical representative
(COTR), 49, 116, 256
Contractor furnished equipment (CFE), 159
Contractor review cycles, 136–39
Black Team, 139
Blue Team, 137
comment sheet, 138
Gold Team, 139

Pink Team, 137
Red Team, 137–39
team purposes, 136
Corporate libraries
cataloging materials, 291
data centers, 291
electronic archiving and backups, 291
intranets, 292
online security, 292
storage media, 292
Cost accounting standards (CAS), 157, 162–63
Cost reimbursement contracts, 93
Costs control, 263–69
business development bonus policy, 266–68
centralization, 264
estimating proposal expense, 264
goal setting, 264
marketing funds, stretching, 268–69
mission statement expansion, 264
outside consultants, 268
proposal expenditures, 263
tracking B&P expenditures, 265–66
Cost volume
automated estimating tools, 162
bottom line prices, 161
confidentiality, 162
direct costs, 161
indirect costs, 161
methodology, 161
ODCs, 161
resource estimates, 161
sales oriented, 162
thoroughness of, 162
Cross-reference matrix. *See* Compliance matrix

Das Gepa-Projekt, 218
Data on Consulting Firms (DACON), 194, 200
Dawnbreaker, 12
DD Form, 254, 385
Defense Contract Administration Service
(DCAS), 385
Defense Contract Audit Agency (DCAA), 385
Defense Technical Information Center
(DTIC), 370
Department of Defense Telephone
Directory, 370

Department of Trade and Industry (DTI) (UK)
  EU, 210
  free information booklets, 212
  market opportunities, 211
  OJ, 211
  small firm definition, 213
  SME, 214
  TED, 211
  Treaty of Rome, 210
Desktop publishing (DTP), 385
Determinations and findings (D&F), 92
DIALOG, 369, 371, 385
Direct contact plans. *See* Call plans
Disclosure statement, 147–48
Discriminators, 385
DoD
  EC/EDI infrastructure, 107
  EC Office, 106
  Electronic Commerce Information Center
    (ECIC), 106–7, 370
Downselecting, 49–51
DR-LINK, 12

EC/EDI, 97–102
  defined, 97
  infrastructure, 107
  resources, 103–6
Economist Intelligence Unit (EIU), 201
EDI Yellow Pages, 109
EI Digest, 378
Eisenhower Award, 32
Electronic Bid Sets (EBS), 109–10
Electronic commerce (EC), 2, 385
Electronic data interchange
    (EDI), 2, 97–102, 385
  CCR, 97
  defined, 97, 385
  Internet vs., 97
Electronic source selection (ESS), 385–86
Electronic submittals, 245–46
  format (.pdf), 246
  WinPET and, 245
Entry Strategies for International Markets, 192
Ethics
  favorable treatment, 64, 70
  gifts, 65
  kickbacks, 64

Euro Info Centres (EIC), 209–10
European Bank for Reconstruction and
    Development (EBRD), 197–200
  CCR, 199
  EDF, 199
  EIB, 198
  EU, 197–98
  Obnova Programme, 198, 199
  OJ, 199
  Phare, 198
  Procurement Policies and Rules, 197
  Tacis, 198
  TED, 199
  Web site address, 197
European Development Fund (EDF), 199
European Investment Bank (EIB), 198
European Procurement Information Network
    (EPIN) (Ireland), 215
European Union (EU), 197–98
EU Small and Medium Sized Enterprise (SME)
    Initiative, 214
Examples
  ACASS ratings graphic, 298
  acquisition process, 91
  ad for small company, 67
  B&P initiation form, 266
  BDG organization, 16
  bid/no bid decision matrix, 122
  bullet draft worksheet, 135
  compliance matrices, 150–52
  direct labor analysis, 265
  disclosure statement for title page, 147
  document exchange between DoD and
    private sector trading partners, 102
  double column proposal page format, 247
  FAA Web page, acquisition management
    system document, 342–47
  features/benefits box, 153
  full kickoff package, 350–61
  international procurement opportunities, 198
  knowledge organization balance sheet, 22
  lessons learned database information
    form, 294
  marketing and proposal life cycle, 118
  marketing information flow, 48
  ODC analysis, 267

organizational element in knowledge-based framework, 306–8
performance scorecard, 59
pre-kickoff proposal products, 123
pre-proposal interview form, 56
primary thematic graphics, 239
project summary format, 288
proposal building blocks, 296
proposal directive package summary sheet, 170
proposal schedule calendar, 185–86
Red Team reviewer's comment sheet, 138
résumé action-oriented words, 285
revenue pipeline, 5
RFP Section M, 77
RFP table of contents, 75
SF254/255 Block 10 Management Section Outline, 331
SF254/255 forms and formats, 316–24
teaming agreement topics, 338
technical volume outline, 153, 168
themes, 131–33
transmittal (cover) letter, 144–45
WBS, 160
workforce questionnaire, 364–66
Executive summary
  company competence, 130, 148
  graphics, use of, 149
  length of, 148
  management strategies, 149
  photographs, use of, 149
  structural components, 148–49
  success stories, 148
  themes, 129, 148
Export-Import Bank of the United States (Ex-Im Bank), 205–6
Export Market Information Centre (EMIC), 214–15
Extranets, 248

Favorable treatment, 64, 70
Federal Acquisition Computer Network (FACNET), 97, 99–100, 108, 110, 386
Federal acquisition process, 87–113
  ANSI X12 standards, 101–2
  CAGE, 107
  CCR, 97

CFR, 88
  competition in government agencies, 93
  contract formation, 90
  contractor selection, 90
  cost reimbursement contracts, 93
  D&Fs, 92
  defined, 90
  DoD EC/EDI infrastructure, 107
  DoD EC Office, 106
  EBS, 109–10
  EC/EDI, 97–102
  EC/EDI resources, 103–6
  ECIC, 106–7
  EDI benefits, 101
  EDI Yellow Pages, 109
  FAA EC, 110
  FACNET, 97, 99–100
  FAR, 88–89
  FASA impact, 96–98, 273
  Federal Register, 89
  firm-fixed-price contracts, 93
  history of, 87–88
  micropurchases, 98
  NDI, 110
  NPR, 94–96
  open competition, 92
  paradigm shift, 93–94
  reinvention teams, 95
  scoring, 91–92
  source selection steps, 89–90
  statutory and regulatory requirements, 88–89
  Third Wave, 94
  TINA, 96
  trading partner information, 98
  Trading Partner Profile, 97
  UCF, 89
  US Air Force model, 90–92
  VANs, 97, 107–9
  weapons systems requirements, 98
  See also Acquisition
Federal Acquisition Reform Act (FARA), 2, 98
Federal Acquisition Regulations (FAR), 386
  8(a) contractors, 25
  competition, 88–89
  debriefings, 140–41

Federal Acquisition Regulations (FAR)
 (continued)
 elaborate proposals, 6
 improper business practices, 64
 information exchange, 62
 oral presentations, 182
 SF254/255, 315–25
Federal Acquisition Streamlining Act
 (FASA), 2, 37, 61, 96–98
Federal marketing and intelligence
 sources, 367–79
 acquisition forecasts, 368
 CBD, 369
 DIALOG, 369, 371
 DoD ECIC, 370
 DoD Telephone Directory, 370
 DTIC, 370
 Federal Marketplace, 372
 Federal Register, 372
 Federal Yellow Book 1997, 372–73
 Freedom of Information requests, 373
Federal Procurement Data Center
 (FPDC), 375–76
Federal Property and Administrative Services
 Act (1949) (FPASA), 386
Federal Register, 89, 111, 372
Federal Yellow Book 1997, 372–73
Firm fixed price (FFP) contracts, 61, 93
Foldout page, 387
Foreign Credit Insurance Association
 (FCIA), 204
Freedom of Information Act
 (FOIA), 272, 386–87
Freelance writers, 249–52
French Committee for External Economic
 Events (CFME), 218–19

Gantt Chart, 387
Ghosting, 387
Global Technology Network (GTN), 226–27
Gold Team review, 139
Government Electronic Marketplace Service
 (GEMS) (Australia), 220
Government furnished equipment (GFE), 159
Government Performance and Results Act
 (GPRA), 387

Government Printing Office (GPO) Style
 Manual, 281
Government Supplies Department
 (GSD), 207–9
Government-wide acquisition contract
 (GWAC) team
 defined, 37
 examples, 37–38
 ITOP, 37
Graphics. See Production graphics
Greatest value approach, 77–78
The Green Book Report, 379
Group of Seven (G7), 193

HijackPro, 240
Human dynamics, 253–61
 administrative support, 255
 BPR, 259
 "business as battlefield" mentality, 259
 change agents, 260
 competitive work ethic, 255–56
 COs, 256
 COTRs, 256
 democracy, role of, 254
 humor, role of, 258
 knowledge identification, 258
 leading by example, 260
 maximizing intellect, 258–60
 nonproductivity, 253
 project performance and proposal
  success, 256, 257
 proposal development, importance
  of, 254–55
 proposal enjoyment, 257–58
 reference verification, 256
 setting expectations, 259
 staff support, 259
 teamwork, 254, 260
 wellness, 261
 See also Knowledge management; Proposal
  manager
Hypermedia, 387

Incumbent, 387
Informational data systems
 CD-ROMs, 296
 intranets, 297
 localized customization, 295

Lotus Notes, 295
  proposal building blocks, 295, 296
Information packaging, 283–98
  boilerplate, 289–90
  CCASS, 297
  CD-ROM scenarios, 296
  company overviews, 284
  corporate libraries, 291–92
  data systems, 295
  electronic format utilization, 283
  electronic proposal toolboxes, 290
  experience summary box, 285
  federal performance appraisal
      systems, 297–98
  format compatibility, 286, 287
  intranet scenarios, 297
  lessons learned databases, 293–95
  Lotus Notes scenarios, 295
  marketing targets, 291
  project descriptions, 287–88
  proposal material storage, 283
  résumés, 284–86
  security, 283–84
  staff and resource management
      integration, 290
  tailoring, 289
Information Technology Omnibus Procurement
    (ITOP) contracts, 37
Integrated logistics support (ILS), 284
Intellectual capital, 19, 27, 300
Intelligence
  acquisition forecasts, 49
  analysis techniques, 49
  call plans, 51, 52–57
  capture plans, 51–52
  company organization, 16
  COs, 49
  COTRs, 49
  downselecting process, 49–51
  gathering, 49–52
  tracking tables, 49, 50
  WinAward, 51
Internal research and development (IR&D), 3
International Bank for Reconstruction and
    Development (IBRD), 230
International Development Association
    (IDA), 230

International marketing leads and information
  media and other organizations, 377–78
  state governments, 377
  U.S. federal government, 377
International proposals, 191–231
  ADB, 200
  APEC, 219–20
  Arabnet, 221–22
  AUMA, 217–18
  Austrade, 220–21
  BABC, 224–25
  benefits to companies, 191
  BOTB, 210
  CCR, 199–200
  CFME, 218, 219
  CTW, 219
  DACON, 194, 200
  Das Gepa-Projekt, 218
  DTI UK, 210–14
  EBRD, 197–200
  EDF, 199
  EIB, 198
  EIC, 209–10
  EIU, 201
  EMIC, 214–15
  EPIN, 215
  EU, 197–98
  Ex-Im Bank, 205–6
  GEMS, 220
  GSD, 207–9
  GTN, 226–27
  host country procurement
      environments, 202–3
  import-export considerations, 203
  in-country partnerships, 202
  information and intelligence sources, 377–79
  Internet-based resources, 206–22
  JETRO, 219
  market planning, 200–201
  MERX, 206–7
  OJ, 199, 211
  OPW, 215
  PIO, 217
  risk assessment, 203
  SIMAP, 216
  SME, 214
  STAT-USA, 222–23

International proposals (continued)
tax liabilities, 231
technological advances and, 191–92
technology transfer, 203
TED, 199, 211
Tender I.N.F.O., 220
terms and conditions, 203–5
THEMiS, 216
Ts & Cs, 203–5
UN procurements, 195–96
USAID, 226–29
USITC, 229
USTDA, 225
World Bank, 192–95
Internet, 36, 113
acquisition forecasts, 49
ADB, 200
APEC, 219
Arabnet, 221
AUMA, 217–18
Austrade, 220
BABC, 224
BOTB, 210
CFME, 218
competitive proposals, 7
Das Gepa-Projekt, 218
DTI UK, 210
EBRD, 197
EBS, 109–10
EDI vs., 97
EIC, 209
EMIC, 214
EPIN, 215
FAA acquisition management system
example, 342–47
Federal Marketplace, 372
Federal Yellow Book 1997, 372
freelance professionals listings, 251–52
GEMS, 220
govcon, 373
graphics, 240
GSD, 207–9
GTN, 226
information packaging, 297
international proposals, 206–22
JETRO, 219
MERX, 206–7

NTDB, 228
OPW, 215
PIO, 217
as PR medium, 68
Pro-Net, 334
RFPs, 73
security, 97
SIMAP, 216
STAT-USA, 222
subcontracting opportunities, 35, 36
Tender I.N.F.O., 220
THEMiS, 216
UNISPHERE, 223–24
USAID, 226
USITC, 229
USTDA, 225
VASs, 108–9
Washington Technology, 376–77
Web pages as advertising, 68
World Bank, 193
Intranets, 248, 297
Invitations for Bid (IFB), 61, 141
IT solutions. *See* Informational data systems

Japan External Trade Organization
(JETRO), 219

Kickbacks, 64
Kickoff meeting, 47, 48, 135
acquisition team members, 169
chairman, 169
competition review, 171
elements, 169–71
intelligence review, 170
management commitment, 170
materials distribution, 169
motivation, 170
staffing review, 171
strengths and weaknesses review, 171
writing assignments, 171
Knowledge management, 19–25, 27, 299–311
balance sheet, 22
benchmarking, 307
business metaphors, 302–3
business model flexibility, 305–9
client redefinition, 305
clients, 24–25
complexity theory, 303

corporate adaptability, 301–2, 303
corporate approaches, 20
CPI, 305
defined, 19, 304
departments, 300
ecological webs, 304
external clients, 24
feedback, 304
information technology expansion, 299–300
intangible values, 22
intellectual capital, 19, 27, 300
internal clients, 24
KOL, 311
language usage, 302
optima, 24, 305
organization charts, 306–8
productivity, 301
products, 22
proposal development, 21–24, 309–10
self-organization, 304
"sharing" business culture, 300
techniques, 20
TQM, 305
work environments, 302
See also Human dynamics; Proposal manager
Knowledge On Line (KOL), 311

Lessons learned databases
constructive introspection, 293
information form sample, 294
management plan accuracy, 293
pattern tracking, 293
Level-of-effort (LOE) contracts, 55
Line of business (LOBs), 15
Lotus Notes, 295

Management volume, 154–61
approach outline, 154–55
CFE, 159
client/contractor interfaces, 156
company capabilities and experience, 158
company facilities, 159
configuration management, 159
corporate organization, 156
cost and time controls, 157
data/information handling, 159
elements of, 154
executive summary, 156

GFE, 159
introduction, 156
management plan, 157
personnel, 158
production deliverability, 161
project organization, 156
quality assurance, 161
related experience, 158
requirements creep, 157
résumés, 158
risk control, 159
scheduling, 157
special resources, 159
staffing plan, 158
subcontracting plan, 158–59
summation, 161
technical performance measurement
    systems, 160
WBS, 159–60
Marketing, 43–70
acquisition forecasts, 49
advertising, 65–69
BBS, 46
best value determination, 69
brand image, 45
buying a solution, 44
call plans, 51, 52–57
capture plans, 51–52
CBD, 46
COTS procurement, 60–61
decentralized, 46
downselecting process, 49–51
ethics, 64–65
FAR, 62, 64
FASA, 61
favorable treatment, 64
federal procurement determinants, 45
FFP, 61, 62
formal kickoff meeting, 47
gifts, 65
ground truth, 48
IFB, 61–62
information flow, 48
intelligence analysis techniques, 49
intelligence gathering, 49–52
key questions, 44–45
kickbacks, 64

Marketing (continued)
    listening, 46–47
    management visibility, 57–58
    performance scorecard, 59
    PM utilization, 58–59
    pre-kickoff activities, 47
    pre-proposal interview form, 56
    pre-RFP sales, 47
    price vs. value, 45
    proposal integration of, 47–49
    Red Team review, 47
    RFC, 62
    RFI, 62
    selling, vs., 43–44
    SF129, 63
    SOQs, 48–49
    SOW, 62
    technical, 55
    tracking tables, 49, 50
    transaction personalization, 45–46
Marketing review processes, 47, 48
Matrix management, 387
Memorandum of understanding (MOU), 32
Mentor-protégé programs, 39–40
MERX, 206–7
Micropurchases, 98, 387
Minority business associations, 373–77
    FPDS, 375–76
    GSA, 375–76
    NTIS, 374
    PASS, 374–75
    Small Business Success, 375
    The Nash & Cibinic Report, 373
    Washington Business Journal, 376
    Washington Technology, 376–77
Minority Enterprise Development (MED)
    program, 7
Minority-owned business, 387–88
Minority/woman business enterprises
    (M/WBE), 32
Mission statement
    defined, 27
    expansion of, 264
    strategic plan relationship, 19
Modular
    contracting, 388
    defined, 388

The Nash & Cibinic Report, 373
National Association of Purchasing Management
    (NAPM), 36
National Partnership for Reinventing
    Government (NPR), 94–96
National Technical Information Service
    (NTIS), 374
National Trade Data Bank (NTDB), 228–29
Nondevelopmental items (NDI), 110, 141
North American Industry Classification System
    (NAICS) codes, 391

Obnova Programme, 198
Office of Public Works (OPW) (Ireland), 215
Office of Small and Disadvantaged Business
    Utilization (OSADBU), 35
Official Journal of the European Communities
    (OJ), 199, 211
Optima, 24, 305
Option years, 389
Oral presentations, 182–83, 389
Other direct costs (ODCs), 161
Out years, 389

The Peck Report, 378
Performance based service contracting
    (PBSC), 389
Performance scorecards, 59
PERT chart, 389
Phare Programme, 198
Pilot Mentor-Protégé Program (PMPP), 39
Pink Team review, 137
Pipeline. See Revenue pipeline
Portable Document Format (.pdf), 246
Portrait/landscape pages, 389
Post-kickoff activities, 172
Pre-kickoff activities
    bullet drafts, 167
    critical supplies identification, 123
    detailed outline, 121
    draft executive summary, 167
    executive summary, 123
    intelligence, 47–48
    kickoff package, 167
    milestone schedule, 121
    planning, 166
    planning meeting, 121–23
    project summaries, 123

related proposals review, 167
résumés, 123
RFP analysis, 166–67
staff procurement, 167
storyboards, 121
technical volume outline, 168
theme development, 121, 167
volume cover design concepts, 123
Preproposal conference, 120, 175
Pre-proposal interview form, 55–56
Pre-Solicitation Announcement (PSA), 11
Private sector marketing leads/information
    EI Digest, 378
    The Green Book Report, 379
    The Peck Report, 378
Private sector solicitation requests
    forms and formats, 84
    marketing style, 83
    point-by-point format, 83
    price importance, 83
    required elements, 84–86
    timing of, 84
Procurement Automated Source System
        (PASS), 374–75
Procurement Information Online (PIO)
        (Germany), 217
Procurement Technical Assistance Centers
        (PTACs), 390
Production graphics
    action captions, 240–41
    AFMC, 245
    configuration control, 241–42
    copyrights, 240
    file backups, 244
    hardware recommendations, 243
    HijackPro, 240
    photo documentation, 240
    publications group role, 242–43
    RAM requirements, 243, 244
    software, 243–44
    throughput capacity, 242
    Web graphics, 240
Profit center, 390
Program Evaluation Review Technique (PERT)
        Chart, 388
Project descriptions
    client project disclosure approval, 288

experience emphasis, 287
    format compatibility, 287
    quantitative narratives, 287
    specificity, 287
    summary format, 288
    updating, 287
Project managers (PMs), 58–59
Pro-Net database, 34, 39, 334
Proposal agility, 188
Proposal building blocks, 295, 296
Proposal components, 143–63
    CAS, 162–63
    compliance matrix, 149–50
    cost volume, 161–63
    disclosure statement, 147–48
    executive summary, 148–49
    front cover, 145–47
    government contract requirements, 162–63
    management volume, 154–61
    narrative body, 150–53
    technical volume, 145–53
    transmittal letter, 144–45
    volumes, 143
Proposal development
    best practices included, 23
    documentation tips, 246–48
    intangible values, 22
    knowledge management, 21
    knowledge products, 21
    market data, 23
    marketing integration in, 47–49
Proposal life cycle, 118–27
    2nd draft, 126
    acquisition teams, 119
    archiving, 126–27
    CBD, 120
    delivery, 183
    direct client contact, 119
    draft review, 126
    final delivery, 126
    final internal bid/no bid
        decision, 120–21, 122
    final production, 126
    Gold Team final review, 126
    government preproposal (bidders)
        conference, 120
    illustrated, 118

Proposal life cycle (continued)
  management and technical input review, 125
  management input submittal to publication
      group, 125
  market intelligence, 119
  pre-kickoff planning meeting, 121–23
  publication group quality check, 126
  Red Team comments integration, 126
  Red Team Review, 126
  revised cost volume preparation, 126
  RFP release, 120
  schedule control, 184–87
  signatures, 126
  staffing information and costs to
      Contracts, 125
  talent management, 119
  technical/management draft preparation, 125
Proposal manager, 173–88
  authors critique, 177
  B&P budget estimates, 176
  bidders conference, 175
  competitive edge, 180–81
  debriefing files, 181
  functions of, 173
  job description, 175–81
  lessons learned review, 187
  meeting preparation, 177
  as motivator, 188
  oral presentations, 182–83
  planning, 175, 178
  project schedule elements, 178–80
  proposal agility, 188
  qualities of, 174
  responsibilities of, 174
  special knowledge requirements, 174
  staff training, 187
  team duties, 175
  themes, 177
  See also Human dynamics; Knowledge
      management
Proposal production
  colored paper, 237
  critical activities, 234–35
  document configuration, 236
  draft retention, 237
  editorial staffing, 235
  freelance/temporary staff, 237

graphics, 239–42
guidebook topic areas, 235
handbook of format and style, 235
internal documentation standards, 235–36
last minute authoring, 238–39
mass storage devices, 236
professionalism, 233
quality layouts, 234
time stamps, 237
version control, 236
Proposals, 115–63, 165–88
  BAFOs, 139–40
  bullet drafts, 133
  choreographic guidelines, 134
  as closing sales presentation, 116–17
  components, 143–63
  contractor review cycles, 136–39
  debriefings, 140–41
  defined, 115–16
  editing, 243
  electronic submittal and evaluation, 245–46
  executive summary, 129–30
  file backups, 244
  freelance writers, 249–52
  key arguments of, 117
  kickoff meeting, 135
  life cycle, 118–27
  management of, 173–88
  marketing life cycle, position in, 118
  as marketing tool, 116
  oral, 139–40
  .pdf format, 246
  planning and organizing, 129–35
  proofreading, 243
  responsiveness, 142
  schedules, 184–87
  storyboards, 133, 134
  team activities, 165–72
  teams, 166
  theme development, 130–33
  writing, 136
Public relations (PR). *See* Advertising

Random access memory (RAM), 243, 244
Red Team review, 47, 126, 139
Reinvention teams, 95
Request for comment (RFC), 62

Request for information (RFI), 62
Requests for proposals (RFPs), 1
    best value, 78
    boilerplate assembly problems, 72
    CDRLs, 78
    compliant submittal, 76
    conflicting requirements, 72
    content influencing, 79–80
    contractual experience, 9
    defined, 71
    earlier contract comparison, 78–79
    evaluation criteria, 79
    federal acquisition process, 87–113
    greatest value approach, 77–78
    non-federal types of, 80–81
    order importance, 72
    Part I-the schedule, 73
    Part II-contract clauses, 73
    Part III-list of documents, exhibits, and other
        attachments, 74
    Part IV-representations and certifications, 74
    performance-based contracting, 78–79
    private sector solicitations, 81
    relationship building, 80
    Section L, 74–76
    Section M (evaluation criteria), 77
    SOW, 72, 78
    table of contents, 75
    word-per-page guidelines, 76
Requirements creep, 157
Responsiveness, 142
Résumés
    action-oriented words, 285
    benefits to projects sections, 285
    "bullet" items, 285
    customization, 284
    editing, 286
    electronic subdirectories, 286
    experience summaries, 284
    ILS, 284
    results emphasis, 285
    RFP requirements, 284
    software compatibility, 286
    tailoring, 284
Revenue pipeline, 4, 5
Roos, Johan, 19, 94, 300, 304
Root, Franklin, 192, 230

The Scientific Principles of Management, 303
Sealed bids. *See* Invitations for Bid (IFB)
SF 33, 390
SF 129, 63, 390–91
SF 254/255, 389
SF 1411, 389
SIC codes, 63
Simplified Acquisition Procedures
        (SAPs), 391
Small and Disadvantaged Business Utilization
        Specialist (SADBUS), 34
Small Business Act (1953), 391
Small Business Administration (SBA), 391–92
    administration, 7
    Small Business Week, 7
Small Business Development Centers
        (SBDCs), 35
Small businesses
    competitive proposals, 1–25
    cross-training, 10
    defined, 391
    disadvantaged (SDBs), 32
    employment statistics, 9
    flexibility and responsiveness, 9–10, 337
    participation by women, 9
Small Business Innovation Research (SBIR)
        program, 10–13
    address, 390
    Bridge Grant, 11
    Dawnbreaker, 12
    DR-LINK, 12
    funding levels, 11, 12
    phases of, 11–13
    typical format, 12–13
Small Business Liaison Officers
        (SBLOs), 34–35
Small Business Success, 375
Small Business Technology Transfer (STTR)
        program, 10, 13
Small Office Solutions (SOS), 41–42
Soat, Dr. Douglas, 259
Sole source acquisition, 392
Source evaluation board (SEB), 53
Source selection, 89
Standard Form (SF) 254/255
        synopses, 1, 315–38
    best-in-class, 27

Standard Form (SF) 254/255
    synopses (continued)
  Block 7, Personnel, 327–29
  Block 8, Project Experience Matrix, 329–30
  Block 10 (evaluation criteria), 330
  Blocks, other, 331–32
  FAR, 315–25
  form preparation overview, 326
  forms and formats examples, 316–24
  Note 24, 326
  strategy of response, 325–27
  structural requirements, 325
  subcontractor participation, 332–34
  teaming agreement building, 334–38
  teaming partner guidelines, 333
Standard Forms (SFs), 389
Standard Industrial Classification
    (SIC), 63, 392
Standards, 111–12
Statement of work (SOW), 62, 72, 78
Statements of qualification (SOQs), 48–49
STAT-USA Globus Information
    System, 222–23
Stet, 392
Storyboards, 121, 133, 134, 145
Strategic and mission planning
  benchmarking, 27
  and company goals, 18
  defined, 18
  as dynamic process, 19
  formulation questions, 18
  functions of, 19
  mission statement, 19
  planned growth patterns, 17–18
Strategic partnering
  alliances, 31
  Eisenhower Award, 32
  GWAC, 37–38
  mentor-protégé programs, 39–40
  MOU, 32
  PMPP, 39
  SBA, 38
  SDBs, 32
  SOS, 40–41
  streamlining authority process, 38–39
  subcontracting, 32–36

Strategic planning and business development
    groups (SP&BDGs). *See* Business
    development groups (BDGs)
Strawman RFP (or proposal), 392
Subcontracting opportunities, 32–36
  B&P costs, 33
  business associations, 36
  conferences, 35–36
  consulting firm utilization, 35
  Internet, 35, 36
  key elements, 33
  LOB agreements, 33
  M/WBE, 32
  NAPM, 36
  OSADBU, 35
  prime contractor relationship, 33
  Pro-Net registration, 34
  SADBUS registration and
    prequalification, 34–35
  SBDC interaction, 35
  SBLOs, 34–35
  small business leveraging, 32
  specific strategies, 34–36
  targeting technical areas of interest, 33
  Thomas Register, 36
  THRUST, 35
  trade associations, 36
  trade shows, 35–36
  TRY US listing, 35
  Web site creation/maintenance, 35
  See also Strategic partnering
Sveiby, Karl E., 19, 311
Systéme d'Information pour les Marchés Publics
    (SIMAP), 216

Tacis Programme, 198
Taylor, Frederick Winslow, 303
Teaming agreements
  elements of, 335
  exclusivity, 335
  legal review, 335
  potential topics, 338
  Pro-Net, 334
  reasons for, 334
Technical leveling, 392
Technical transfusion, 392
Technical volume, 145–53

artwork, 146–47
    back cover design, 147
    elements of, 145
    front cover, 145–47
    logo, use of, 146
    narrative body, 150–53
    storyboards, 145
Technology transfer, 203
Tender I.N.F.O. (Electronic Tendering and
        Procurement Network), 220
Tenders Electronic Daily (TED), 199, 211
Terms and conditions (Ts & Cs)
    FCIA, 204
    in international proposals, 203
    line items, 204–5
    risk awareness, 203
Themes, 392–93
    development, 130–33
    example, 131–33
    Executive summary, 129, 148
    proposal manager, 177
THEMiS, 216
The Thomas Register of American
        Manufacturers, 36
THRUST program, 35
Tiger Team, 393
Toffler, Alvin and Heidi, 93–94
Total quality management (TQM), 305
Trading Partner Profile, 97
Transfer ratios, 9
Transmittal letter, 144–45
Truth in Negotiations (TINA), 96, 162
TRY US National Minority Business
        Directory, 35

UN Development Businesses
        (UNDB), 193, 194
Uniform Contract Format (UCF), 89
Unisphere Institute, 223–24
United Nations procurements, 195–96
    application address, 196
    application requirements, 196
    related subagencies, 195
    UN's supplier roster, 195
U.S. Agency for International Development
        (USAID), 60, 226–29
    function of, 226

GTN, 226–27
    Internet address, 226
    NTDB, 228–29
    regional business outreach offices, 227–28
    related agencies, 226–29
    USITC, 229
U.S. Air Force Material Command
        (AFMC), 245
U.S. General Services Administration
        (GSA), 375–76
U.S. Trade and Development Agency
        (USTDA), 225

Value-added networks (VANs), 97, 107–9
Value-added services (VASs), 108–9
Virtual proposal centers, 248–49

Walsh-Healey Act (1936), 393
War room, 393
Washington Business Journal, 376
Washington Technology, 376–77
Web-based resources. *See* Internet
WinAward market managing system, 51
Windows Proposal Evaluation Tool
        (WinPET), 245
Woman-owned business (WOB), 393
Work breakdown structure (WBS), 159–60
World Bank, 192–95
    DACON, 194
    described, 230
    G7, 193
    IBRD, 230
    IDA, 230
    Internet, 193
    role of, 195
    scope of, 192–93
    UNDB, 193, 194
World Wide Web. *See* Internet
Writing, 136, 271–82
    ACASS, 273
    accuracy, 273
    acronyms and abbreviations, 278
    action captions, 278–79
    active voice, 274
    brevity, 273
    clarity, 273
    competition, separation from, 272
    editing checklist, 280–81

Writing (continued)
  enhancement methods, 279–80
  evaluator guiding, 275–78
  facts usage, 277
  features as benefits, 277
  feature translation, 274
  FOIA, 272
  freelance writers and, 249–52
  general to specific, 276
  government recognized standards, 281
  GPO Style Manual, 281
  guidance sources, 282

  hi-impact descriptors, 276
  knowledge-based sales documents, 271
  marketing intelligence, 274
  material utilization, 279
  packaging technical solutions, 272
  reading aloud, 280
  RFP terminology, 278
  tangible client benefits, 276–77
  technical data dumps, 274
  thinking graphically, 278
  useful verbiage, 275

Recent Titles in the Artech House Technology Management and Professional Development Library

Bruce Elbert, Series Editor

Designing the Networked Enterprise, Igor Hawryszkiewycz

Evaluation of R&D Processes: Effectiveness Through Measurements, Lynn W. Ellis

Decision Making for Technology Excutives: Using Multiple Perspectives to Improve Performance, Harold A. Linstone

Introduction to Information-Based High-Tech Services, Eric Viardot

Introduction to Innovation and Technology Transfer, Ian Cooke, Paul Mayes

Managing Engineers and Technical Employees: How to Attract, Motivate, and Retain Excellent People, Douglas M. Soat

Managing Virtual Teams: Practical Techniques for High-Technology Project Managers, Martha Haywood

Successful Marketing Strategy for High-Tech Firms, Second Edition, Eric Viardot

Successful Proposal Strategies for Small Businesses: Winning Government, Private Sector, and International Contracts, Second Edition, Robert S. Frey

The New High-Tech Manager: Six Rules for Success in Changing Times, Kenneth Durham and Bruce Kennedy

For further information on these and other Artech House titles, including previously considered out-of-print books now available through our In-Print-Forever® (IPF®) program, contact:

Artech House
685 Canton Street
Norwood, MA 02062
Phone: 781-769-9750
Fax: 781-769-6334
e-mail: artech@artechhouse.com

Artech House
46 Gillingham Street
London SW1V 1AH UK
Phone: +44 (0)171-596-8750
Fax: +44 (0)171-630-0166
e-mail: artech-uk@artechhouse.com

Find us on the World Wide Web at:
www.artechhouse.com